METHODS IN MOLECULAR BIOLOGY™

Series Editor
John M. Walker
School of Life Sciences
University of Hertfordshire
Hatfield, Hertfordshire, AL10 9AB, UK

For further volumes:
http://www.springer.com/series/7651

High-Throughput Phenotyping in Plants

Methods and Protocols

Edited by

Jennifer Normanly

University of Massachusetts, Amherst, MA, USA

Editor
Jennifer Normanly
University of Massachusetts
Amherst, MA, USA

Please note that additional material for this book can be downloaded from http://extras.springer.com

ISSN 1064-3745 ISSN 1940-6029 (electronic)
ISBN 978-1-61779-994-5 ISBN 978-1-61779-995-2 (eBook)
DOI 10.1007/978-1-61779-995-2
Springer New York Heidelberg Dordrecht London

Library of Congress Control Number: 2012942930

Printed on acid-free paper

Humana Press is a brand of Springer
Springer is part of Springer Science+Business Media (www.springer.com)

Preface

Genetic approaches to understanding plant growth and development have always benefitted from screens that are simple, quantitative, and fast. Visual screens and morphometric analysis have yielded a plethora of interesting mutants and traits that have provided insight into complex regulatory pathways. Still, many genes within any given plant genome remain undefined. The premise underlying this book is that the higher the resolution of the phenotype analysis the more likely that new genes and complex interactions will be revealed. Recent advances in automation and highly sensitive analytical techniques have substantially expanded the plant biologist's toolbox with which to screen for mutants and traits and identify new genes. There are now centers and institutes dedicated to high-throughput phenotyping of plants, and it has been the subject of at least three international conferences to date. As always, the value of the data obtained through high-throughput phenotyping methods depends upon the experimental design, which is discussed here.

The methods described in this book can be generally classified as either quantitative profiling of cellular components, ranging from ions to small molecule metabolites and nuclear DNA, or image capture that ranges in resolution from chlorophyll fluorescence from leaves and time-lapse images of seedling shoots and roots to individual plants within a population at a field site. The sort of high-throughput analytical analysis described in these chapters will be relevant to plant researchers who rely on phenotype analysis to define gene function and characterize genome responses to the environment; this includes biochemists, molecular geneticists, ecologists, evolutionary biologists, and population geneticists. As robotics, computing, and imaging technologies all continue to advance at a rapid rate, the list of quantifiable assays that can be carried out in high-throughput and at high resolution will continue to expand, providing more tools to understand plant growth and development.

Amherst, MA *Jennifer Normanly*

Contents

Contributors

LOUISE I. AHL • *Department of Plant Biology and Biotechnology, University of Copenhagen, Frederiksberg, Denmark*
BETTINA BERGER • *The Plant Accelerator, University of Adelaide, Urrbrae, SA, Australia*
JOSEPH A. BERRY • *Department of Global Ecology, Carnegie Institution of Washington, Stanford, CA, USA*
MAGDALENA BEZANILLA • *Department of Biology, University of Massachusetts, Amherst, MA, USA*
CARLO BICCHI • *Dipartimento di Scienza e Tecnologia del Farmaco, University of Turin, Turin, Italy*
REDDAIAH BODANAPU • *School of Life Sciences, University of Hyderabad, Hyderabad, India*
JUSTIN BOREVITZ • *Department of Ecology and Evolution, University of Chicago, Chicago, IL, USA*
TIM BROWN • *Time Science, Salt Lake City, UT, USA*
WEN-PING CHEN • *Department of Horticultural Science and Microbial and Plant Genomics Institute, University of Minnesota-Twin Cities, St. Paul, MN, USA*
JOANNE CHORY • *Plant Biology Laboratory and Howard Hughes Medical Institute, Salk Institute for Biological Studies, La Jolla, CA, USA*
MADS H. CLAUSEN • *Department of Plant Biology and Biotechnology, University of Copenhagen, Frederiksberg, Denmark*
JERRY D. COHEN • *Department of Horticultural Science and Microbial and Plant Genomics Institute, University of Minnesota-Twin Cities, St. Paul, MN, USA*
BENJAMIN J. COLE • *Plant Biology Laboratory, Salk Institute for Biological Studies, La Jolla, CA, USA; Division of Biological Sciences, University of California – San Diego, La Jolla, CA, USA*
BAS DE REGT • *The Plant Accelerator, University of Adelaide, Urrbrae, SA, Australia*
JACK EGELUND • *Department of Plant Biology and Biotechnology, University of Copenhagen, Frederiksberg, Denmark*
JONATAN U. FANGEL • *Department of Plant Biology and Biotechnology, University of Copenhagen, Frederiksberg, Denmark*
JOACHIM FISAHN • *Max Planck Institute of Molecular Plant Physiology, Potsdam, Germany*
DAVID W. GALBRAITH • *School of Plant Sciences, University of Arizona, Tucson, AZ, USA*
PETER GEIGENBERGER • *Department Biologie I, Ludwig-Maximilians-Universität München, Planegg-Martinsried, Germany*
SAMUEL P. HAZEN • *Department of Biology, University of Massachusetts, Amherst, MA, USA*
ADRIAN D. HEGEMAN • *Department of Horticultural and Microbial and Plant Genomics Institute, Department of Plant Biology, University of Minnesota-Twin Cities, St. Paul, MN, USA*
ÉVA HIDEG • *Institute of Biology, Faculty of Sciences, University of Pécs, Pécs, Hungary*
DIRK HINCHA • *Max Planck Institute for Molecular Plant Physiology, Potsdam-Golm, Germany*

Denis Klimov • *Monterey Bay Aquarium Research Institute, Moss Landing, CA, USA*
Karin Köhl • *Max Planck Institute for Molecular Plant Physiology, Potsdam-Golm, Germany*
Mikiko Kojima • *RIKEN Plant Science Center, Tsurumi, Yokohama, Japan*
Zbigniew S. Kolber • *University of California, Santa Cruz, Institute of Marine Sciences, Santa Cruz, CA, USA*
Joachim Kopka • *Max Planck Institute for Molecular Plant Physiology, Potsdam-Golm, Germany*
Georgina M. Lambert • *School of Plant Sciences, University of Arizona, Tucson, AZ, USA*
Scott J. Lee • *Plant Biology Graduate Program, University of Massachusetts, Amherst, MA, USA*
Susan B. Leschine • *Department of Microbiology, University of Massachusetts, Amherst, MA, USA*
Kaleb Lowe • *Kansas Lipidomics Research Center, Division of Biology, Kansas State University, Manhattan, KS, USA*
Massimo Maffei • *Unità di Fisiologia Vegetale, Dipartimento di Biologia Vegetale, University of Turin, Turin, Italy*
Petra Majer • *Institute of Plant Physiology, Biological Research Centre, Szeged, Hungary*
Will I. Menzel • *Department of Horticultural Science and Microbial and Plant Genomics Institute, University of Minnesota-Twin Cities, St. Paul, MN, USA*
Nina Noah • *Department of Ecology and Evolution, University of Chicago, Chicago, IL, USA*
C. Barry Osmond • *Plant Sciences Division, Research School of Biology, Australian National University, Canberra, Australia*
Whitney Panneton • *Department of Ecology and Evolution, University of Chicago, Chicago, IL, USA*
Henriette L. Pedersen • *Department of Plant Biology and Biotechnology, University of Copenhagen, Frederiksberg, Denmark*
Roland Pieruschka • *Forschungszentrum Jülich, IBG: 2 Plant Sciences, Jülich, Germany*
Charles A. Price • *School of Plant Biology, University of Western Australia, Crawley, Perth, Australia*
Uwe Rascher • *Forschungszentrum Jülich, IBG: 2 Plant Sciences, Jülich, Germany*
Mary R. Roth • *Kansas Lipidomics Research Center, Division of Biology, Kansas State University, Manhattan, KS, USA*
Maja Gro Rydahl • *Department of Plant Biology and Biotechnology, University of Copenhagen, Frederiksberg, Denmark*
Hitoshi Sakakibara • *RIKEN Plant Science Center, Tsurumi, Yokohama, Japan*
Armando Asuncion Salmean • *Department of Plant Biology and Biotechnology, University of Copenhagen, Frederiksberg, Denmark*
Thilani Samarakoon • *Kansas Lipidomics Research Center, Division of Biology, Kansas State University, Manhattan, KS, USA*
László Sass • *Institute of Plant Physiology, Biological Research Centre, Szeged, Hungary*
Christian Schudoma • *Max Planck Institute for Molecular Plant Physiology, Potsdam-Golm, Germany*
Javier Seravalli • *Redox Biology Center and Department of Biochemistry, University of Nebraska-Lincoln, Lincoln, NE, USA*
Rameshwar Sharma • *School of Life Sciences, University of Hyderabad, Hyderabad, India*

SUNITHA SHIVA • *Kansas Lipidomics Research Center, Division of Biology, Kansas State University, Manhattan, KS, USA*
HEIKE SPRENGER • *Max Planck Institute for Molecular Plant Physiology, Potsdam-Golm, Germany*
YELLAMARAJU SREELAKSHMI • *School of Life Sciences, University of Hyderabad, Hyderabad, India*
MATTHIAS STEINFATH • *Max Planck Institute for Molecular Plant Physiology, Potsdam-Golm, Germany*
PAMELA TAMURA • *Kansas Lipidomics Research Center, Division of Biology, Kansas State University, Manhattan, KS, USA*
MARK TESTER • *The Plant Accelerator, Australian Centre for Plant Functional Genomics, University of Adelaide, Urrbrae, SA, Australia*
YING TU • *Department of Chemical Biology and Therapeutics, St. Jude Children's Research Hospital, Memphis, TN, USA*
JOOST T. VAN DONGEN • *Max Planck Institute for Molecular Plant Physiology, Potsdam-Golm, Germany*
RAJU NAIK VANKUDAVATH • *School of Life Sciences, University of Hyderabad, Bio-Medical Informatics Centre and National Institute of Nutrition, Hyderabad, India*
SILVIA VIDAL-MELGOSA • *Department of Plant Biology and Biotechnology, University of Copenhagen, Frederiksberg, Denmark*
DIRK WALTHER • *Max Planck Institute for Molecular Plant Physiology, Potsdam-Golm, Germany*
THOMAS A. WARNICK • *Department of Microbiology, University of Massachusetts, Amherst, MA, USA*
RUTH WELTI • *Kansas Lipidomics Research Center, Division of Biology, Kansas State University, Manhattan, KS, USA*
WILLIAM G.T. WILLATS • *Department of Plant Biology and Biotechnology, University of Copenhagen, Frederiksberg, Denmark*
SHU-ZON WU • *Plant Biology Graduate Program, University of Massachusetts, Amherst, MA, USA*
BING YAN • *School of Chemistry and Chemical Engineering, Shandong University, Jinan, China*
NIMA YAZDANBAKHSH • *Max Planck Institute of Molecular Plant Physiology, Potsdam, Germany*
CHRISTOPHER ZIMMERMANN • *Time Science, Salt Lake City, UT, USA*
ELLEN ZUTHER • *Max Planck Institute for Molecular Plant Physiology, Potsdam-Golm, Germany*

Chapter 1

Image-Based Analysis of Light-Grown Seedling Hypocotyls in Arabidopsis

Benjamin J. Cole and Joanne Chory

Abstract

Time-resolved hypocotyl length measurements in seedlings have the potential to greatly aid genetic studies looking at light, hormone, and circadian regulation of cell expansion. Recently, several computer-based tools have been developed to quantify hypocotyl length during photomorphogenesis and early seedling development. Here we detail a method for quantifying Arabidopsis seedling hypocotyls in an image-based assay, focusing on light-grown seedlings responding to shade conditions.

Key words: Hypocotyl, Image processing, Shade avoidance, HyDE, Arabidopsis

1. Introduction

Hypocotyls serve multiple functions in development. Hypocotyl structure is very simple, consisting of files of up to 20 cells, which grow almost exclusively by cell expansion (as opposed to cell division) (1). When germinated under soil, hypocotyls expand to push the undeveloped photosynthetic organs up to the surface where they can intercept light and undergo photomorphogenesis (2). During photosynthetic growth, they connect the root system with the cotyledons and leaves of the plant, allowing nutrients, minerals, and signaling hormones to be transported to their sites of action, and their elongation is under circadian control (3). Arabidopsis hypocotyls also elongate in response to proximity stress, generated by the threat or onset of foliar shade, to position photosynthetic organs optimally (4). This shade condition is most often associated with an altered quality of light, namely a decreased red to far-red light ratio (5, 6). Thus, hypocotyl length serves as an incredibly

Jennifer Normanly (ed.), *High-Throughput Phenotyping in Plants: Methods and Protocols*, Methods in Molecular Biology, vol. 918, DOI 10.1007/978-1-61779-995-2_1, © Springer Science+Business Media, LLC 2012

useful readout for many aspects of light signaling, as this parameter tends to be inversely proportional to the amount of light available. The high responsiveness to the light environment and the simplistic nature of the hypocotyl has made it an obvious subject of many mutant screens aimed at identifying important molecular contributors to light signaling (7–9). It is thus useful to extend these studies of hypocotyl length to dynamic studies of growth rates and kinetics. To aid these studies, new assays have been developed (10–12) that can accurately quantify length and growth rate of hypocotyls in time-lapse images under various light conditions. Image-based assays are advantageous, being non-invasive, sensitive (with the help of high-resolution camera technology), and relatively easy to perform. Here, we describe a protocol for growing, imaging, and quantifying hypocotyl length in a time-resolved assay of Columbia seedlings encountering shade conditions.

2. Materials

It is important to keep all solutions sterile, as fungal or bacterial growth can affect growth dynamics and image quality.

2.1. Seedling Growth and Imaging Components

1. Solid plant growth medium: 0.8% agar, 1/2 Linsmaier and Skoog (LS) nutrients, 2.5 mM MES. Prepare by adding MES powder to 1/2 LS solution, pH to 5.7 with 5 M KOH, and then add the agar powder. Autoclave solution for 20 min.
2. ~100 sterilized, stratified Col-0 seeds.
3. Sterile pipettes for plating seeds.
4. Circular petri dishes and semicircular sterile molds (fitting snugly into the petri dishes, such that half of the volume is occluded).
5. Long-term seedling growth chamber (Percival) with two to four cool white fluorescent bulbs and two incandescent light bulbs. Total fluence rate should be approximately 50 μmol/m^2 s, and the temperature is kept at 20°C.
6. Spectroradiometer.
7. Dedicated imaging growth chamber (Percival) with a three channel LED light source (blue, red, far-red), whose light conditions are variable, but adjusted to match the long-term seedling growth chamber conditions (calibrate with a spectroradiometer).
8. IR LED backlight source (>800 nm, see Note 1).
9. High-resolution video camera (>1 megapixel, see Note 2).
10. Sample vessel and mount (see Note 3).
11. Macro video zoom lens (see Note 4).

2.2. Image Processing and Analysis

1. Camera Control Software (see Note 5).
2. Hypocotyl Analysis Software (see Note 6).
3. ImageJ installed with the MultiStackReg plugin (http://www.stanford.edu/~bbusse/work/downloads.html).

3. Methods

3.1. Seedling Growth and Imaging

1. Prepare solid growth medium on which to plate seeds (pouring molten agar solution into plate fitted with mold; see Note 3).
2. Plate seeds onto sample vessel, and cover each seed with a thin layer (small drop) of molten agar solution to ensure root submergence and support for apical growth.
3. Grow seedlings in diurnal growth conditions in long-term seedling growth chamber. For this assay, we use long-day (16 h light/8 h dark) growth conditions to keep hypocotyl length short initially (see Note 3) and keep the temperature at 20°C. Typically, hypocotyls are at their most dynamic developmental stage 4–7 days post-germination under these conditions.
4. On day 5 (or 1–2 days earlier or later, depending on the assay), transfer the seedlings to the dedicated imaging chamber, and mount the seedling vessels in front of the camera.
5. Begin video image capture. Parameters should be set using the control software such that an image of a single seedling is taken every 5–10 min. This can be accomplished by adjusting the frame rate and the number of frames between image capture events (see Note 5).
6. Image seedling growth for 2 h under normal (simulated white-light) conditions to establish a base line.
7. After 2 h, increase the intensity of the far-red light channel (this can be programmed into the LED chamber). Continue this treatment for up to 36 h.

3.2. Image Processing and Analysis

1. Convert all images into 8-bit grayscale TIFF images using ImageJ (or a similar image processing program).
2. For each image stack (all time series images associated with a single seedling), align images using a non-seedling reference point (e.g., the agar surface adjacent to the seedling). ImageJ plugins are available for this task (see MultiStackReg).
3. Crop the aligned image such that only hypocotyl information is contained at the very bottom of the image (see Fig. 1). Some of the basal hypocotyl can be eliminated as dynamic length measurements need only new growth. Cotyledons can be cropped out, as this information is not important for determining hypocotyl length.

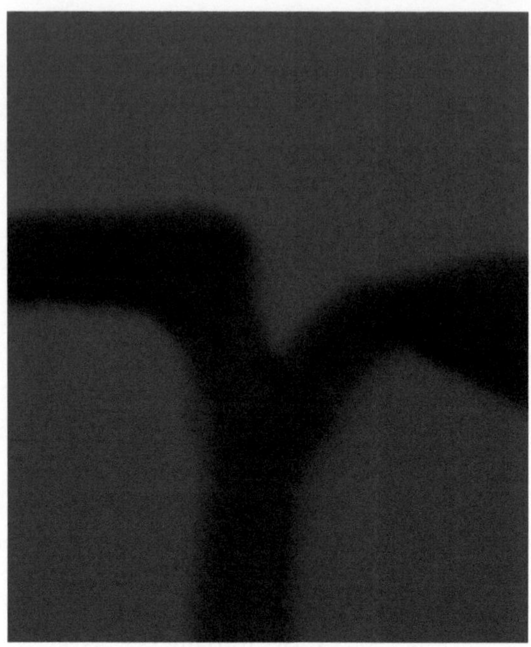

Fig. 1. An ideally cropped seedling image for measuring hypocotyl growth. Note how only the hypocotyl is touching the very bottom of the image, important for automated measurement software. Resolution and zoom should be high enough such that hypocotyl detection is accurate.

4. Measure hypocotyl length in the stack using image analysis software (see Note 6).
5. Manually inspect output from the software, taking care to eliminate growth curves that do not accurately reflect the hypocotyl length seen in the image stack. This can be done by overlaying the midline information with the original set of images. Eliminate stacks where the line deviates from the hypocotyl termination point.
6. Report results in units of millimeters of new hypocotyl growth.

4. Notes

1. To enable imaging in darkness, it is necessary to have a light source that emits long wavelength infrared light (>800 nm), which plants are "blind" to, yet can still be detected on a CCD camera. The LED source we use emits at 880 nm, and we also installed a 790 nm cutoff filter on the camera lens such that this is the only light source allowed into the CCD detector.

2. Here, we use a 1,392 × 1,050 pixel CCD camera, capable of imaging at least one frame per second. Any preexisting filters on the camera itself that would prevent detection of long wavelengths were removed.
3. It is important that seedlings are oriented vertically (for normal growth experiments; special conditions may be necessary for gravitropism studies, for example). To accomplish this for light-grown seedlings, we use a circular petri dish half-filled with solid growth medium (facilitated by a semicircular plexiglass mold during pouring) and plate seeds on the ledge formed. In this way, the areal portion of the seedling is allowed to grow unobstructed in air while the roots are allowed to penetrate through the growth medium. We have found that if seedlings are grown on a full agar plate (no ledge created), frictional forces generated as the seedling grows upward tend to impede growth, or cause the hypocotyl and root system to move downward (complicating downstream image processing algorithms). Furthermore, we add a thin layer of "top-agar" to the seeds, ensuring that they have enough support for vertical growth as young seedlings. It also helps if the hypocotyls are relatively short (3–5 mm) at the time of imaging, although this is not always possible (e.g., imaging a photoreceptor mutant or under short-day conditions). This sample vessel (the petri dish) must then be mounted vertically to prevent vibration and displacement. In our case, we had a custom sample mount fabricated. Alternatively, if higher-throughput imaging is desired, individual seedlings can be grown in cylindrical tubes (e.g., USA Scientific, 1.1 mL tubes). This eliminates problems associated with seedling orientation, as the tubes can be freely rotated. Special considerations are needed for preparation of the agar surface, and controlling condensation when growing seedlings in these tubes: we use 0.1–10 μL pipette tips (upside-down) to mold the surface in the cylindrical tubes into a short pedestal (minimizing optical obstruction from the meniscus formed), and punch a small hole in the tube cap to allow gas exchange and to minimize humidity.
4. A lens capable of delivering in-focus images at >100 pixels/mm is helpful. The higher the zoom factor on the lens, the more pixels a seedling image will occupy, increasing the accuracy of downstream image processing algorithms.
5. Software is necessary for controlling the camera (notifying it when to take and save images). Often this software is provided by the camera's manufacturer, but in some cases, manual programming is necessary. If using a motion-control device in combination with video imaging (to increase throughput), custom-designed control software must be engineered such that image capture is coordinated with motion. Take care that

the scale of the image is accurately determined (measure how many pixels per millimeter the image represents).

6. To reduce error associated with (sometimes subjective) measurements of hypocotyl length, computer-based software is needed. If you are designing your own tool for measurement, it is important to find some distinguishing feature of the start and end points of the structure to be measured. For measuring hypocotyl length in stacks described above, the start point is fixed (the cropped bottom of the hypocotyl image). The end point is the only parameter that must be determined for accurate measurement. HyDE is one software-based tool we developed to automate measurements, removing bias. This tool can be downloaded from http://cactus.salk.edu/hyde. This program uses information at the shoot-apical meristem to determine where the hypocotyl ends, and traces a midline from the hypocotyl base up to this point. Once identified, the program calculates the length of this midline, and reports these values for each image successfully processed. For measuring hypocotyl length in etiolated seedlings, the reader is directed to HYPOTrace (see ref. 12) (http://phytomorph.wisc.edu/software/hypotrace.php). These tools as well as other software for phenotyping are also being made publicly available through the iPlant collaborative project (http://www.iplantcollaborative.org).

Acknowledgments

This chapter was previously published in the Ph.D. thesis of Benjamin J. Cole, Rapid and dynamic growth of Arabidopsis seedlings in response to changes in light quality: a live imaging study, University of California, San Diego, 2011.

References

1. Gendreau E, Traas J, Desnos T, Grandjean O, Caboche M, Höfte H (1997) Cellular basis of hypocotyl growth in *Arabidopsis thaliana*. Plant Physiol 114:295–305
2. Chen M, Chory J, Fankhauser C (2004) Light signal transduction in higher plants. Annu Rev Genet 38:87–117
3. Nozue K, Covington MF, Duek PD, Lorrain S, Fankhauser C, Harmer SL, Maloof JN (2007) Rhythmic growth explained by coincidence between internal and external cues. Nature 448:358–361
4. Franklin KA (2008) Shade avoidance. New Phytol 179:930–944
5. Kasperbauer MJ (1987) Far-red light reflection from green leaves and effects on phytochrome-mediated assimilate partitioning under field conditions. Plant Physiol 85:350–354
6. Ballaré CL, Scopel AL, Sánchez RA (1990) Far-red radiation reflected from adjacent leaves: an early signal of competition in plant canopies. Science 247:329–332
7. Tao Y, Ferrer J-L, Ljung K, Pojer F, Hong F, Long JA, Li L, Moreno JE, Bowman ME, Ivans

LJ, Cheng Y, Lim J, Zhao Y, Ballaré CL, Sandberg G, Noel JP, Chory J (2008) Rapid synthesis of auxin via a new tryptophan-dependent pathway is required for shade avoidance in plants. Cell 133:164–176

8. Fairchild CD, Schumaker MA, Quail PH (2000) HFR1 encodes an atypical bHLH protein that acts in phytochrome A signal transduction. Genes Dev 14:2377–2391
9. Nagatani A, Reed JW, Chory J (1993) Isolation and initial characterization of Arabidopsis mutants that are deficient in phytochrome A. Plant Physiol 102:269–277
10. Cole B, Kay SA, Chory J (2011) Automated analysis of hypocotyl growth dynamics during shade avoidance in Arabidopsis. Plant J 65: 991–1000
11. Miller ND, Parks BM, Spalding EP (2007) Computer-vision analysis of seedling responses to light and gravity. Plant J 52:374–381
12. Wang L, Uilecan IV, Assadi AH, Kozmik CA, Spalding EP (2009) HYPOTrace: image analysis software for measuring hypocotyl growth and shape demonstrated on Arabidopsis seedlings undergoing photomorphogenesis. Plant Physiol 149:1632–1637

Chapter 2

High-Throughput Phenotyping of Plant Shoots

Bettina Berger, Bas de Regt, and Mark Tester

Abstract

Advances in automated plant handling and image acquisition now make it possible to use digital imaging for the high-throughput phenotyping of plants. Various traits can be extracted from individual images. However, the potential of this technology lies in the acquisition of time series. Since whole shoot imaging is nondestructive, plants can now be monitored throughout their lifecycle, and dynamic traits such as plant growth and development can be captured and quantified. The technique is applicable to a wide range of plants and research areas and makes high-throughput screens possible, reducing the time and labor needed for the phenotypic characterization of plants.

Key words: Plant imaging, Growth analysis, Leaf area, Shoot morphology

1. Introduction

The remarkable progress in plant genetics over recent years has made increasingly apparent that plant phenotyping is lagging behind and has become the rate-limiting step in plant science and the generation of improved crop varieties. Traditionally, whole shoot phenotyping involves techniques such as visual assessment of plants, manual measurement of height and leaf dimensions, or destructive sampling to determine biomass accumulation, making it a time-consuming and labor-intensive process. High-throughput phenotyping protocols are therefore needed and, as with genetics, this will be a technology-driven process.

The ability to capture and store information in images is not new and has been used for a long time. Automated plant handling and imaging systems have rendered plant shoot phenotyping high-throughput. Using digital imaging as a means of shoot

Jennifer Normanly (ed.), *High-Throughput Phenotyping in Plants: Methods and Protocols*, Methods in Molecular Biology, vol. 918, DOI 10.1007/978-1-61779-995-2_2,

phenotyping has several advantages. (1) Whole shoot imaging is nondestructive and noninvasive, making it possible to image the same plant throughout the course of its lifecycle to measure dynamic traits such as growth; (2) it is possible to determine several traits within a single image, thereby increasing the information captured; (3) digital images can be stored and reanalyzed if there are improvements in image processing or different research questions arise; (4) morphological parameters or leaf symptom measurements derived from images are quantitative rather than arbitrary units subject to human assessment; (5) imaging can extend beyond the range of visible light and allows the analysis of traits that are invisible to the human eye.

Nevertheless, the phenotypic traits amenable to high-throughput imaging protocols need to fulfill certain criteria. Capturing the trait reliably in the images and extracting it in an automated manner through image processing are critical. Not all features of a plant shoot obvious to the researcher, such as individual stems of a wheat plant, can easily be identified through image processing. Also, some traits might be subject to circadian rhythms, such as leaf angles or leaf temperature, and the respective protocols to measure those traits need to incorporate a suitable time window for imaging.

The traits measured by imaging will obviously depend on the research question at hand, and it is beyond the scope of this chapter to present an exhaustive list. We will therefore focus on the use of digital color imaging to measure growth dynamics, a trait important for many areas, such as abiotic stress or nutrient use efficiency.

2. Materials

2.1. Seed Treatment

1. Uniformly sized seeds (see Note 1).
2. 70% (v/v) ethanol.
3. 3% (v/v) sodium hypochlorite (see Note 2).
4. Alternatively, Thiram or similar fungicides.

2.2. Growth in Potting Mix

2.2.1. Measurement of Field Capacity of Potting Mix

1. Sintered glass funnel.
2. 1.3-m silicon or clear plastic tubing with diameter to fit the funnel outlet.
3. Retort stand and clamp.
4. Large beaker or bucket as water reservoir.

2.2.2. Pot Preparation and Plant Growth in Potting Mix

1. Plastic pots with a capacity of about 3 L (see Note 3).
2. If the application of nutrients or water to the bottom of the pot is necessary, draining pots should be placed in saucers that enclose the bottom third of the pot.
3. Potting mix (see Note 4).

2.3. Biological Validation of Shoot Imaging for Biomass Measurements

1. Leaf area meter (e.g., LI-3100C, LI-COR, USA).
2. Drying oven.
3. Analytical balance.

2.4. Image Acquisition

1. Industry grade digital color camera with automated software control (e.g., LemnaTec 3D Scanalyzer system, LemnaTec GmbH, Germany).
2. Automated setup to move plants to the camera or vice versa. If manual systems are used, experiments are usually limited to about 150–200 plants per experiment.
3. Adequate computer hardware for image storage (see Note 5).
4. Adequate illumination equipment.
5. Optional: A color reference card and/or ruler for calibration purposes (e.g., RHS Colour Chart; ColorChecker, X-Rite, USA).

2.5. Image Analysis

1. Adequate computer hardware for high-throughput image processing.
2. Image analysis software package, included with imaging system, e.g., LemnaGrid (LemnaTec GmbH, Germany) and/or standalone software such as MATLAB (Mathworks, USA), Halcon (MVTec Software GmbH, Germany), or Labview (National Instruments, USA). An open source alternative is ImageJ (http://rsbweb.nih.gov/ij).

3. Methods

3.1. Seed Treatment

1. Surface sterilize uniformly sized seeds for 1 min in 70% (v/v) ethanol followed by 5 min in 3% (v/v) sodium hypochlorite.
2. Rinse the seeds several times in deionized water (see Note 6).

   Or

3. Surface coat the seeds with Thiram following the manufacturer's instructions (see Note 6).

3.2. Growth in Potting Mix

3.2.1. Measurement of Field Capacity of Potting Mix

When working in pots, it is important to carefully consider the watering regime to avoid waterlogging and hypoxia (1). Many experiments will adjust watering to "water holding capacity" or "pot capacity," which is the volumetric water content of a free-draining pot. However, this value greatly depends on the height of the pot and might often result in hypoxia, especially with fine potting mixes or field soil. In our experiments, we measure "field capacity," defined as the volumetric water content of the potting mix or soil at 1 m suction.

The setup described here to measure this parameter is comparable to the one shown in Fig. 2 of Passioura (1).

1. Attach the silicon tubing to the funnel outlet.
2. Mount the funnel with tubing on a retort stand about 1 m above the water reservoir (see Note 7).
3. Add about 2 L of water to the water reservoir below the funnel.
4. Fill the funnel and silicon tube with water ensuring that all air bubbles are removed.
5. Add the soil/potting mix to be tested into the funnel and let it settle. About half to two-thirds of the funnel should be filled with soil.
6. Once the water has drained to just above the soil level, cover the funnel with clingfilm to avoid evaporation from the surface.
7. To ensure hydraulic conductivity, there should be no air bubbles present between the filter plate, tubing, and water reservoir.
8. Adjust the position of the filter to obtain a height of 1 m from the sintered filter plate down to the water level in the reservoir.
9. Let the soil/potting mix equilibrate for several days up to one week, ensuring that no air bubbles form.
10. Take out the wet soil from the funnel and record the wet weight (WW).
11. Dry the soil in an oven at 105°C until constant weight is reached.
12. Record the dry weight (DW).
13. The volumetric field capacity is given by the equation (WW – DW)/DW.

3.2.2. Plant Growth in Potting Mix

The following protocol describes growth of plants under well-watered conditions with complete fertilizer present in the potting mix. If experiments for nutrient use efficiency are performed, a fertilizer free potting mix should be used, and nutrients should be supplied through fertilizer solutions with a defined nutrient composition. In the case of drought experiments, the required watering level for the low watering regimes can be determined through establishing a soil water retention curve, using a pressure plate apparatus (2) or through measuring pre-dawn leaf water potential with a pressure bomb (3).

1. Fill a pot to about 4 cm below the rim after gentle tapping and then weigh it.
2. Use the same weight to fill up all remaining pots.
3. Include several spare pots to monitor water evaporation from the soil during the experiment and at least two pots to determine the oven dry weight of the soil.
4. Once all pots are filled, add enough water for germination.
5. Plant three to four seeds per pot, about 1 cm deep and cover them with soil.
6. Use the soil dry weight to calculate the target weight of a pot at field capacity as determined by Subheading 3.2.1.
7. Adjust the watering level of each pot to field capacity about 2–3 times per week and record the water use.
8. Once the seedlings are about established, thin out to one seedling per pot.
9. Image the plants daily or every second day during the period important for phenotypic measurements.

3.3. Biological Validation of Imaging for Shoot Biomass Measurements

We found a good correlation of the plant area measured from three images (two images from the side at 90° rotation and one image from the top) and shoot biomass for a variety of plants including wheat, barley, sorghum, and tomato. However, this might not be the case for all plant types and certainly not for the whole lifecycle of the plant. It is therefore necessary to establish a calibration for the specific plant type analyzed and the developmental stages of the plant critical for phenotyping.

1. Grow several replicates of plants to the desired growth stage under the same conditions used for the phenotyping experiments (see Note 8).
2. Image the plants immediately prior to destructive harvest (see Note 9).
3. Harvest the shoot and measure the shoot fresh weight. If individual organs, such as leaf and stem, can be differentiated in the images, measure them separately.
4. Measure the leaf area with a leaf area meter (see Note 10).
5. Dry the shoot or separated shoot organs in a drying oven until constant weight is reached.
6. Measure the shoot dry weight or the dry weight of the individual organs.

7. Establish a calibration curve for the projected shoot area extracted from the images (see below) and shoot area or shoot biomass. If the images of the shoot can be differentiated into individual organs, take the different biomass for those organs into account when establishing the calibration curve; e.g., the same pixel area of stem may account for more shoot biomass than the same pixel area of leaf.

3.4. Image Acquisition to Monitor Plant Growth

How images are acquired will greatly depend on the hard- and software available to the researcher and the trait to be measured. There are complete systems available from LemnaTec (LemnaTec GmbH, Germany) that combine plant handling, imaging hardware, and the control software. Other institutes might have the capability to build their own automated in-house solutions (4, 5) or use a fairly simple camera setup and manual handling of plants. We will therefore only present aspects of image acquisition that are generally applicable and important for any type of setup.

1. The aim of any imaging setup should always be to obtain the best possible image of the plants for measuring the trait of interest. Image acquisition should be done as consistently as possible. This will greatly facilitate the image analysis and ideally allow the generation of automated image analysis algorithms that require minimum user input.
2. In general, there are two methods for image acquisition.
   (a) The plants are stationary and the camera is moved to the plant. This is most commonly used for plants with a simple architecture, such as, Arabidopsis, where a single image from the top often provides sufficient data.
   (b) The plants are moved to a stationary camera setup. This is an advantage for plants with a complex morphology, such as wheat and barley, where images from several angles will greatly increase the quality of data obtained through imaging. In addition, the imaging environment, such as background and illumination, is easier to control.
3. Illumination conditions should be as uniform as possible, both over time and throughout the field of view. It is important to preheat the lamps until constant illumination is reached before the first images are taken. Hunter et al. (6) give detailed information on how to achieve optimal lighting and avoid shadows and reflections.
4. Use of a color card and ruler allows calibration of the imaging setup. If both are present in an image, it is possible to normalize the recorded colors and calibrate for the zoom factor used. This allows comparisons between different imaging setups that differ in lighting conditions and the cameras used.

5. The imaging background should be chosen carefully to facilitate the identification of the plant in subsequent analysis. Backgrounds, such as white or blue, are preferable, since the green of the plant will be easy to differentiate.
6. Green and gray should be avoided as pot colors. White, blue, and black are suitable for most plant types and white has the advantage of keeping the soil cooler than darker colors. Materials with a flat finish reduce undesired reflections.
7. The soil surface can become challenging in the image analysis, since sandy or drying soils can have very similar colors to senescent leaves. Colored plastic mulch or white gravel on the surface can reduce this problem and have the further advantage of reducing water loss from the soil surface.
8. Many plants, especially wheat and barley, will need some sort of support when grown in pots, such as carnation frames. Again, they should not be green and if metal they need to be tested to determine if they can be easily eliminated in the image analysis. In some cases, it might be easier to get color-coated frames to avoid problems during the automated image analysis.
9. When choosing the exposure for the images, it is generally better to have a lower exposure. Overexposure will lead to white spots and thus a loss of color information that cannot be compensated for by image analysis.
10. The file format for storing the images should not lead to loss of image information (e.g., JPG or BMP). PNG and TIFF are the commonly used formats and do not lead to loss of information through compression.

3.5. Image Analysis to Measure Projected Shoot Area

Since plant imaging allows daily recordings, simple image analyses such as plant size measurement yield valuable information about plant growth and performance. Nevertheless, basic image analysis also requires the use of specialized software, computing infrastructure and database management if it is to be performed at high-throughput.

Depending on the software solution used, different levels of prior knowledge in image analysis and programming are necessary to develop image analysis algorithms, and collaboration with scientists experienced in that area is advisable.

MATLAB (MathWorks, Massachusetts, USA) is possibly the most commonly used and powerful software to develop image analysis algorithms and offers solutions for automated image acquisition. Halcon (MVTec Software GmbH, Germany) is a fairly comprehensive application for image analysis, and it is compatible with common programming languages such as C, C#, and .NET. ImageJ (http://rsbweb.nih.gov/ij) presents a Java-based solution for image analysis that is open source, so it is easily accessible. However, all three software programs require a certain amount of programming skills to write and implement analysis algorithms. The built-in image analysis solution of LemnaTec setups, LemnaGrid, is designed to allow researchers without prior programming knowledge to create algorithms for image analysis through drag-and-drop software where individual operators can be connected to create a processing pipeline. Unfortunately, algorithms can only be shared among LemnaTec users and the functionalities are not as comprehensive as those of specialized image analysis software.

Since the specific algorithms will depend on the software used and the imaging setup, we will only discuss general steps common to digital image processing (7) that are necessary to measure the size of the plant and to perform subsequent growth analysis.

1. *Image retrieval.* Recorded images need to be loaded into the software from a database or storage folder. Images may need to be cropped or a region of interest (ROI) may need to be set to shorten the computing time and/or to remove unnecessary parts of the image that can become a source of noise.
2. *Image preprocessing.* The application of filters to minimize noise or increase sharpness can improve the outcome of the subsequent analysis steps. However, there is a possibility of losing information that cannot be retrieved in later steps. If thresholding is used to make a binary image in the next step, the color image needs to be converted into a grayscale image by transforming the 3D RGB color information into a single channel.
3. *Image segmentation.* The next step is the segmentation of the image into objects of interest and objects that will later be discarded, such as the background, pot, support frames, or soil. Depending on the composition of the image, there are several options to produce a binary image. Classification by color with a supervised nearest neighbor algorithm or thresholding of a grayscale image is commonly used. In both instances, the result is a binary image, where pixels that belong to the object of interest are set to a value of 1, all others to 0.
4. *Noise reduction.* Morphological operations such as erosion-dilation steps or filling holes can be used to correct for unavoidable imperfections in the binary image that result from noise from image acquisition or difficulties in distinguishing between parts of the object and background that have similar colors.
5. *Image composition.* Leaves can often become fragmented in earlier steps due to curling of the leaves, and the individual fragments need to be merged to create one single object, the plant.

6. *Image description.* Features of the identified object, such as area, height, width, convex hull, or compactness, are quantified. The features mostly consist of mathematical characteristics calculated from the object.

7. *Color classification.* The identified object, the plant, can now be extracted from the original RGB image. Based on the color information of the original image, the leaves can be subdivided according to their color and the respective areas quantified using supervised nearest neighbor color classification. This can be used to quantify necrotic or senescent leaf area. A similar approach that uses the color information of the plant to determine the chlorophyll content is presented in chapter 6, this volume.

3.6. Basic Plant Growth Analysis

The following protocol describes basic measurements of several growth parameters. For more detailed plant growth analyses, refer to the excellent publications by Hunt (8, 9). All steps presented here assume a linear correlation between plant biomass and the projected shoot area measured from the images. If this is not the case, the calibration established in Subheading 3.3 should be used to convert the measured projected shoot area to estimated biomass or leaf area.

1. Increase in shoot area (A) over time (t). For a first evaluation of the data, plot the shoot area for individual plants or treatment groups over time. This will allow a visual assessment of treatment or genotype effects and the identification of biological outliers (entire growth curve is affected) or technical outliers from the imaging process (generally only individual points of the growth series are affected). Most plant species have a sigmoid growth curve when imaged from seedling stage to early reproductive stage, consistent with other measuring techniques. Once leaves start to senesce during seed ripening, this will obviously result in a decrease in projected leaf area, which is then no longer a good indicator of plant biomass. It is possible to overcome this technical challenge by using the color information of the leaves to differentiate between green and senescent leaf areas if experiments need to extend over the whole growth cycle; however, this needs to be tested for each plant species.

2. Use the data of shoot area over time to generate a growth model through curve fitting. Growth models, such as higher order polynomials or cubic splines that make no prior assumption about the data, are preferable. Higher order polynomials can be generated with basic spread sheet software, such as Microsoft Excel (Microsoft Cooperation, USA). Spline curves generally need statistical software packages.

3. Use the growth model to compute the absolute growth rate of the plants, which is the first derivative (dA/dt) of the growth model. The absolute growth rate will reveal how much area the plant gained per day at any time during the experiment. If plants were imaged over most of the lifecycle, the absolute growth rate will show an increase during early growth, reaching a maximum when plants shift from vegetative to reproductive growth and a subsequent decline as plants mature. The time interval for plants to reach maximum absolute growth can be regarded as a trait. Certain stress treatments, such as drought or salinity can alter the length of the interval, indicating altered plant development.
4. Relative growth rate ($dA/dt \cdot 1/A$). In addition to the absolute growth rate, a growth model can be used to calculate the relative growth rate (RGR) at any given time. The RGR is generally highest for young seedlings and then declines gradually. Since RGR is independent of plant size, it allows comparison of plants and varieties with fairly different growth habits. Analysis of RGR over time can reveal when genotype or treatment effects become apparent.
5. Leaf area duration (LAD). The expression of leaf area duration was used by Watson in 1947 (10) for the integral of the leaf area over the entire lifecycle and was described as the "whole opportunity for assimilation" of the plant. Using the previously developed growth model, it is possible to calculate LAD for the entire experiment or certain intervals relevant to the treatment. LAD will give a measure of the leaf area and its persistence over the chosen period.
6. Morphological measurements. As for total size of the plant, morphological measurements are most powerful when considered over the whole growth period, rather than just at a single time point. There are numerous morphological parameters that can be extracted and quantified from images. Most obvious is probably height and width of the plant. Another frequently used measurement is compactness. This is defined as the ratio of the plant area to the convex hull, the area that entirely encloses the plant. Compactness can be a very useful measure to describe the morphology of fairly rigid plants such as Arabidopsis. However, it is prone to noise for grassy plants such as wheat or barley, where leaves are highly flexible and do not stay in the same position. A simple alternative in this case is to calculate the ratio of leaf area to plant height. This value will increase with increasing tiller number and can show clear differences between control and stressed plants. Another option would be to divide the side view images into several segments, e.g., horizontal sections in 10 cm intervals extending both above and below pot level. Quantifying the percentage of leaf area in those segments will allow a description of the leaf denseness at various heights.

4. Notes

1. We have used the described methods for numerous species including wheat, barley, maize, sorghum, tomato, and chickpea. Since the assay presented here is based on growth analysis, it is extremely important that the seeds and seedlings used are as uniform as possible. If sufficient seed is available, one should always plant excess amounts to be able to select for evenly sized seedlings. If it is known that the lines being used germinate at different rates, the sowing should be staggered to have evenly sized seedlings at the start of the experiment.
2. The sodium hypochlorite solution can be prepared using a household product such as Domestos®, when taking into account the lower active concentration of Cl^- compared to a lab grade solution.
3. The color of the pot should allow an easy distinction from the plants in the image-processing step, preferably white or blue. Black is possible, but it leads to an increased soil temperature. Standard green nursery pots should not be used.
4. The choice of potting mix will obviously depend on the experiment. Some might require controlled nutrient application and should therefore be free of fertilizer. If the pots are placed on an automated conveyor system, the substrate should not be too loose (such as pure sand) since it might shift through the movement on the belt and damage the root system. The clay content should not be too high since there is the potential for compaction on the conveyor belt and consequently root anoxia.
5. We generally take three images per plant (two from the side at 90° rotation and one from the top) at about 15–20 time points throughout an experiment. With a file size of about 4 MB, this amounts to 4 MB × 3 images × 20 time points = 240 MB per plant. Even a smaller scale experiment with 200 plants will therefore need 47 GB of storage.
6. Seed treatment might not be necessary, depending on the source of the seed. However, fungal infections of young seedlings can influence the growth rate and their sensitivity to certain stress treatments.
7. If no large retort stand is available, a smaller one can be placed on a table with the water reservoir on the ground.
8. The growth conditions can influence parameters such as leaf thickness and will consequently also influence the correlation between leaf area and biomass.
9. The number of images taken per plant will depend on the shoot morphology and the desired throughput. We found that

three images (two from the side and one from the top) are sufficient for most plants. Plants such as Arabidopsis generally require only a single image from the top.

10. If no leaf area meter is available, a simple flat-bed scanner can be used. However, this requires mounting the leaves on paper and extracting the leaf size from the acquired images. It is generally more labor intensive and not suitable for a large sample number.

References

1. Passioura JB (2006) The perils of pot experiments. Funct Plant Biol 33:1075–1079
2. Klute A (1986) Water retention: laboratory methods. In: Klute A (ed) Methods of soil analysis, Part I, 2nd edn. Agronomy Monograph 9, American Society of Agronomy and Soil Science Society of America, Madison
3. Scholander PF, Hemmingsen EA, Hammel HT et al (1964) Hydrostatic pressure and osmotic potential in leaves of mangroves and some other plants. Proc Natl Acad Sci USA. doi:10.1073/pnas.52.1.119-125
4. Granier C, Aguirrezabal L, Chenu K et al (2006) Phenopsis, an automated platform for reproducible phenotyping of plant responses to soil water deficit in *Arabidopsis thaliana* permitted the identification of an accession with low sensitivity to soil water deficit. New Phytol 169:623–635
5. Jansen M, Gilmer F, Biskup B et al (2009) Simultaneous phenotyping of leaf growth and chlorophyll fluorescence via growscreen fluoro allows detection of stress tolerance in *Arabidopsis thaliana* and other rosette plants. Funct Plant Biol 36:902–914
6. Hunter F, Biver S, Fuqua P (2007) Light-science & magic: an introduction to photographic lighting. Focal Press, Elsevier, Oxford, UK
7. Gonzalez RC, Woods RE (2006) Digital image processing, 3rd edn. Prentice-Hall, Upper Saddle River
8. Hunt R (1978) Plant growth analysis. Edward Arnold, London
9. Hunt R, Causton DR, Shipley B, Askew AP (2002) A modern tool for classical plant growth analysis. Ann Bot 90:485–488
10. Watson DJ (1947) Comparative physiological studies on the growth of field crops. Ann Bot 11:41–76

Chapter 3

High-Throughput Phenotyping of Root Growth Dynamics

Nima Yazdanbakhsh and Joachim Fisahn

Abstract

Plant organ phenotyping by noninvasive video imaging techniques provides a powerful tool to assess physiological traits, circadian and diurnal rhythms, and biomass production. In particular, growth of individual plant organs is known to exhibit a high plasticity and occurs as a result of the interaction between various endogenous and environmental processes. Thus, any investigation aiming to unravel mechanisms that determine plant or organ growth has to accurately control and document the environmental growth conditions. Here we describe challenges in establishing a recently developed plant root monitoring platform (PlaRoM) specially suited for noninvasive high-throughput plant growth analysis with highest emphasis on the detailed documentation of capture time, as well as light and temperature conditions. Furthermore, we discuss the experimental procedure for measuring root elongation kinetics and key points that must be considered in such measurements. PlaRoM consists of a robotized imaging platform enclosed in a custom designed phytochamber and a root extension profiling software application. This platform has been developed for multi-parallel recordings of root growth phenotypes of up to 50 individual seedlings over several days, with high spatial and temporal resolution. Two Petri dishes are mounted on a vertical sample stage in a custom designed phytochamber that provides exact temperature control. A computer-controlled positioning unit moves these Petri dishes in small increments and enables continuous screening of the surface under a binocular microscope. Detection of the root tip is achieved by applying thresholds on image pixel data and verifying the neighbourhood for each dark pixel. The growth parameters are visualized as position over time or growth rate over time graphs and averaged over consecutive days, light–dark periods and 24 h day periods. This setup enables the investigation of root extension profiles of different genotypes in various growth conditions (e.g., light protocol, temperature, growth media) and is especially suited for the detection of diurnal or circadian growth rhythms.

Key words: Phenotyping, Root growth dynamics, Video imaging, Screening robot, Computerized video image analysis, Diurnal rhythm, Petri dish culture

1. Introduction

High-throughput quantification of visible plant phenotypes is attracting considerable attention as a tool to characterize gene function, circadian and diurnal rhythms, geomagnetic variations,

Jennifer Normanly (ed.), *High-Throughput Phenotyping in Plants: Methods and Protocols*, Methods in Molecular Biology, vol. 918, DOI 10.1007/978-1-61779-995-2_3, © Springer Science+Business Media, LLC 2012

improvements of plant growth performance, and biomass production (1). Previously, several methods that provide characterization of a few plant growth parameters have been developed; however, they are limited by the amount of detectable replica. Initial simple approaches in the study of growth patterns in plants involved the use of rulers or digital calipers to determine the displacement of marked sectors. More recently, automated technologies using digital image processing have evolved to monitor organ development in a noninvasive manner (2–4). Most of these tools are appropriate to address very specific questions related to the complexity of whole-plant, organ, or segmental growth. Because of this specification, only a small proportion of the dynamic and architectural parameters have been extracted from the sampled images (5, 6).

Recent platforms for noninvasive analysis of root growth provide accurate and reliable results with high temporal or spatial resolution, but they are time-consuming, often targeted, have a low throughput, and do not control/capture the environmental conditions. The effect of environmental stimuli on biological processes, especially growth and development, is well established, but only scarcely included in the interpretation of plant growth experiments. Thus, it is difficult to reveal reliable patterns and detect small changes in root elongation from currently available data. We here describe construction details and operating protocols for a recently developed plant root monitoring platform (PlaRoM) that enables high-throughput monitoring of growing seedlings with high spatial and temporal resolution, detection of root elongation profiles, and documentation of light and temperature conditions. The unprecedented accuracy provided by the automated image processing software application makes it especially suited for investigation of dynamics of root elongation rate, the detection of circadian, ultradian, and diurnal rhythms in root elongation.

2. Materials

2.1. Growth Medium Containing Petri Dishes

1. Growth medium: 11 g Murashige and Skoog (MS), 2.5 g MES, 4.5L VE-water, 7 g select agar, pH 5.7 adjust with KOH (see Note 1).
2. 13 cm × 13 cm rectangular Petri dishes (see Fig. 1).

2.2. Seed Sterilisation and Germination

1. Sterilization solution: 10 % sodium hypochlorite, 0.1 % surfactant (Triton X 100).

Fig. 1. Petri dish with growing seedlings that has been used for detection of root growth kinetics using PlaRoM. Rectangular Petri dish filled with solid agar medium. Arabidopsis seedlings are placed in a row 3 cm from the top on the surface of the medium. After 1 day in a commercial phytochamber, plates will be moved to the vertical sample stage attached to the screening robot. Root monitoring will start after 2 days of acclimation.

2.3. Automatic Root Tip Detection: Hardware Components

High-throughput characterization of growth requires robotized equipment (7, 8) for imaging, and adequate software for capturing and processing these records. Since plant and organ growth is affected by both endogenous processes and environmental stimuli, obtaining robust, reproducible, and reliable growth data requires tight control, recording, and documentation of the environmental parameters that the organism is exposed to. To meet these requirements we developed a plant root monitoring platform (PlaRoM) that enables high-resolution monitoring of up to 50 seedlings growing in two rectangular Petri dishes. In the following sections we will describe this setup and explain the key functionalities of this platform, as well as documentation of growth conditions in detail (see Figs. 2 and 3).

1. A custom designed phytochamber houses the detection unit of the measuring head (see Fig. 2a). This phytochamber provides temperature control during the entire measurement by a digital thermometer and thermostat (232 DTT, B&B Electronics, Ottawa, IL, USA) that regulates a cooling

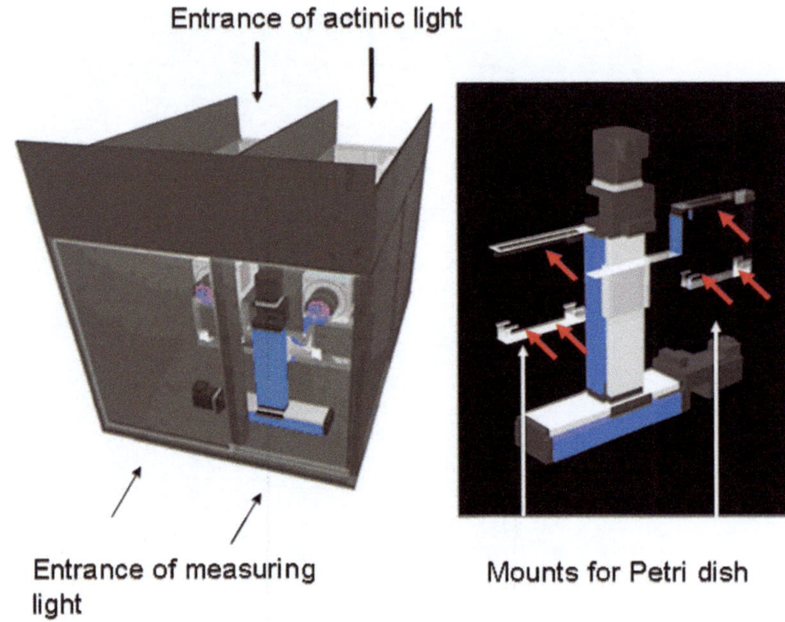

Fig. 2. Hardware components of the imaging platform. A custom designed phytochamber encloses the measuring head. The 110 cm × 72 cm × 58 cm Perspex shield was built with special emphasis on the accessibility to the microscope focusing dials and alignment of the Petri dishes with the optical axis. With the front and top sides consisting of transparent material, the imaging and actinic light sources can be placed outside the chamber to avoid any heat production inside the phytochamber. The robot arm consists of two perpendicularly arranged linear stages and holds the sample stage. The dual sample stage holds two Petri dishes on each side of the vertical axis in front of the microscope camera units. *Red arrows* indicate the position of levelling screws to provide correct alignment of the Petri dishes with the optical axis.

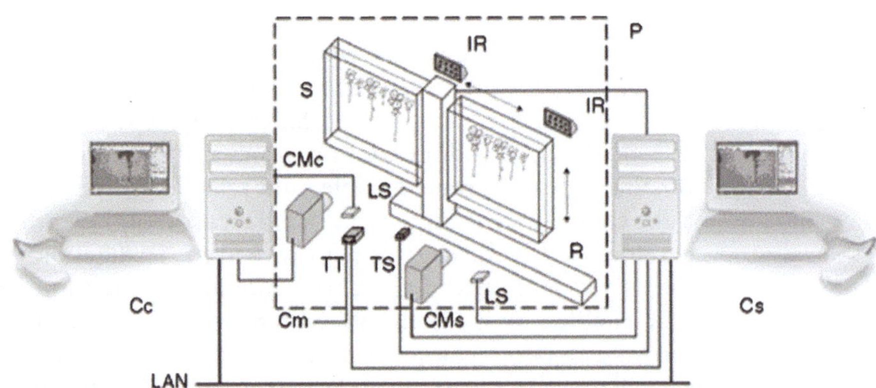

Fig. 3. Hardware components of the imaging platform. Schematic illustration of the electronic and optical components of the platform. A computer-controlled robot arm (R) mounts two Petri dishes in front of the two microscope camera units (CMs, CMc). Imaging light is provided by infrared light sources (IR). This measuring head is enclosed in the temperature-controlled phytochamber (*dashed line* labelled with P) shown in Fig. 2. Each camera-microscope is connected to an individual microcomputer (Cs; Cc). These two microcomputers communicate through the LAN (local area network) in a client/server mode. A light sensor (LS) is connected to each computer that monitors the actinic light at the level of the two Petri dishes inside the phytochamber. Accurate temperature control is provided by the server via a thermometer (*TS* temperature sensor) and a thermostat (*TT* thermometer and thermostat).

device (Waeco, ECO, Pilsen, Czech Republic). Temperature accuracy provided by this setup is better than 0.5 °C (see Note 2). The controlling software application designed for this platform sets the temperature according to the requirements of the experiment through the digital thermometer and the thermostat and records the temperature at the beginning of each screening round. Establishment of local temperature gradients within the phytochamber is avoided by the operation of three fans (Papst 3412 NLE, Conrad Electronics, Berlin, Germany) that permanently circulate the air inside the chamber. For more accurate documentation of the temperature within the measuring head, a digital thermometer (USB-Temp, TFA Dostmann GmbH, Wertheim, Germany) is used. USB-Temp, the software application provided by the manufacturer of this device, records the temperature every 5 s.

2. With the front and top layers of the phytochamber consisting of transparent Perspex, the imaging and photosynthetic light sources are placed outside the chamber to avoid any heat production inside. The actinic photon flux density at the surface of the leaves of the seedlings is typically 70–90 μmol/m^2 s. Actinic light is provided by ten fluorescent light tubes (5 Philips TL-D 18, 830 and 5 Philips TL-D 18, 840; Growland, Hamburg, Germany). A handset interval timer is used to control the photoperiod and thus the switching of the actinic lights. Two light sensors (Luxmeter, LX-1108, Voltcraft, Hirschau, Germany) are placed inside the phytochamber to continuously record the photosynthetic photon flux density at the level of the Petri dishes. Incident photon flux densities are directly transmitted to the computer through a serial port and stored in a document file throughout the entire measurement.

   Central to the multi-parallel recording of many roots is the screening robot (see Fig. 2b). This robot is located in the front part of the phytochamber (see Fig. 2a). The robot consists of two perpendicularly arranged linear stages (Pico LPT 60, Feinmess Dresden GmbH, Dresden, Germany) controlled by MDrive programmable stepper control units (Schneider Electric Motion, Marlborough, CT, USA). Two hundred steps per cycle of the stepper motor driving the linear stage lead to a 0.1 in. movement along the linear axis, which provides the robot arm a 1.27 μm spatial accuracy and high reproducibility. For accurate and reproducible positioning, both ends of each linear stage are equipped with Hall effect end switches (TLE4905G, Feinmess, Dresden GmbH, Dresden, Germany). Upon arrival of the moving stage at the terminal switch position, a signal is transmitted to the MDrive control unit. The corresponding position of the moving stage can then be defined as zero via a code residing in the MDrive controller. In particular, our present MDrive application consists of a machine code algorithm developed in the MDrive language that is executed at power on, moves the travelling stage to the stepper motor margin, and sets the stop position (defined by the Hall switch) as origin in the robot arm coordinates system. Furthermore, this code determines the maximum movement velocity, acceleration, and deceleration values. In detail, the acceleration of the individual linear stages is set two times higher than that of the deceleration to obtain a smooth movement and less vibrations of the sample stage when the robot stops its movement.

3. Stepwise movement of the robot arm, controlled by the imaging software application, changes the position of the sample stage in a plane. This automated motion enables screening of a user-defined area in the robot arm coordinate system over the surface of each Petri dish for several days.
4. The sample stage consists of two custom designed pockets that mount the Petri dishes (see Fig. 2b).
5. Each pocket contains three adjustment screws that provide the correct alignment of the Petri dishes with the optical axis and the focus of the microscope camera. To obtain the optimal focal orientation of each Petri dish, prior to each measurement the operator moves the robot arm to each corner of the Petri dish and adjusts these micro-positioning screws.
6. Live stream images of Petri dish segments are viewed by two horizontally placed binocular microscopes (Leica MZ6, Leica Microsystems GmbH, Wetzlar, Germany) each equipped with an additional camera port (see Fig. 2a). A CCD-camera (Panasonic Colour CCTV Camera, WV-CP210/G, Matsushita Communication Industrial Co. Ltd, Yokohama, Japan) is connected to each of these camera ports. Microscope and cameras are installed inside the phytochamber to establish an optical axis between measuring light, Petri dishes, and camera.
7. Measuring light is provided by two infrared light sources (Infra-Red Illuminator CE 7710, Jenn Huey Enterprise Co. Ltd, Taipeh, Taiwan) that are placed outside in front of the phytochamber (see Fig. 2a). Due to the supply of infrared measuring light the seedlings can be monitored by the microscope camera unit in absence of actinic light. To exclude the entry of actinic light into the camera, an infrared filter (Infrarot RG780 E25, Heliopan, Gräfelding, Germany) is installed between each camera and microscope unit.
8. The microscope camera is connected to the BNC connector (video in) of a Picolo PCI capture board (Picolo series, Euresys Inc., Itasca, IL, USA, see Fig. 3) installed in a microcomputer outside the phytochamber.

9. The entire setup enables continuous monitoring of a 4.58 mm × 3.33 mm view frame area on the surface of the Petri dish through a 768 × 576 pixel channel. As a result, time-lapse records with 5.96 × 5.76 μm/pixel resolution are obtained. Alternative image resolutions can be obtained by switching the objective lenses of the microscope.
10. To enable fast image acquisition of both Petri dishes that are screened by the robot arm, two microcomputers, each equipped with a Picolo frame grabber (see Fig. 3), are used which were communicating via a LAN connection.

2.4. Software Control of the Imaging Platform

1. Movement of the screening robot is under control of a custom designed software application that resides in the server computer (Cs, see Fig. 4).
2. The program sets the screening area limits in the robot coordinate system, moves the sample stage in front of the camera-microscope units, captures time-lapse records, reads the temperature, and stores the entire data set (see Note 3).
3. The motion control software application advances the position of the sample stage every 6 s. In particular, screening the user-defined area starts from left to right. Upon arrival at the right margin the robot arm moves the Petri dishes vertically down by a preset number of steps and screening continues from right to left. To improve the data quality of root tip positions that will be detected from images to sub-pixel level, three time-lapse records are retrieved with 800 ms intervals before each upcoming step of the robot arm.
4. At the start of each measurement, the operator provides a unique name for the actual measurement. This name includes the growth medium (e.g., MS002s for MS medium containing 0.02 % sucrose), genotype and number of seedlings (Col23, e.g., 23 seedlings of Arabidopsis Col-0), and sowing date (e.g., 110615 for 15 June 2011). The recording software application automatically creates a folder with this name together with the measurement date-time (hence, if the measurement starts on 1 July 2011 at 10:50:25, the folder name will be: MS002s_Col23_110615 01.07.2011 10:50:25) and populates it with subfolders named according to the *x*–*y* coordinates of the screening area on the Petri dish. Obtained image files are subsequently stored in corresponding folders labelled by a text string comprised of the defined nomenclature of the measurement (e.g., MS002s_Col23_110615 in the example above), coordinates of the viewing frame (e.g., 20–20 for the robot arm position (20, 20)), image index from that position (e.g., 12 for the images taken in the 12th screening round), image repeat index (a value between 1 and 3, denoting the images taken

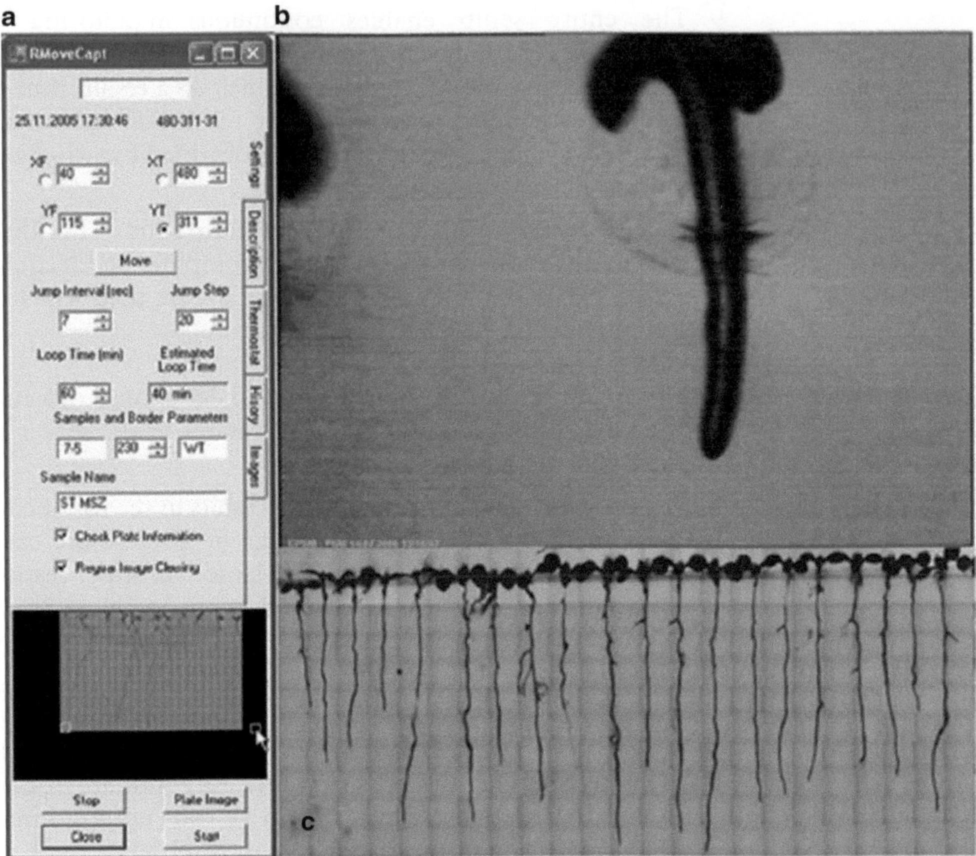

Fig. 4. Control of robot arm movement during image acquisition. (**a**) User interface of the PlaRoM software application subunit. The stepwise movement of the robot arm enables screening the [(XF, YF), (XT, YT)] defined area over the surface of the Petri dish. Once the radio button behind any of the *x* or *y* values is selected, a click on "move" advances the sample stage to the corresponding (*x* or *y*) position. Each movement is performed after an at least 5 s time interval (defined in "jump interval") and positions the sample stage 20,000 microsteps (in robot arm coordinate system, defined as the "jump step" value × 1,000) in horizontal and 70 % of this value in vertical direction. Depending on the time needed for screening of the defined area (presented in "estimated loop time"), the "loop time" value can be defined by the user. This value sets the time gap between consecutive screenings. In comparative studies several wild-type and mutant seedlings are placed on the same plate. Hence, time-lapse records taken from the area of the Petri dish containing wild type and mutants are labelled with the ecotype name. Names of the two genotypes and the robot arm coordinate corresponding to their border are provided by the user in "samples and border parameters." Each measurement is labelled by a name provided by the user in "sample name." The sowing date and the description of the growth medium are inserted by the user in the "description" tab page. The user can set the desired temperature and read the current temperature status inside the phytochamber in the "thermostat" tab page. An image composed of individually screened segments over the surface of the Petri dish is presented in the lower panel. A mouse click on any part of this composed image moves the robot arm to the corresponding position. (**b**) A typical time-lapse record captured by the imaging platform. Due to the magnification set by the binocular microscope, the 768 × 576 pixel image covers an area of 4.58 mm × 3.33 mm. (**c**) An example of the combined image produced by the imaging software application. This image presents the growth status of all seedlings in the screened area of the Petri dish and is stored after each screening round.

from the same position with 800 ms interval), and capture time. Thus, the record name "MS002s_Col23_110615_20–20_12_1 01.07.2011 18:57:23.png" documents the entire measurement information of the first image captured at 6:57:23 p.m. on 1 July 2011 from the view frame with position (20, 20) in robot arm coordinates, during the 12th screening round of a Petri dish which contains 23 Col-0 seedlings planted on 15 June 2011 on MS medium containing 0.02 % sucrose. As described above, this record will be stored in a folder named "20–20" under the measurement folder "MS002s_Col23_110615 01.07.2011 10:50:25" (see Note 4).

5. A resized copy of one of the records in each position is used to assemble an image of the whole plate, which is completed by the end of each screening round.
6. In addition, light and temperature data are recorded at the end of each completed screen.
7. Screening of two third of the plate surface for an hour produces 1,800 records that when stored in .jpg or .png format will occupy ca. 105 MB computer storage.

3. Methods

3.1. Growth Medium Containing Petri Dishes

Pour liquid medium in 13 cm × 13 cm rectangular Petri dishes (Fig. 1) until it forms a layer of ca. 8–10 mm thickness. Assure that the Petri dish will stay in a horizontal position until the medium solidifies. Noteworthy, for Petri dishes that contain the experimental seedlings, the thickness of the medium has to be uniform throughout the entire dish. This uniformity in nutrient layer thickness is highly important. Due to the high magnification used for imaging, the distance between root and camera should remain constant. All steps have to be performed under sterile conditions (see Note 5).

3.2. Seed Sterilization and Germination

1. Surface sterilize seeds for 20 min with sterilization solution.
2. Rinse seeds five times with sterile water.
3. Plate seeds on agar medium in Petri dish.
4. Keep the Petri dish containing seeds in a horizontal position for 4 days for stratification. Subsequently, place Petri dishes in vertical orientation in a phytotron (21 °C, constant day and night temperature, 100 $\mu mol^{-2}\ s^{-1}$ photon flux density).
5. Adjust photoperiod and temperature according to the requirements of the scheduled experiment.

6. After 9 days select seedlings that have developed roots of at least 1 cm length and transfer these seedlings to a new Petri dish filled with growth medium (Subheading 2.1).
7. Distribute the selected seedlings equally on the surface of the medium along a row that is 3 cm from the upper border (see Fig. 1).
8. Place these Petri dishes for another day in the phytotron in a vertical position before transferring them into the measuring unit.
9. Assure that the light and temperature conditions in the measuring unit are set to the requirements of the scheduled experiment.
10. Transfer Petri dishes that contain the experimental seedlings to the measuring unit at least two days prior to the measurement. Performing steps 8–10 diminishes the effect of a change in environment on root elongation behavior and provides reliable growth rates.

A general work flow of the entire method to obtain a high number of root growth kinetics is depicted in Fig. 5. Images of growing roots are captured by the screening platform described above. Image stacks of individual roots are selected by the operator, and root tip detection is automatically performed. Since the positions of the points identified as the root tip can be followed in time, growth rates are calculated from the displacement of the root tip at consecutive time points. Statistical analysis of these values and averaging them over several time periods provide the root extension profiles, which are plotted over time (see Fig. 5). Subsequently we will describe these steps in detail.

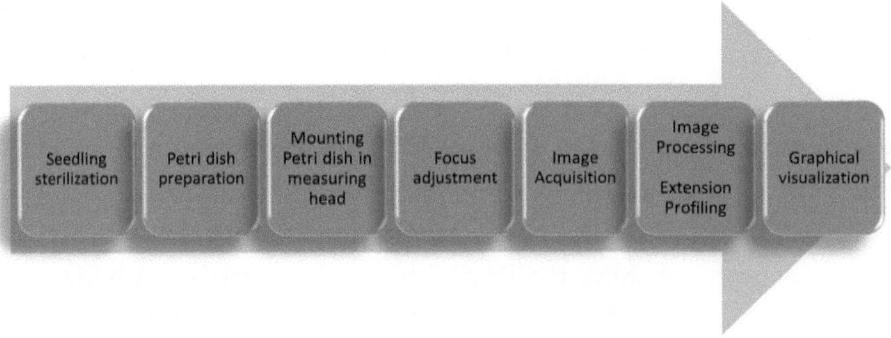

Fig. 5. Work flow of automated root growth profiling. The method uses digital images to produce plots of total root elongation rates versus time. Image acquisition provides high-resolution time-lapse records from growing seedlings. The image processing module applies a novel root tip detection algorithm to multiple images and reveals the coordinates of the root tip position. Growth rates are calculated from the displacement of the root tip at consecutive time points. Statistical analysis of these values and averaging them over several time periods provide the root extension profiles, which are plotted over time.

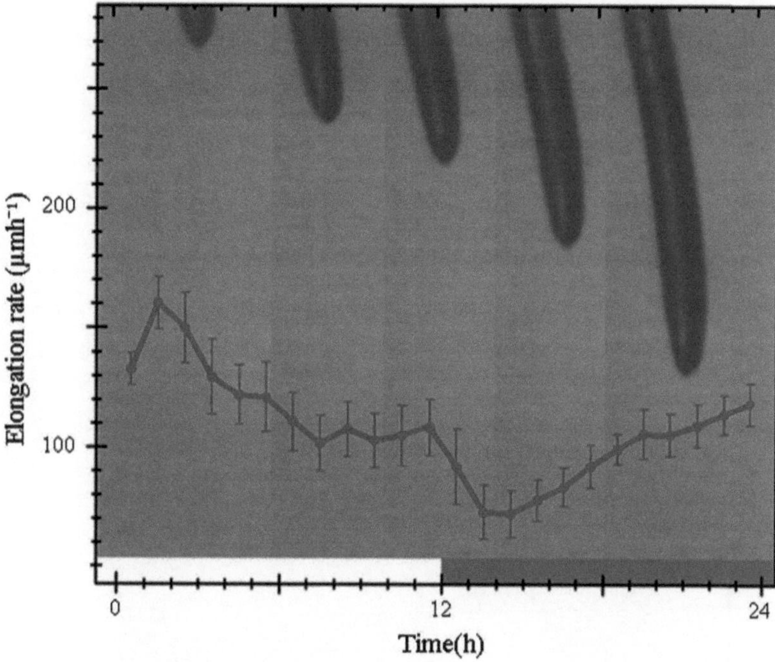

Fig. 6. Total root elongation is calculated from the displacement of the root tip. Tip displacement of a root growing on a surface describes a curve in a plane. This curve can be approximated by connecting a finite number of points using line segments. Dividing the length of consecutive line segments by the respective time interval provides the rate of root elongation in the corresponding period. *Trace*: a stack of images was used to calculate root elongation rates based on root tip displacement (*upper panel*). Measuring the displacement of the root tip in this image stack enables the root elongation rates to be calculated.

3.3. Automatic Root Tip Detection

1. The tip displacement of a root growing on the surface of a solid plate over time describes a curve in a plane (see Fig. 6). This curve can be approximated by connecting a finite number of points using line segments to create a polygonal path. It is straightforward to calculate the distance between two successive points by using the theorem of Pythagoras in Euclidean space. If the curve is not already a polygonal path, a better approximation to the curve can be obtained by decreasing the time interval between images collected during a root growth experiment.

2. Quantification of huge root image stacks requires automated root tip detection. Tip detection starts by applying thresholds on the red, green, and blue values of image pixel data (see Fig. 7a–d). As a result, each image pixel whose red, green, and blue values are below the applied thresholds will be replaced with a black point (for which red = green = blue = 0) (Fig. 7c). Subsequently, the image will be screened from right to left and bottom to top. The root tip is defined as a black pixel ($p(x, y)$)

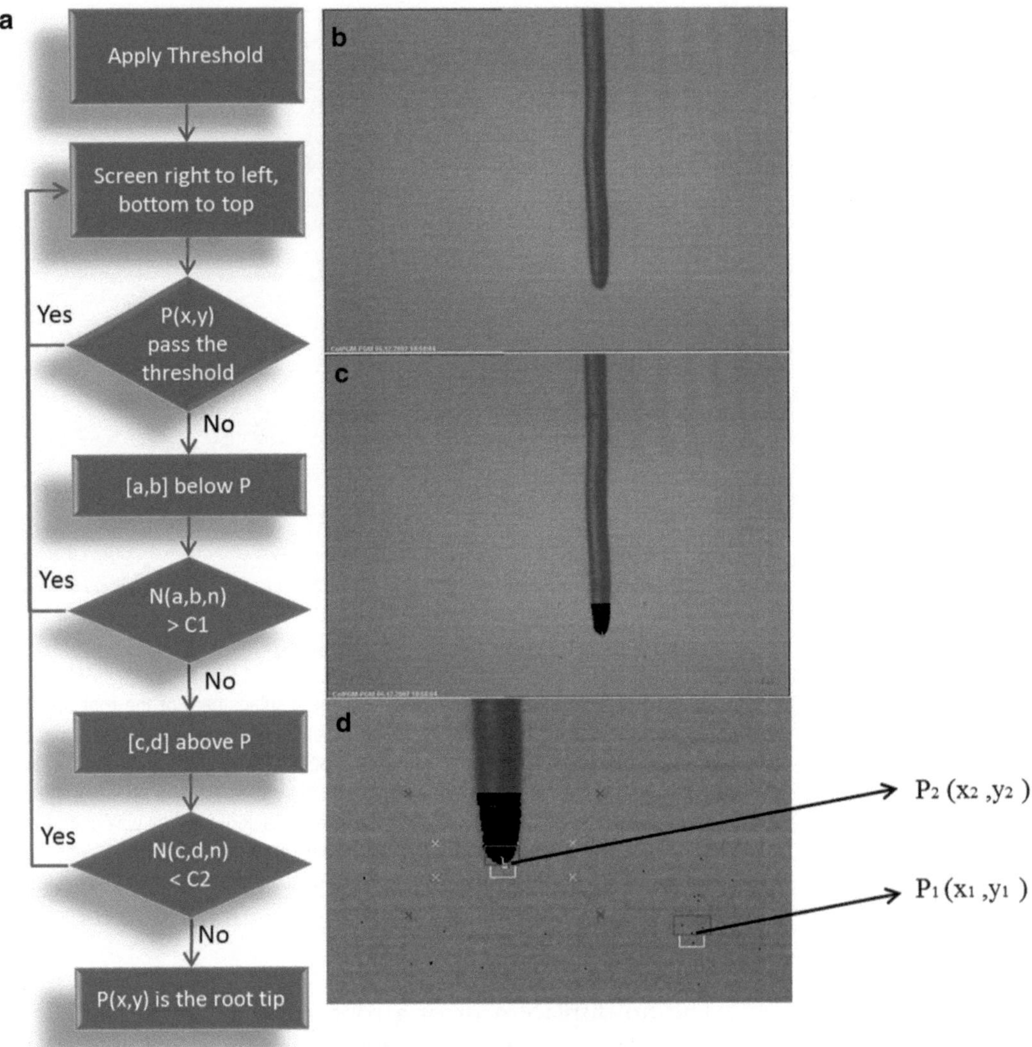

Fig. 7. Root tip detection algorithm. Detection of the root tip starts by applying thresholds on color values of the bitmap image pixels. Screening the whole image or the area of interest, below each pixel [*P*(*x*, *y*)] whose color value does not succeed the threshold, a rectangle (2*a* pixels in width and *b* pixels in height) is defined. If the number of dark pixels in this region exceeds a user-defined value (C1), screening the image will continue. Only if this condition is not satisfied, a second rectangle (2*c* in width and *d* in height) is defined above *P*. If the number of dark pixels in this region exceeds, a second user-defined threshold (C2), *P*(*x*, *y*) is reported as the root tip. (**a**) Flowchart of the root tip detection algorithm; (**b**) a typical time-lapse record before applying thresholds; (**c**) bitmap obtained after applying the threshold on pixel color values; (**d**) *P*(*x*, *y*) is defined as the root tip when the number of dark pixels in a rectangle below *P* (*lower rectangle*) is smaller than C1 and the number of dark pixels in the rectangle defined above *P* (*upper rectangle*) exceeds C2.

detected during this screen, that is characterized by very few black pixels in an area below it, but many black pixels in the area above it. To solve this identification procedure, the image processing algorithm defines one rectangle below and one above each candidate pixel. The first rectangle is defined below the candidate pixel ($p(x, y)$) and has a height of b and width of $2a$. Therefore, this rectangle is determined by its diagonal corners $(x-a, y)$ and $(x+a, y+b)$ (Fig. 7d, yellow rectangle). Successively, the number of black pixels in the lower area will be counted. Only if the sum falls below a certain threshold value (C1 in Fig. 7a), the second rectangle will be defined above the candidate pixel. This upper rectangle has a height of d and width of $2c$ and will be located in the image coordinate system denoted by its corner coordinates $(x-c, y)$ and $(x+c, y-d)$ (Fig. 7d, red rectangle). In the next step, the number of black pixels in the upper rectangle will be counted. The candidate pixel is defined as the root tip if the number of black pixels exceeds a second threshold (C2 in Fig. 7a). To give the detection algorithm further flexibility, neighboring pixels can also be considered in counting the black pixels within both rectangles. Consideration of neighboring pixels requires higher thresholds (C1 and C2 in Fig. 7a), thus provides more efficient root tip detection especially for low-quality images (out of focus or low contrast). Among successive records the root tip can only progress over a small distance from the previously detected position. In particular, the root tip can be expected not to move upwards or far left or right. Hence, to speed up the image processing procedure, after successful detection of the root tip, the software application performs the root tip detection algorithm only on a user-defined area of interest (AOI) around the previously detected root tip position.

3. The root tip detection algorithm successfully calculates coordinates of root tips when individuals are growing with sufficient spatial separation. However, when roots grow right behind each other, or if a dark object exists in the background of the image, this algorithm is susceptible to failure. To detect the root tip correctly in these conditions, image subtraction should be activated by the user. Image subtraction will highlight the differences between two images by removing the static parts. To perform image subtraction, the image processing software applies the threshold on pixel color values (of the whole image or only the AOI region) of the current record (with index n) and a previously recorded image from the same position (with record index of $n-i$). Subsequently, all black pixels detected in record $n-i$ will be removed from the current image. After image subtraction the root tip detection algorithm can calculate the tip coordinates of the desired root. Furthermore, applying image subtraction allows lateral root growth to be studied from stages of initial development.
4. The sequential image processing module applies the root tip detection algorithm to a stack of images (see Fig. 8). In detail, the user defines the root tip detection parameters and proposes

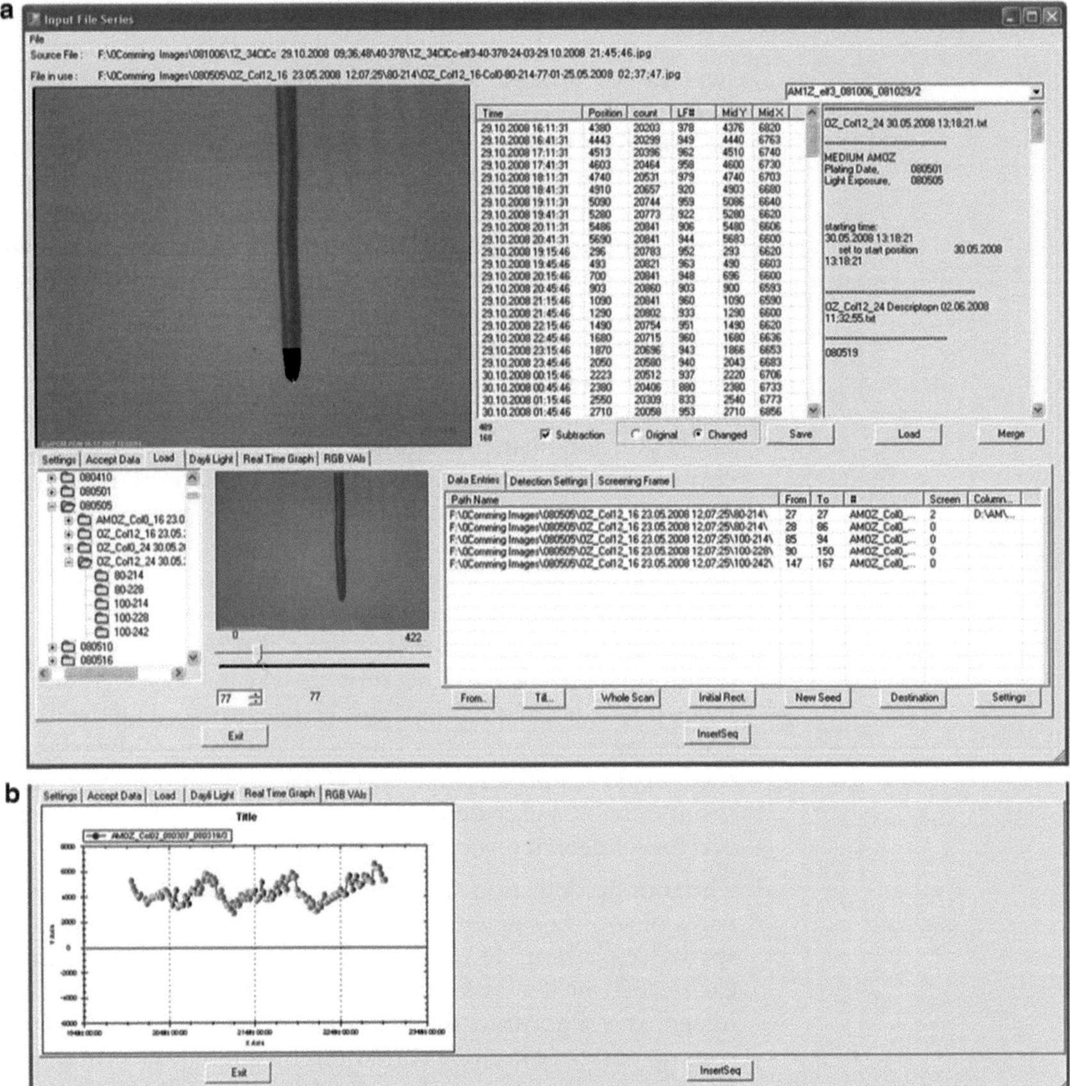

Fig. 8. The multiple image analyser applies the root tip detection algorithm to a stack of time-lapse records. (**a**) A stack of images is provided by the user, which includes the index numbers of time-lapse records as well as the desired root tip detection parameters. Upon image processing the resulting bitmap is presented and the detected root tip coordinates are listed. (**b**) To monitor the precision of selected root tip detection parameters, this application also calculates the root extension rates and plots them online during image processing on the display.

a list of the indices of images showing the root tip. The image processing module displays the detection parameters and image indices in three lists (Fig. 8a). For a measurement applied over 4 days with time-lapse records captured every 30 min, the image stack contains at least 576 records. Upon stack selection and program activation, the root tip coordinates of the detected root tip as well as the corresponding image capture time will be displayed in a list (see Fig. 8a). To monitor the precision of

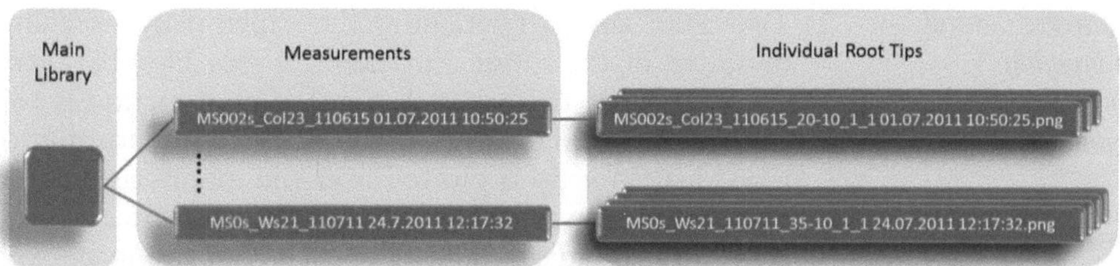

Fig. 9. Library structure used in this software application. Detected coordinates as well as times of dawn and dusk for each root tip are stored in the individual root tip .XML library file. For all individuals of the same genotype, filenames and storage addresses of these libraries are stored as a new entry in the genotype library. The same information regarding the genotype libraries is stored as a new entry in the main library of the software application. All entries are accessible through the main form of the software application.

selected root tip detection parameters, this application also calculates the root extension rates and plots them online during image processing on the display (see Fig. 8b). After analysis of all records of an individual root, the user defines a suitable nomenclature for saving specified data.

5. The root tip positions obtained are stored in XML data library structures that possess three hierarchical levels (see Fig. 9). On the lowest level the individual root tip libraries contain the detected root tip coordinates together with the record capturing time. These files are labelled by the name assigned to the respective individual (e.g., "MS002s_Col23_110615_110701_Col_2" for the second Col0 root detected by the measurement started on 1 July 2011 using the Petri dish explained in Subheading 2.4). Libraries of the higher hierarchical levels, namely the main library of the software application and the genotype libraries, store the filename and storage address of the genotypes and individual root tip libraries, respectively. Content of the main library is presented in the graphical user interface of this software application. Here the user can select a single entry or groups of individuals and apply colors and symbols to each set. Calculation of the obtained profiles enables their growth patterns to be compared.
6. Calculation and visualization of root elongation profiles enables comparison of extension patterns. The root extension profiling software application analyzes the time-lapse records and provides specified visualization of root extension profiles. A statistical package is included in the PlaRoM imaging software application that provides user-defined averaging and normalization of various combinations of data, together with the standard error. In addition to calculation of absolute elongation rates, this package normalizes each elongation rate value to the daily mean value of the same individual on the same solar day period.

3.4. Software Control of the Imaging Platform: Start-Up Procedure

1. Two Petri dishes containing the experimental seedlings are mounted in the robot arm cassettes (see Fig. 2b). Screening areas are set in the robot dialog, the photoperiod is adjusted, and the recording started. Subsequently, the position of each individual root tip is plotted over the monitoring period (see Fig. 10a). Growth rates per half hour time interval are calculated from the displacement of each root tip, in both x and y directions, at successive measuring points. These quantities are then averaged among individuals within each plate. Further data processing enables the display of additional growth response characteristics (see Fig. 10a–g). Figure 10c displays the average daily growth rate of individual seedlings growing in Petri dish a (red trace) and b (blue trace). It is noteworthy that root growth rates remained almost constant during the three-day measurement period. In particular, not only the absolute elongation rates, but also the kinetics of root elongation detected in the two separate Petri dishes exhibited highly reproducible characteristics. Figure 10d provides the average rates of growth during each light and dark period. Averaged extension rates detected in the dark were higher compared to the light period. To substantiate the detected time constants the averaged 24 h growth pattern of seedlings can be calculated (see Fig. 10e).
2. Individual seedlings exhibit slight differences in the mean absolute growth rate. All growth rate values in Fig. 10b–e show averaged absolute elongation rates of roots (see Notes 6–8). As indicated by the standard error, these averages are affected by the different mean elongation rates of the individual roots. Normalizing growth rates of each individual seedling to the median of each day removes this source of noise. These normalized growth rates averaged in time and in 24 h growth profiles are depicted in Fig. 10f, g, respectively. In parallel to the absolute elongation rates, the red and blue traces present identical patterns. Therefore, these graphs clearly demonstrate identical results obtained from simultaneous measurement of two Petri dishes, and the existence of reproducible diurnal rhythms in *Arabidopsis thaliana* root elongation rate.
3. From these calculations it was demonstrated that root elongation exhibited a diurnal rhythm with highest extension rates occurring 1.5–2 h after the beginning of the light period and the lowest elongation rates several hours prior to the onset of dusk.

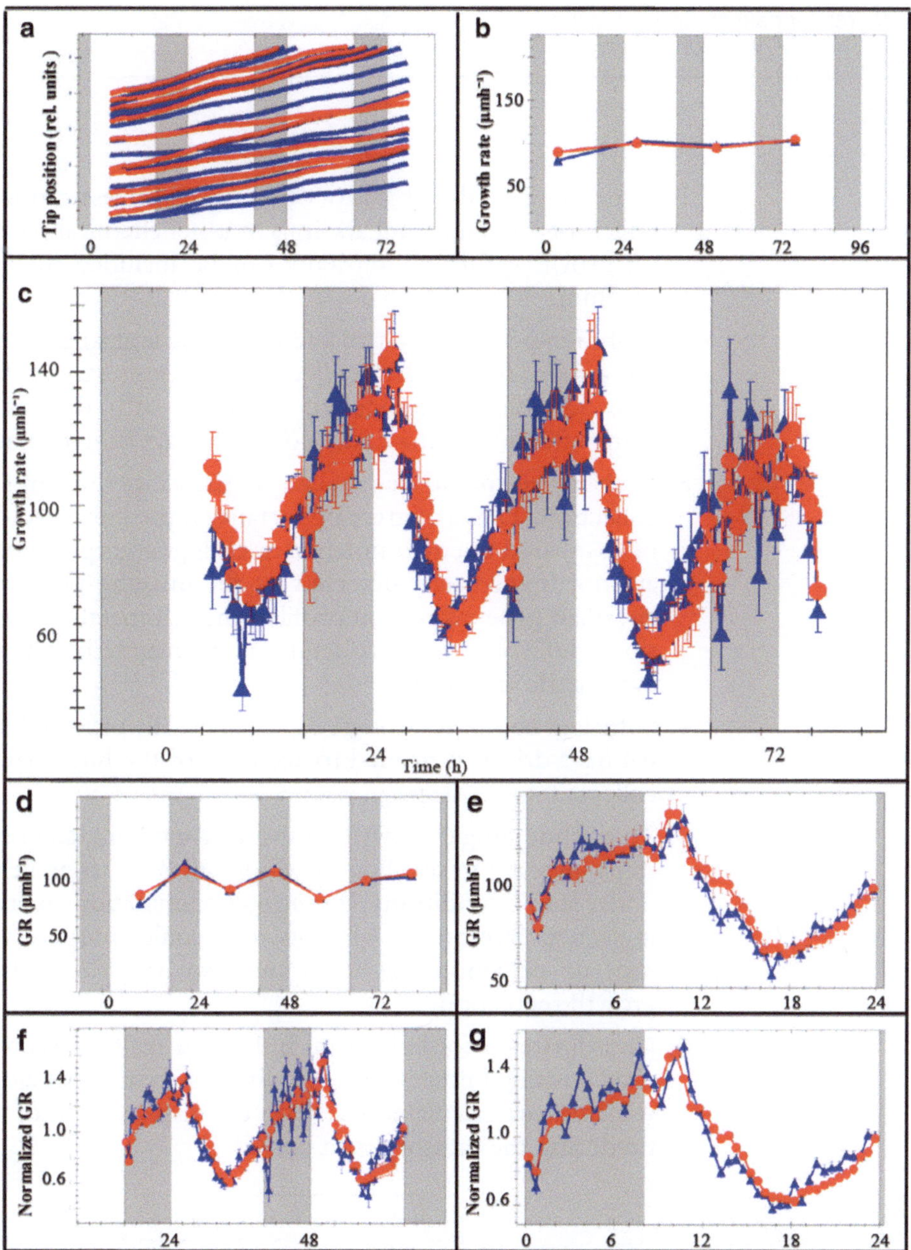

Fig. 10. Root growth profile of Col0 seedlings growing in 16 h photoperiod. 13-Day-old seedlings growing in 21 °C in two Petri dishes ($n=10$, *red dots*; $n=16$, *blue dots*) were monitored simultaneously for 72 h. (**a**) Positions of the root tips plotted over the time represent almost *straight lines*. (**b**) Averaged growth rate of seedlings growing on the same Petri dish grouped together during the measuring time period. (**c**) Average of the extension rate of all individuals of a group, each averaged over consecutive days. It should be noted that as the measurement had not been started or finished at midnight (solar time), the changed values for first and last days should not be considered. (**d**) Time course of average extension rate. Average of the extension rate of all individuals of a group, each averaged over consecutive light or dark periods, plotted over time. This curve reveals that the extension rate of the seedlings averaged in light period represent slightly lower values than the average at night. (**e**) Average of extension rates of all individuals of a group, each averaged over a 24 h period. This plot presents a 24 h rhythm of the extension rate. (**f**) Average of normalized extension rates of each individual of a group over time. This curve presents a similar pattern to (**b**) but due to the normalization algorithm, it starts from midnight and ends at midnight (solar time). (**g**) Average of normalized extension rates of all individuals of a group, each averaged over a 24 h period. Normalized curves present similar patterns as (**b**) and (**e**) but with smaller error bars. Values are the average extension rate of all ten seedlings and error bars represent the SE. *GR* growth rate.

4. Notes

1. Depending on the objective of the study (investigations of the effects of exogenous sugars, ions or other chemical compounds on growth), further nutrients can be included in this basic medium.
2. Tight temperature control and recording is absolutely essential for a reproducible data collection. Previous studies failed to provide sufficient temperature control with the result that many effects were masked by daily changes in temperature.
3. Each image should be labelled by the exact time and date of collection. This is of great importance if the measured data will be compared to environmental or endogenous parameters that are not immediately detected by the imaging platform, e.g., circadian gene expression profiles, diurnal metabolite modulations, lunisolar tidal acceleration, geomagnetic activity, and barometric variations.
4. Sufficient computer storage capacity, usually provided by external hard drives, is needed to account for the huge sizes of the collected image stacks.
5. To exclude fungal growth on the agar medium, all preparative steps involved in seedling development have to be performed under sterile conditions. If fungal contamination emerges during kinetic recordings of roots, this could interfere with the root tip detection algorithm and also with the endogenous growth regulation.
6. Coordinates recorded for an individual root tip represent its pixel position in the image coordinate system. Root elongation velocities are calculated based on the displacement of the root tip during the measurement period:

$$v_{(t)} = \frac{\sqrt{(x_{t+1} - x_t)^2 + (y_{t+1} - y_t)^2}}{t_{t+1} - t_t},$$

where $v_{(t)}$ is the root velocity at time t, and x and y are the root tip coordinates at times t and $t+1$. The elongation rate of each individual root tip will be calculated as long as the root tip is visible within the same viewing frame position. To obtain a continuous elongation plot, when a root is about to leave the viewing frame, at least for one time point the root tip should be visible in two records. For example, if the root tip is elongating downwards and leaving the viewing frame of the robot arm position (20, 20), the record captured on the same screening round from the viewing frame right beneath [position (20, 34) of the robot arm coordinating system assuming 14 units downwards movement] should display the root tip. To perform this, stepwise movement of the robot arm is set in a way that the top 10 % region of each record overlaps with the record captured in the viewing frame above. Similarly, two overlapping regions of 5 % are also defined for left and right margins of each viewing frame.

7. Considering different individuals of the same genotype being measured, extension rates can be averaged. Figure 10b depicts the averaged elongation rate ($\overline{v}_{(t)}$) of seedlings growing under a 16/8 h light–dark period:

$$\overline{v}_{(t)} = \frac{\sum_{s=1}^{n} v_{(t)}}{n}$$

where ($\overline{v}_{(t)}$) is the average extension rate at time t, $v_{(t)}$ is the elongation rate of an individual at time t, s is the seedling index and n is the total number of individuals being averaged together.

8. Elongation rates are further averaged over consecutive days ($\overline{v}_{(d)}$) or periods of light and dark ($\overline{v}_{(1)}$), leading to daily averaged (see Fig. 10c) or light–dark period averaged elongation rates (see Fig. 10d).

$$\overline{v}(d) = \frac{\sum_{s=1}^{n} \sum_{t=t_d}^{t_{d+1}-1} v_{(t)}}{n(t_{d+1} - t_d)}$$

where ($\overline{v}_{(d)}$) is the averaged elongation rate of selected individuals during day d, s is the seedling index, td is the first measured time point on day d, $v_{(t)}$ is the root extension of an individual at time t, and $n(td+1 - td)$ is the total number of time points recorded for all s individuals on day d.

Acknowledgments

Development of the described platform was supported by the Max Planck Society and by a contract to N.Y. We appreciate the permissions of Oxford University Press and CSIRO Publishing to reproduce some parts of our figures that were originally published in Annals of Botany and Functional Plant Biology.

References

1. Yazdanbakhsh N, Fisahn J (2010) Analysis of *Arabidopsis thaliana* root growth kinetics with high temporal and spatial resolution. Ann Bot 105:783–791
2. Yazdanbakhsh N, Fisahn J (2009) High-throughput phenotyping of root growth dynamics, lateral root formation, root architecture and root hair development enabled by PlaRoM. Funct Plant Biol 36:938–946
3. Yazdanbakhsh N, Sulpice R, Graf A, Stitt M, Fisahn J (2011) Circadian control of root elongation and C partitioning in *Arabidopsis thaliana*. Plant Cell Environ 34:877–894
4. Yazdanbakhsh N, Fisahn J (2011) Stable diurnal growth rhythms modulate root elongation of *Arabidopsis thaliana*. Plant Root 5:17–23
5. Yazdanbakhsh N, Fisahn J (2011) Mutations in leaf starch metabolism modulate the diurnal root growth profiles of *Arabidopsis thaliana*. Plant Signal Behav 6:1–4
6. Iijima M, Matsushita N (2011) A circadian and an ultradian rhythm are both evident in root growth of rice. J Plant Physiol. doi:10.1016/j.jplph.2011.06.005
7. Yazdanbakhsh N, Fisahn J (2007) Investigation of plant root elongation by screening the surface of a Petri dish. In: Arabnia HR (ed) Proceedings of the 2007 international conference on image processing, computer vision and pattern recognition. CSREA Press
8. Yazdanbakhsh N, Fisahn J (2007) Development of a robot based platform applied to simultaneous root growth profiling of seedlings growing in a Petri dish. In: Aggarwal A, Yager R, Sandberg IW (eds) Proceedings of the 8th WSEAS international conference on mathematics and computers in biology and chemistry (MCBC'07). World Scientific and Engineering Academy and Society Press

Chapter 4

LEAF GUI: Analyzing the Geometry of Veins and Areoles Using Image Segmentation Algorithms

Charles A. Price

Abstract

The Leaf Extraction and Analysis Framework Graphical User Interface (LEAF GUI) software is designed for biologists who wish to analyze the structure of vessel bundles (veins) in leaves. The software enables users to extract descriptive statistics on the dimensions and positions of leaf veins and areoles by utilizing a series of thresholding, cleaning, and segmentation algorithms applied to images of leaf veins. The resulting statistics for the dimensions of individual veins and the areoles they surround can then be used to evaluate numerous hypotheses regarding the structure and function of leaf veins.

Key words: Leaf vein, Vessel bundle, Vein density, Vein network, Network geometry, Areole, Xylem, Phloem, Image segmentation, Graphical User Interface

1. Introduction

The development of tools to facilitate the quantification of leaf vein geometry has implications for many areas of plant biology. For example, several authors have speculated that differential vein investment is an unrecognized source of variability underlying the leaf economics spectrum (1, 2). In addition, recent studies have shown that leaves are a major hydraulic bottleneck in plants (3), and thus measures of the geometry of veins will help motivate attempts to model patterns of conductance (4, 5). Network structure can influence photosynthesis via hydraulic efficiency, with recent work implicating vein density as a good predictor of photosynthetic rates (6, 7). Leaf vein patterning is also associated with whole leaf shape suggestive of shared developmental pathways (8). Unfortunately to date, quantifying the dimensions of leaf veins has relied primarily on slow, interactive "point-and-click" approaches.

Jennifer Normanly (ed.), *High-Throughput Phenotyping in Plants: Methods and Protocols*, Methods in Molecular Biology, vol. 918, DOI 10.1007/978-1-61779-995-2_4, © Springer Science+Business Media, LLC 2012

To further our ability to rapidly measure the dimensions of veins in leaves, I have developed (with coauthors) the Leaf Extraction and Analysis Framework Graphical User Interface (LEAF GUI), a platform independent software package, free for academic use. The software was developed in the Matlab programming environment, and contains a series of algorithms designed to threshold, clean, and segment leaf vein images in preparation for subsequent measurement. Extensive descriptions of the software and methodology can be found in the associated journal article (9), in the instruction manual, and in a series of tutorial videos all of which are available at www.leafgui.org.

The procedure that a user might take to process a leaf image and automatically measure structure within the leaf network can be separated into five major steps: (1) setting the scale of the leaf image, (2) image cropping, (3) thresholding, (4) binary image cleaning and processing, (5) network and areole feature extraction. The tools invoked in this series of steps can be seen in Fig. 1 and are described below.

There is significant variability in the methods used to obtain vein images, and in their quality, resolution, and content (see Note 1). To deal with this variability, the software is designed to give users substantial control throughout the entire process. In the

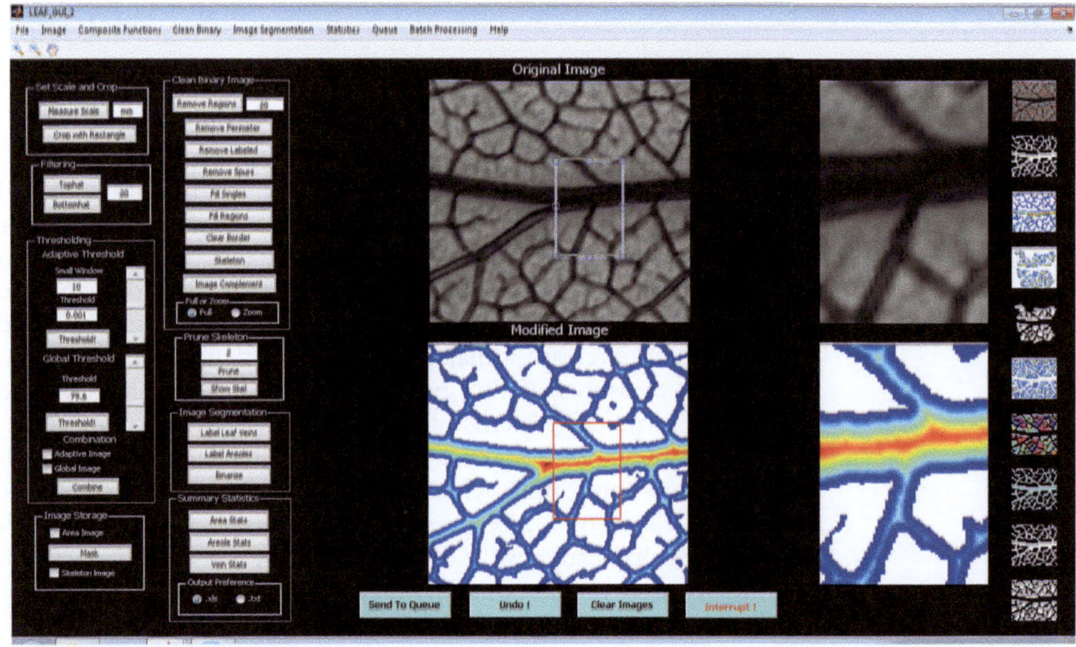

Fig. 1. A screenshot of the LEAF GUI software interface: a series of interactive panels on the left allows a user to process images. The original image (a zoomed in portion of a leaf in this example) is in the *Original Image* window with the results of varying image processing steps in the *Modified Image* window. A movable, resizable zoom window is visible in both images with the area of interest to the right. A queue is visible on the far right, which allows the user to store up to ten images as they are working.

following I present a few examples of the steps for using LEAF GUI. The first two examples are for entire leaf images that differ in type and quality. The third example covers batch processing, which is available at present for binary images only. Italicized words throughout this chapter refer to specific components of the LEAF GUI software (Fig. 1).

2. Materials

1. Leaf vein image: in order to resolve all orders of leaf veins, various clearing methods are usually employed (10, 11), and I mention them only briefly here. A common approach is to place leaves in a 5 % NaOH solution for days to weeks followed by staining with Safranin O. Other researchers have explored the use of X-ray images (12). We have provided sample leaf images for use in the following protocols that are downloadable from www.leafgui.org.
2. LEAF GUI software: available free online at www.leafgui.org. Executable files for Windows, Mac, and Linux.

3. Methods

3.1. Example 1 (Ficus religiosa)

The first example is a scan of a chemically cleared *Ficus religiosa* leaf (Image 1, downloadable from www.leafgui.org). The image is relatively high in quality requiring minimal cleaning subsequent to image thresholding and thus demonstrates the types of results one can obtain from images of this type.

3.2. Setting the Scale and Cropping the Image

1. Open the image either using File -> Open Image, or ctrl+O.
2. For this image, we will assume the scale is known in advance, so enter 11.78 in the Measure Scale box and type "Enter" on the keyboard. For situations in which the scale is not known, see Subheading 3.7.
3. As the image is of moderately large size, we will work from a cropped portion for the purposes of this example. Click on the *Crop with Rectangle* button under the *Set Scale and Crop* panel.
4. Select a small square region near the middle of the leaf including some of the midrib and secondary veins. I'd suggest a region approximately 5–10 % of the entire leaf image. When finished resizing, type "Enter" on the keyboard and the cropped region will now fill both the *Original Image* and *Modified Image* windows.

3.3. Thresholding and Masking

1. Set the value of the *Global Threshold* to about 18 using the slider and click *Threshold!*
2. To obtain an image in which the veins are white (1's) and the background is black (0's), click *Image Complement* under the *Clean Binary Image* panel.

3.4. Cleaning the Binary Image

1. There are small disconnected fragments throughout the image, some of which are veins that were not connected to the whole network, and some of which are simply image noise. Set the *Remove Regions* value to 10 and click *Remove Regions.*
2. Click *Fill Singles* once.
3. Click *Remove Spurs* once.

3.5. Image Segmentation

1. Click on the *Label Leaf Veins* button under the *Image Segmentation* panel. A color image appears in which all continuously connected white regions are assigned the same color. As can be seen, there are disconnected bits here and there, but the vein network is largely connected.
2. Click the *Undo* button to return the binary image to the *Modified Image* window.
3. Click the *Label Areoles* button. Here all contiguous black regions (areoles) are colored with different colors. The colors are chosen randomly, and due to the finite number of colors available, some adjacent regions have the same color giving them the appearance that they are connected when they are not.

3.6. Summary Statistics

1. Check the *.xls* radio button under the *Output Preference* subpanel.
2. Click the *Vein Stats* button. An Excel spreadsheet will open up with both summary statistics for the entire network, and measurements for each individual vein segment.
3. Note that in this subimage, some of the areoles overlap the border and are thus incomplete. To remove these border regions, click on *Image Complement*, then *Clear Border.* Click on *Image Complement* again to return to white veins on a black background.
4. Click the *Areole Stats* button. A popup box will open asking if the image is surrounded by a background, click *No.* An Excel spreadsheet will open up shortly with measurement statistics for all of the areoles (contiguous regions inside the leaf vein perimeters).

3.7. Example 2, Image 2 (Gualtheria oppositifoila)

For the second example, we will use a copy of an image obtained from the national cleared leaf collection maintained by the Natural History Museum at the Smithsonian. The image is of moderate to poor quality and thus demonstrates the types of results one can obtain from images of this type (Image 2, downloadable from www.leafgui.org).

3.8. Setting the Scale and Cropping the Image

1. Open the image either using File -> Open Image, or ctrl+O.
2. Set the scale by entering the length of the scale bar in mm (50 here), in the small box to the right of the *Measure Scale* button. Then click the *Measure Scale* button. As you mouse over the *Original Image*, crosshairs will appear. Click on one end of the scale bar, then drag the bar so that its other end lines up with the other end of the scale bar.
3. Double click the scale bar. The image scale (in pixels per mm) should now appear in the same box where you entered the length of the scale bar. The value here should be about 23 or so.
4. Click the *Crop with Rectangle* button.
5. Mouse over the *Original Image* and crosshairs will appear. Click and drag a rectangle that just encloses the leaf.
6. Hit Enter on the keyboard. The cropped image will now appear in both the *Original Image* and *Modified Image* windows.

3.9. Thresholding and Masking

1. Enter a value of 10 for the *Small Window* size (this is the default value).
2. Enter a value of 0.001 for the *Threshold* value (this is the default value).
3. Click *Threshold!*. A binary representation of the *Original Image* will now appear in the *Modified Image* panel.
4. Check the *Adaptive Image* checkbox. This stores a copy of the image for subsequent combining. A copy of the *Modified Image* should now appear in the *Queue.*
5. Set the value of the *Global Threshold* to about 90 either by using the slider.
6. Check the *Global Image* checkbox. As with the *Adaptive Image* checkbox, this stores an image for combining and a copy of the image will also now appear in the *Queue.*
7. Set the *Global Threshold* value at around 180 using the slider and click *Threshold!*
8. Within the *Clean Binary* Image panel, set the *Remove Regions* value to 1000 and click *Remove Regions.*
9. Under the *Image Storage* panel, check the *Area Image* box. This will store a copy of the *Modified Image* for masking and removing of perimeter. A copy of the mask image will also be sent to the *Queue.*
10. Click *Combine* under the *Thresholding* panel.

11. Click *Mask* under the *Image Storage* panel. There is now a binary representation of the vein image in the *Modified Image* window. Note that some of the veins fail to connect, particularly along the midrib. This is due in part to a small lighter band that occurs along the midrib within the original image. Higher quality images will help to remove this source of error.

3.10. Cleaning the Binary Image

1. Thresholding can cause the leaf margin to appear a bit thicker than what it actually is. To correct this, click *Remove Perimeter* once.
2. There are abundant small disconnected fragments throughout the image, some of which are veins that were not connected to the whole network, and some of which are simply image noise. Set the *Remove Regions* value to 10 and click *Remove Regions.*
3. Click *Fill Singles* once.
4. Click *Remove Spurs* once.
5. Send a copy of this image to the *Queue* for subsequent use.

3.11. Pruning the Skeleton

1. Use the default setting of 2 in the *Prune Skeleton* text box and click *Prune* (see Notes 2–5).
2. Click *Show Skel.*
3. Under the *Image Storage* panel, check the *Skeleton Image* check box. This stores a copy of the skeletonized and pruned image for subsequent use by the vein measurement algorithms. A copy of the image is also sent to the *Queue.*

3.12. Image Segmentation

1. Click on the *Label Leaf Veins* button under the *Image Segmentation* panel. A color image appears in which all continuously connected white regions are assigned the same color. As can be seen, there are disconnected bits here and there, but the vein network is largely connected.
2. Click the *Undo* button to return the binary image to the *Modified Image* window.
3. Click the *Label Areoles* button. Here all contiguous black regions (areoles) are colored with different colors. The colors are chosen randomly, and again, due to the finite number of colors available, some adjacent regions have the same color giving them the appearance that they are connected when they are not. Note that due to the poor resolution in the original image, many adjacent areoles are in fact connected here. An image of this quality would thus be unsuitable for analyzing areole dimensions.

3.13. Summary Statistics

1. Send the image in Queue position 2 back to the Modified Image window by clicking Image 2 under the Queue drop down menu.

2. Check the *.xls* radio button under the *Output Preference* subpanel.
3. Click the *Areole Stats* button. A popup box will open asking if the image is surrounded by a background, click *Yes*. An Excel spreadsheet will open up shortly with measurement statistics for all of the areoles (contiguous regions inside the leaf perimeter).
4. Click the *Vein Stats* button. An Excel spreadsheet will open up with both summary statistics for the entire network, and measurements for each individual vein segment (see Note 4).

3.14. Example 3, Batch Processing

Due to the high variability in the quality and type of initial images, at present only a limited *Batch Processing* option is available in the LEAF GUI. This approach requires that binary images of leaves have been created and no further processing is required save the removal of unwanted regions and/or pruning.

1. Click on the *Batch Processing* dropdown menu and click *Batch*.
2. Enter the path to the folder where the binary images are stored. The folder should be empty of other file types if possible.
3. Enter the known *Scale* in pixels per millimeter.
4. Enter the size of the contiguous regions to be removed in the *Remove Regions* window.
5. Enter the critical value for the *Prune* function.
6. Check the *Areole Stats* checkbox if you would like statistics on the areole dimensions. If the images have a background, i.e., entire leaves surrounded background, check the *Background Present* box.
7. Check the *Vein Stats* checkbox if you would like statistics on the vein dimensions.
8. Click the *Execute Batch Processing* button. Text files corresponding to the individual vein statistics, summary vein statistics and areole statistics will be iteratively created in the same folder where the images are stored.

4. Notes

1. LEAF GUI is designed for images where there is good contrast between veins and image background. If the background is too similar in intensity to the veins themselves, the different thresholding approaches will not be able to segment the images well, and there will be substantial noise in the resulting binary images.

2. If images are particularly high resolution, the skeletonization algorithms may create false *Edges*, particularly in the larger veins. There are several approaches within the LEAF GUI to reduce this source of error. The primary approach is to utilize the *Prune Skeleton* function. The Prune Skeleton panel allows users to remove spurious edges based on the distance from the tips to the nearest areole (*tip distance*), the distance from the base to the nearest areole (*base distance*), their length (*edge length*), and a *critical factor* that is set interactively by the user. By increasing the critical factor, the user removes more and more edges. Other approaches including using: *Dilate Image*, *Erode Image*, *Fill Regions*, *Remove Spurs*, *Diagonal Fill*, *Close Image*, *Bridge*, *Thin* and *Fill* and many of these functions on subsections of images with the *Full or Zoom* option; however, all of these methods should be used with caution.
3. As the results from the software are dependent on the quality of the images, a little extra effort beforehand to improve image quality will help significantly in improving the software's performance.
4. Some of the algorithms, in particular the statistics and pruning algorithms can take a substantial amount of time (minutes) to complete. The amount of time increases exponentially with the size of the leaf. Using LEAF GUI on a computer with a relatively slow processor is not a good idea, and may produce an "out of memory" error if working with large leaves. We suggest installing LEAF GUI on a faster machine if possible.
5. Almost all of the functions in LEAF GUI are interruptible, and if a new function is invoked too soon, the previous function may not have had time to finish. Thus, to avoid confusing results, patience is advised.

References

1. Niinemets U, Portsmuth A, Tena D, Tobias M, Matesanz S, Valladares F (2007) Do we underestimate the importance of leaf size in plant economics? Disproportional scaling of support costs within the spectrum of leaf physiognomy. Ann Bot 100:283–303
2. Niklas KJ, Cobb ED, Niinemets Ü, Reich PB, Sellin A, Shipley B, Wright IJ (2007) "Diminishing returns" in the scaling of functional leaf traits across and within species. Proc Natl Acad Sci USA 104:8891–8896
3. Sack L, Holbrook NM (2006) Leaf hydraulics. Annu Rev Plant Biol 57:361–381
4. Cochard H, Nardini A, Coll L (2004) Hydraulic architecture of leaf blades: where is the main resistance? Plant Cell Environ 27:1257–1267
5. Price CA, Wing SL, Weitz JS (2012) Scaling and structure of dicotyledonous leaf venation networks. Ecol Lett 15:87–95
6. Brodribb TJ, Feild TS, Jordan GJ (2007) Leaf maximum photosynthetic rate and venation are linked by hydraulics. Plant Physiol 144:1890–1898
7. Sack L, Frole K (2006) Leaf structural diversity is related to hydraulic capacity in tropical rain forest trees. Ecology 87:483–491

8. Dengler N, Kang J (2001) Vascular patterning and leaf shape. Curr Opin Plant Biol 4:50–56
9. Price CA, Symonova O, Mileyko Y, Hilley T, Weitz JS (2010) Leaf extraction and analysis framework graphical user interface: segmenting and analyzing the structure of leaf veins and areoles. Plant Physiol 155:236–245
10. Bates JC (1931) A method for clearing leaves. Am Nat 65:288
11. Klucking EP (1989) Leaf venation patterns: vol. 4 Melastomataceae. J. Cramer. Berlin
12. Wing SL (1992) High resolution leaf x-radiography in systematics and paleobotany. Am J Bot 79:1320–1324

Chapter 5

Remote Chlorophyll Fluorescence Measurements with the Laser-Induced Fluorescence Transient Approach

Roland Pieruschka, Denis Klimov, Joseph A. Berry, C. Barry Osmond, Uwe Rascher, and Zbigniew S. Kolber

Abstract

The interaction of plants with their environment is very dynamic. Studying the underlying processes is important for understanding and modeling plant response to changing environmental conditions. Photosynthesis varies largely between different plants and at different locations within a canopy of a single plant. Thus, continuous and spatially distributed monitoring is necessary to assess the dynamic response of photosynthesis to the environment. Limited scale of observation with portable instrumentation makes it difficult to examine large numbers of plants under different environmental conditions. We report here on the application of a recently developed technique, laser-induced fluorescence transient (LIFT), for continuous remote measurement of photosynthetic efficiency of selected leaves at a distance of up to 50 m. The ability to make continuous, automatic, and remote measurements of photosynthetic efficiency of leaves with the LIFT provides a new approach for studying the interaction of plants with the environment and may become an important tool in phenotyping photosynthetic properties in field applications.

Key words: Chlorophyll fluorescence, Photosynthesis, Remote sensing, Quantum yield

1. Introduction

Chlorophyll fluorescence analyses have become one of the most powerful techniques to nondestructively quantify photosynthetic efficiency and non-photochemical energy dissipation (1). The most commonly used technique to quantify photosynthetic efficiency from chlorophyll fluorescence yield is the pulse amplitude modulated (PAM) approach, which relies on the application of saturating light flashes (2, 3). Short duration of approximately 1 s and high intensity (>5,000 μmol photons/m^2 s) flashes are applied to transiently close all PSII reaction centers and induce a maximum

Jennifer Normanly (ed.), *High-Throughput Phenotyping in Plants: Methods and Protocols*, Methods in Molecular Biology, vol. 918, DOI 10.1007/978-1-61779-995-2_5,

fluorescence emission. Comparison of this maximum fluorescence level (F'_m) with the steady-state fluorescence (F) in light gives information on the effective quantum yield ($\Delta F F_m'^{-1}$, with $\Delta F = F'_m - F$) and, in the absence of light, on the potential quantum yield ($F_v F_m^{-1}$, with $F_v = F_m - F$). Additionally relating the F_m of dark-adapted leaves to the F'_m in light allows the estimation of non-photochemical quenching mechanisms ($NPQ = (F_m - F'_m) / F'_m$, which is related to the energy conversion into heat (4). These measurements have been widely used to study photosynthesis and to detect and quantify different stress levels that are related to and affect the photosynthetic apparatus (1). Recently, the saturation pulse approach has been included in phenotyping platforms to screen a large number of genotypes for the photosynthetic performance under various environmental conditions (5, 6). However, since the quantitative application of saturating pulses is only practical in close proximity to the photosynthetically active tissue, this approach can hardly be used to measure photosynthetic performance beyond the level of a single leaf or a small plant (4). Thus, the photosynthetic screening experiments have been mostly limited to small rosettes of *Arabidopsis thaliana* (5, 6). Additionally, the use of saturating light pulses may lead to photoinactivation of the photosystem II when applied over an extended time period, for example, to derive diurnal courses (7).

The laser-induced fluorescence transient (LIFT) approach uses low-intensity pulses instead of the brute force of a saturating pulse and is based on the principle of fast repetition rate fluorometry (FRRF) (8). The approach employs low-intensity laser light to manipulate the level of photosynthetic activity and to measure the corresponding fluorescence transient, which is interpolated to a maximum fluorescence level using a fluorescence model (9). The maximum fluorescence level is then related to the minimum fluorescence in the same way as the saturating pulse method (9–11). The LIFT approach correlates well with the saturating pulse method and has been used to monitor photosynthetic efficiency of plants in a controlled environment at Biosphere II Laboratories (9, 10), to scan tree canopies (12), and to detect cold stress impact on photosynthesis (11). The LIFT system can be used for measurement of photosynthetic efficiency from a distance between 5 and 50 m. Currently, a portable benchtop device has been developed and tested for measurements ranging from few meters to centimeters (Kolber and Osmond, unpublished results). The FRR fluorometry with a LIFT or a benchtop device may allow a very flexible way for rapid and quantitative measurement of chlorophyll fluorescence over a range of scales in the lab, greenhouse, and under field conditions. Here we focus on the presentation of the LIFT apparatus.

2. Materials

The LIFT fluorometer has been developed for remote measurement of photosynthetic status and biochemical parameters of terrestrial vegetation from a distance of 5–50 m. A collimated laser beam is used as an excitation source and the emitted fluorescence signal is collected by a telescope and detected by a photodiode.

2.1. Laser Source and Optical Collimator

The main laser light source with collimated output used in the LIFT fluorometer is based on semiconductor laser diodes, which are widely used in industry and science in areas ranging from speed detection to bar code scanners to diagnostics and remote sensing. For safety reasons the laser light system of the LIFT is divided into two subsystems: (1) the embedded laser diode (Boston Laser 1F5257D, Binghamton, NY) with fiber output, (2) the optical collimator mounted on the top of the telescope to send the beam to the target (Fig. 1). The laser has an excitation wavelength of 660 nm and operates in a pulsed mode with up to 2,000 flashes with individually controlled length and time intervals. The collimated laser beam produces a beam of approximately 9 cm in diameter with peak optical power of 125 W/m^2 (approximately 680 μM

Fig. 1. Schematic diagram of the LIFT instrument. The Cassegrian telescope with motorized pan and tilt allows remote targeting of the instrument from a terminal connected to the Internet.

photons/m^2 s). The beam divergence of the collimator is 1 mrad and the beam reaches a diameter of 12 cm at a distance of 50 m, which approximately doubles the beam area at the target. It is estimated that measurements beyond 50 m do not provide sufficient excitation power for reliable excitation of a fluorescence transient and are not recommended. The focal length of the optical system of the telescope is 5 m and determines the minimum distance to the target.

2.2. Chlorophyll Fluorescence Detection

The emitted fluorescence signal is collected by a 250 mm Cassegrian telescope (Meade 200 LX GPS, Meade Corporation, Irvine, CA) filtered with a 690 nm, 10 nm bandwidth filter and detected by a cooled large area avalanche photodiode (LAPD 639-70-72-631 Advanced Photonix, Inc., Camarillo, CA). The detected signal is amplified by a variable, 0–96 dB gain preamplifier (AD 675, Analog Devices, Norwood, MA) and digitized at 4 MHz.

2.3. Auxiliary Equipment

The system can be run with a standard computer. A video-camera (W500FL, Sony Corporation, NY) is mounted coaxially on the body of the telescope (Fig. 1), and it is accessible via a remote video server (Axis 2400, Axis Communications Corporation, Chelmsford, MA). The telescope has a remotely controlled pan (360°) and tilt (−45 to +60°) and can target any leaves within its field of view. For field experiments the equipment is enclosed in a weather proof and temperature-controlled enclosure.

3. Methods

3.1. Calibration of the LIFT

Chlorophyll fluorescence measurements with the LIFT were performed on leaves enclosed in a commercial gas exchange system (such as MPH 1000, Campbell Scientific Inc., Logan, UT, USA) under non-photorespiratory conditions (2% O_2), constant [CO_2] of at 400 μmol CO_2/mol, leaf temperature of 25°C, and photosynthetic photon flux density (PPFD) ranging between 50 and 1,200 μmol/m^2 s. LIFT-based electron transport rates (ETR_{LIFT}) are calculated as $ETR_{LIFT} = \Delta F F'^{-1}_{mLIFT} \times PPFD \times 0.85 \times 0.5$, with 0.85 as an estimate of absorbed light and 0.5 accounts for the partitioning of light between photosystems I and II (2). Gas exchange-based electron transport rates (ETR_A) are taken as a reference and the calculation has been previously described (13, 14). The relation of ETR_A vs. ETR_{LIFT} is empirically corrected by multiplying $\Delta F F'^{-1}_{mLIFT}$ with a factor ϕ to obtain an ETR_A vs. ETR_{LIFT} slope of 1. The correction factor ϕ was found to be constant for several species with $\phi = 1.628$ (11) (see Note 1).

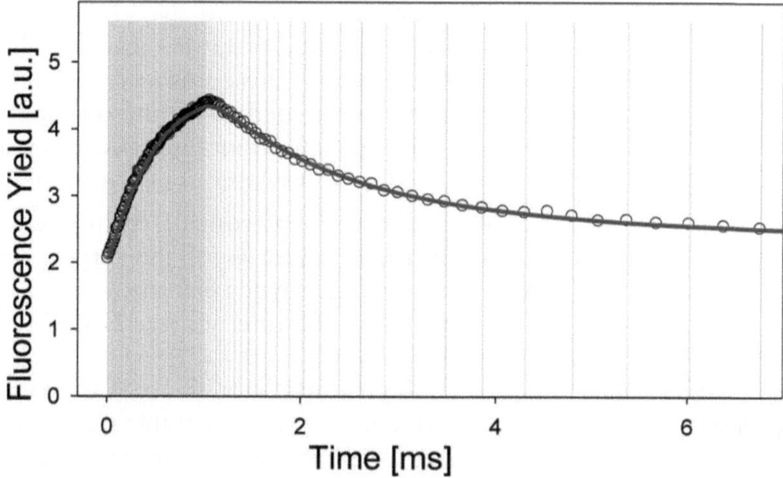

Fig. 2. Principle of the LIFT method. The excitation power is modulated by changing the frequency of the excitation flashes (*gray bars*). The fluorescence signal (*open symbols*) transiently increases when the rate of stable charge separation exceeds the capacity of the photosynthetic electron transport (high frequency flashes) and decreases when the rate of stable charge separation is lower than the rate of photosynthetic electron transport. The *solid line* is the numerical fit of the fluorescence transient.

3.2. Experimental Procedure

1. After starting the LIFT program, a GUI written in C++, an excitation protocol can be loaded and a graphic representation of the excitation protocol is displayed on the screen. The protocol can be modified if necessary by determining the duration of the laser pulse and the intervals in between for both the high and low duty cycles. The modified file can be stored and loaded accordingly. For field conditions with high background illumination empirical tests resulted in a pulse train of 100 flashlets with 5 μs duration and intervals of 5 μs during the high duty cycle; 40 flashlets with 5 μs duration and exponentially increasing intervals during the low duty cycle (Fig. 2). In order to improve the signal-to-noise ratio, 30–50 pulse trains are averaged before calculating the maximum (F'_{mLIFT}) and steady-state fluorescence (F_{LIFT}) by numerically fitting the fluorescence transient to a theoretical model.
2. The system can be operated in a manual or automated mode. In the manual mode the telescope can be set to aim at any target within the field of view and activated manually. The fluorescence signal is displayed on the monitor and the gain can be adjusted to avoid oversaturation of the detector and enable fluorescence acquisition within the dynamic range of the detector.
3. Before starting the automated mode the coordinate system of the telescope is used to determine the position of the targeted leaves in the canopy. Currently 12 coordinates can be defined.

They are stored, and the LIFT system recurrently returns to the predefined positions to assess fluorescence. The acquired fluorescence transient is recorded and displayed on the screen, in the automated mode the transient is fitted automatically to derive the important fluorescence parameters, these parameters can also be displayed for each of the targets.

4. The instrument control, including selection of the excitation protocol, selection of the target, data acquisition, and data processing, can be performed remotely, from any terminal connected to the Internet (see Note 2).

3.3. Model and Data Analysis

1. At high duty cycle the rate of stable charge separation, the reduction of the first stable electron acceptor Q_A, exceeds the rate of photosynthetic electron transport, and Q_A becomes progressively reduced causing a transient increase of the fluorescence yield (Fig. 2). At low duty cycle the rate of Q_A reduction is lower than the photosynthetic electron transport and the fluorescence decreases (Fig. 2). The dynamics of the Q_A redox processes are determined by the rate of charge separation (R_{CS}) and the rate of photosynthetic electron transport (R_{ETR}):

$$\frac{\partial[Q_A^-]}{\partial t} = R_{CS} - R_{ETR} \quad (1)$$

2. The R_{CS} depends on the excitation power ($i(t)$), the ambient irradiance, (E), the concentration of PSII reaction centers, (n), the functional absorption cross section of PSII, σ_{PSII}, and the reduction level of $[Q_A^-]$:

$$R_{CS} = (i(t) + E)\sigma_{PSII}(n - [Q_A^-]). \quad (2)$$

The R_{ETR} is controlled by the kinetics of the photosynthetic electron transport (κ_{ETR}), which depends on the time constant of the Q_A^- reoxidation (τ_{Q_A}), which is usually expressed as a sum of two exponential decays:

$$R_{ETR} = [Q_A^-]\kappa_{ETR} = [Q_A^-]\frac{1}{\tau_{Q_A}}. \quad (3)$$

The product of $(i(t) + E)\sigma_{PSII}$ presents the rate of stable charge separation in open reaction centers and transient changes in $[Q_A^-]$ level can be described as:

$$\frac{\partial[Q_A^-]}{\partial t} = (i(t) + E)\sigma_{PSII}(n - [Q_A^-]) - [Q_A^-]\frac{1}{\tau_{Q_A}}. \quad (4)$$

Dividing the Eq. 4 by the concentration of reaction centers results in:

$$\frac{\partial[Q_A^-]}{\partial tn} = (i(t) + E)\sigma_{PSII}\left(1 - \frac{[Q_A^-]}{n}\right) - \frac{[Q_A^-]}{n}\frac{1}{\tau_{Q_A}}. \quad (5)$$

By expressing $\frac{[Q_A^-]}{n} = C$ the level of PSII reduction in Eq. 5 can be rewritten as:

$$\frac{\partial}{\partial t}C = (i(t) + E)\sigma_{PSII}(1 - C) - C\frac{1}{\tau_{Q_A}}. \quad (6)$$

In the absence of the LIFT excitation signal ($i(t) = 0$), C operates at a steady state under the given illumination (C_E) and C_E is defined by:

$$C_E = \frac{E\sigma_{PSII}}{E\sigma_{PSII} + \frac{1}{\tau_{Q_A}}}. \quad (7)$$

The additional change $[Q_A^-]$ induced by the LIFT excitation $i(t)$ can be described as:

$$\frac{\partial}{\partial t}C_i = [i(t)\sigma_{PSII}(1 - C_E)(1 - C_i) - C_i\frac{1}{(1 - C_E)\tau_{Q_A}}. \quad (8)$$

This allows the LIFT-induced fluorescence to be expressed as:

$$f(t) = F + (F_m' - F)C_i \quad (9)$$

with F_m' as the fluorescence level at fully reduced $[Q_A^-]$ and F as the fluorescence level at the background illumination.

3. By numerically integrating Eq. 8 for the duration of the excitation protocol, substituting C_i to Eq. 9 and, fitting that equation into the measured fluorescence transient F, F_m' (F_m in darkness), τ_{Q_A} and σ_{PSII} can be calculated. For more details on the model and fitting procedure, see (8–10).

4. Notes

1. The current LIFT apparatus has been developed to prove the concept of active remote monitoring of chlorophyll fluorescence of plants under field conditions. The apparatus

was successfully tested in the Biosphere II laboratories (see refs. 9, 10), in the lab (see ref. 11) and under field conditions to monitor photosynthetic performance of arrange of different species (see refs. 11, 12, 15). The LIFT has demonstrated that it can reliably detect photosynthetic dynamics in a dynamic environment and assess different stress levels. However, the system is still work in progress and the next generation instrument has to be modified with measurement of the incident light intensity on the target leaves.

2. In general the approach to actively measure chlorophyll fluorescence with the FRRF protocol has the potential to be used over a range of scales, which in contrast to the saturating pulse approach may provide large flexibility in quantitative measurement of fluorescence. Depending on the optical system the chlorophyll fluorescence can be measured from distances of up to 50 m such as in the described LIFT system. Recently, a benchtop device has been tested for measurements in a range of a meter to few centimeters providing a way for detailed mechanistic studies of PSII dynamics or fast screening systems (Kolber and Osmond, unpublished data). Potentially the detector of such a system can be substituted by a high-speed camera and provide fluorescence images in future applications. Additionally the combination of passive measurements of reflectance spectra of plants or canopies with FRRF measurements may provide a very powerful tool for a very detailed assessment of the interaction of plants with the environment.

Acknowledgments

We are very grateful to numerous colleagues at the Biosphere II Laboratories, Carnegie Institution for Science and, Forschungszentrum Jülich for the extensive support in developing and testing the LIFT apparatus. RP was supported by Marie Curie Outgoing International Fellowships (Nr: 041060-LIFT).

References

1. Papageorgiou GC, Govindjee (2005) Chlorophyll a fluorescence: a signature of photosynthesis. Springer, Heidelberg
2. Genty B, Briantais JM, Baker NR (1989) The relationship between the quantum yield of photosynthetic electron transport and quenching of chlorophyll fluorescence. Biochim Biophys Acta 990:87–92
3. Schreiber U, Bilger W (1993) Progress in chlorophyll fluorescence research: major developments during the past years in retrospect. Prog Bot 54:151–173
4. Maxwell K, Johnson GN (2000) Chlorophyll fluorescence—a practical guide. J Exp Bot 51:659–668
5. Badger MR, Fallahi H, Kaines S et al (2009) Chlorophyll fluorescence screening of *Arabidopsis thaliana* for CO(2) sensitive photorespiration and photoinhibition mutants. Funct Plant Biol 36:867–873

6. Jansen M, Gilmer F, Biskup B et al (2009) Simultaneous phenotyping of leaf growth and chlorophyll fluorescence via GROWSCREEN FLUORO allows detection of stress tolerance in *Arabidopsis thaliana* and other rosette plants. Funct Plant Biol 36:902–914
7. Apostol S, Briantais JM, Moise N et al (2001) Photoinactivation of the photosynthetic electron transport chain by accumulation of over-saturating light pulses given to dark adapted pea leaves. Photosynth Res 67:215–227
8. Kolber Z, Prasil O, Falkowski PG (1998) Measurements of variable chlorophyll fluorescence using fast repetition rate techniques: defining methodology and experimental protocols. Biochim Biophys Acta 1367:88–106
9. Kolber Z, Klimov D, Ananyev G et al (2005) Measuring photosynthetic parameters at a distance: laser induced fluorescence transient (LIFT) method for remote measurement of photosynthesis in terrestrial vegetation. Photosynth Res 84:121–129
10. Ananyev G, Kolber Z, Klimov D et al (2005) Remote sensing of heterogeneity in photosynthetic efficiency, electron transport and dissipation of excess light in Populus deltoides stands under ambient and elevated CO_2 concentrations, and in a tropical forest canopy, using a new laser-induced fluorescence transient device. Global Change Biol 11:1195–1206
11. Pieruschka R, Klimov D, Kolber Z et al (2010) Continuous measurements of the effects of cold stress on photochemical efficiency using laser induced fluorescence transient (LIFT) approach. Funct Plant Biol 37:395–402
12. Rascher U, Pieruschka R (2008) Spatio-temporal variations of photosynthesis: the potential of optical remote sensing to better understand and scale light use efficiency and stresses of plant ecosystems. Prec Agric 9: 355–366
13. Laisk A, Loreto F (1996) Determining photosynthetic parameters from leaf CO_2 exchange and chlorophyll fluorescence. Plant Physiol 110:903–912
14. Peterson RB, Havir EA (2004) The multiphasic nature of nonphotochemical quenching: implications for assessment of photosynthetic electron transport based on chlorophyll fluorescence. Photosynth Res 82:95–107
15. Pieruschka R, Rascher U, Klimov D et al (2009) Optical remote sensing and laser induced fluorescence transients (LIFT) to quantify the spatio-temporal functionality of plant canopies. Nova Acta Leopoldina 96: 49–62

Chapter 6

Leaf Hue Measurements: A High-Throughput Screening of Chlorophyll Content

László Sass, Petra Majer, and Éva Hideg

Abstract

Computer analysis of digital photographic images provides fast, high-throughput screening of leaf pigmentation. Pixel-by-pixel conversion of red, green, blue (RGB) parameters to hue, saturation, value (HSV) showed that Hue values were proportional to total chlorophyll, offering an alternative to photometric analysis of leaf extracts. This is demonstrated using tobacco leaves with various chlorophyll contents due to senescence but shows the possibility of applications in studies of stress conditions accompanied by chlorophyll loss.

Key words: Chlorophyll fluorescence, High-throughput screening, Leaf Hue, Photochemical yield, RGB color model, HSV color model

1. Introduction

Chlorophyll loss is characteristic to advanced stages of a number of biotic and abiotic stress conditions. For example, high light conditions result in photobleaching, and infections cause chlorosis or necrosis. Developmental conditions such as senescence also result in selective catabolism of chlorophyll and thus changes in leaf coloration (1). Therefore chlorophyll content of leaves is a valuable marker of plant health status (2–4). Conventional methods of chlorophyll determination are based on spectrophotometric measurements of pigment extracts (5, 6). These, however, are not adequate for testing large numbers of leaf samples. Analysis of photographic or fluorescence images offer high-throughput alternatives. Moreover, these noninvasive methods allow testing the same leaf repeatedly at various stages of development or stress response (7–13).

Jennifer Normanly (ed.), *High-Throughput Phenotyping in Plants: Methods and Protocols*, Methods in Molecular Biology, vol. 918, DOI 10.1007/978-1-61779-995-2_6,

In this chapter, we demonstrate the use of a photographic method and show that the Hue parameter of leaf color offers a fast, high-throughput initial screening system for measuring leaf chlorophyll content. In an earlier publication, we used senescing leaves as a model system to validate this method and also showed that Hue values were not only proportional to total chlorophyll but also offered a good estimate of the photochemical yield of photosystem (PS) II in leaves that lost less than 80% chlorophyll (14).

2. Materials

1. Select and arrange plant material for imaging. Whole plants, excised leaves or leaf cuttings are equally suitable for photographic analysis.
2. Arrange on a flat surface with a white background (see Note 1).
3. For the experiment illustrated here, we collected fully expanded leaves from greenhouse grown (22°C, natural light conditions) tobacco (*Nicotiana tabacum* cv. Petit Havana SR1) plants with different color resulting from different phases of senescence; 0.6 cm diameter disks were cut and arranged on white filter paper.
4. Digital camera: Olympus C-7070WZ (Olympus Hungary KFT, Budapest, Hungary).
5. Software for image analysis: MATLABs (version 2008b) with the Image Processing Toolbox™ (The MathWorks Inc., Natick, MA, USA).

3. Methods

The image used as an example (Fig. 1a) was taken with a digital camera with the following camera settings: aperture of f/7, shutter speed of 1/250 s, ISO 400, resolution: 3,072 × 2,304, quality: SHQ, format: jpeg (see Note 2). There are several commercially available software programs for image analysis. We use MATLAB.

3.1. Apply White Balance Correction

1. Create a background image (see Note 3).
2. Compensate image for white balance, dividing the original picture pixel by pixel with the background image. The resulting white compensated image is shown in Fig. 1b.

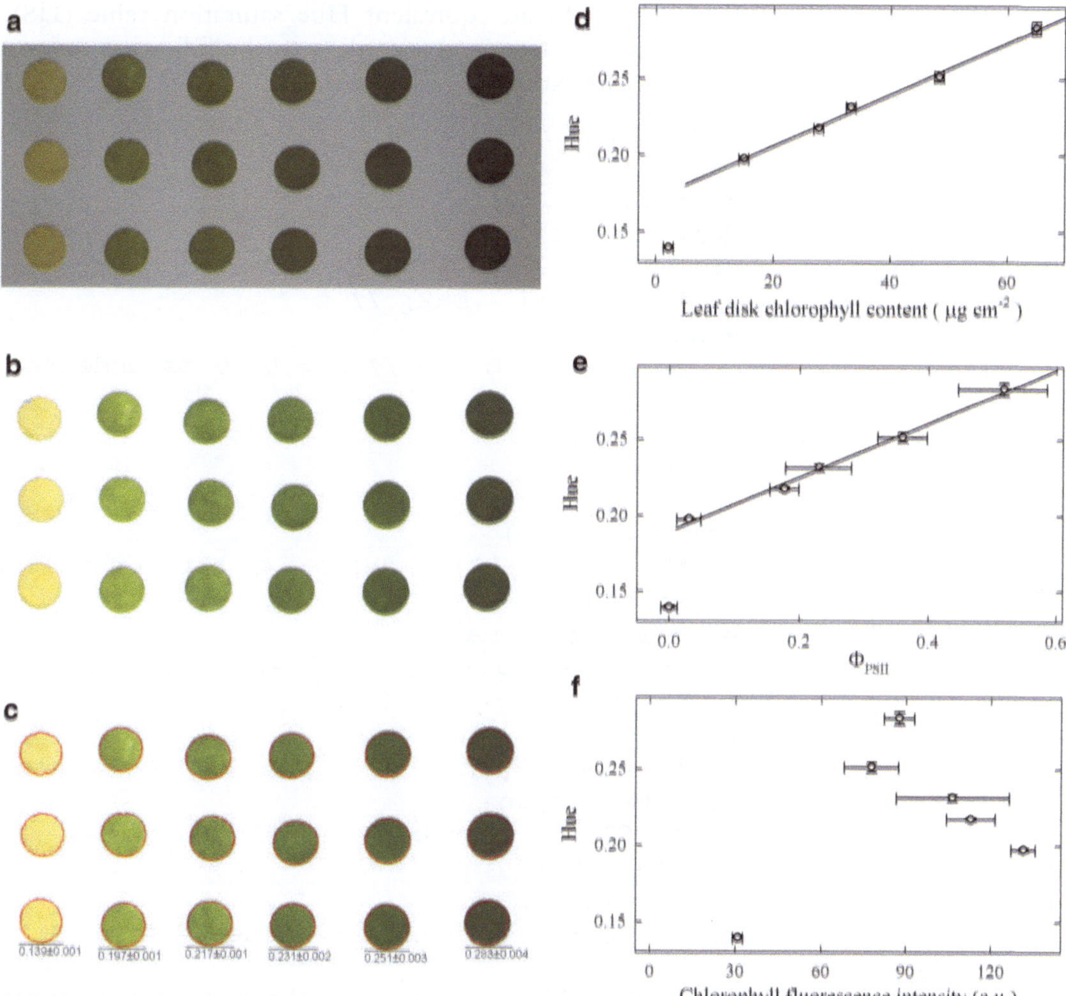

Fig. 1. Using the Hue parameter to represent leaf chlorophyll content. (**a**) Digital image of 12 leaf disks cut from tobacco leaves. Disks were arranged in columns according to various phases of leaf senescence, the first column representing advanced senescence and strong color change and the last column representing non-senescent mature green leaves. (**b**) The same image after white compensation. (**c**) Hue parameters were calculated first for each leaf disk, averaged from all pixels in the *red encircled areas*, then mean ± standard deviation characteristic for each column were calculated, using the average Hue of the three disks. (**d**) Hue and chlorophyll content data of leaf disks in the image (*open circles*) and linear least squares fit (*solid line*, $R^2 = 0.995$) of all data with the exception of the one point with very low chlorophyll. (**e**) Hue and photochemical quantum yield (Φ_{PSII}) data of leaf disks in the image (*open circles*) and linear least squares fit (*solid line*, $R^2 = 0.993$) of all data with the exception of the one point with $\Phi_{PSII} = 0$. (**f**) Hue and chlorophyll fluorescence intensity data of leaf disks in the image (*open circles*).

3.2. Select Areas of Interest and Convert Red, Green, Blue to Hue Saturation Value

1. Each pixel of an image has red, green, blue (RGB) parameters, where R, G, and B are between 0.0 and 1.0, with 0.0 being the least amount, and 1.0 being the greatest amount of that color.

2. For each RGB, an equivalent Hue saturation value (HSV) color can be determined according to the following series of formulas (see Note 4):

```
min = MIN( r, g, b );
max = MAX( r, g, b );
*v = max; // v
delta = max - min;
if( max != 0 )
   *s = delta / max; // s
else {
   // r = g = b = 0 // s = 0, v is undefined
   *s = 0;
   *h = -1;
   return;
}
if( r == max )
   *h = ( g - b ) / delta; // between yellow
& magenta
else if( g == max )
   *h = 2 + ( b - r ) / delta; // between cyan
& yellow
else
   *h = 4 + ( r - g ) / delta; // between
magenta & cyan
*h *= 60; // degrees
if( *h < 0 )
   *h += 360;
*h /= 360; // converts h values 0 to 1
```

3.3. Use Hue Parameters to Characterize Leaves

In our example, Hue parameters are shown as mean ± standard deviation characteristic to each column of the leaf matrix in Fig. 1c.

3.4. Hue Values Are Good Markers of Total Chlorophyll Content

After photography and photosynthesis measurements (see below), leaf disks were ground in 80% acetone and chlorophyll content was determined by photometric measurements as described (5). Figure 1d shows a good positive correlation between Hue and chlorophyll content for leaf disks which lost less than 80% of their chlorophyll.

3.5. Hue May Also Give an Estimate of Photochemical Yield of Photosynthesis

In our example, photochemical yield was determined from the same set of leaf disks with an Imaging PAM fluorometer (MAXI version, Heinz Walz GmbH, Effeltrich, Germany).

1. Keep leaf disks in darkness for 15 min to relax non-photochemical quenching.

2. Expose leaf disks to 55 μmol/m²/s blue actinic light for 2 min.
3. Measure minimal (F) and maximal (F'_m) fluorescence yields before (F) and after (F'_m) a saturating pulse, while keeping the blue actinic illumination on.
4. Calculate the effective photochemical yield of photosystem II in the illuminated samples as $\Phi_{PSII} = (F'_m - F) / F'_m$, according to Schreiber (15) and Genty et al. (16).

The graph in Fig. 1e contains Hue and Φ_{PSII} data points from the leaf disks shown in Fig. 1a–c. A good linear correlation between Hue and Φ_{PSII} was found for leaf disks, which had photochemical activity (i.e., all, except the disks in the first column). Although data presented here as well as results published earlier suggest that the Hue parameter gives a good approximation of leaf photosystem II photochemistry (14), Hue cannot generally substitute more accurate chlorophyll fluorescence yield-based measurements (see Note 6).

Chlorophyll fluorescence, which is emitted by bulk chlorophyll and can be measured with fluorescence microscopy, is not to be confused with the above discussed, photosynthesis-related chlorophyll fluorescence yield parameters, and is not correlated with Hue (Fig. 1f) or chlorophyll content (data not shown) and cannot be used for evaluating leaf color.

4. Notes

1. It is advisable that samples are photographed on a flat surface. Leaves on plants can be temporarily supported from underneath.
2. Use identical light conditions when photographing a series of samples. To ensure this, ambient light or illumination from above is best. In our experience, photography with a flash may result in reflections on leaf surface, which disturbs data evaluation. Omitting flash illumination and using ambient laboratory light may result in relatively dark backgrounds (as illustrated in Fig. 1a), but does not disturb picture analysis if corrections are applied.
3. There are several ways to compensate images for white balance. The one illustrated here uses the assumption that changes in background color are continuous, i.e., background of areas under the leaf disks can be interpolated from surrounding pixels.

(a) Select areas of interest and create a mask image. This is shown in Fig. 2a, background pixels are in black, areas of interest are in white.

(b) Fill up the masked areas in red, green, and blue planes (pixel matrices). In this process, the original red pixel composition of the image (Fig. 2b) is changed so pixels of the areas of interests are replaced by red pixel values interpolated from those at the perimeter (borders) of the area of interest (Fig. 2e). The same process is repeated for green (Fig. 2c, f) and blue (Fig. 2d, g) color planes.

(c) Superpose the filled RGB planes to get the background (Fig. 2h). This process can be used to make objects in an image seem to disappear as they are replaced with values that blend in with the background area. This background image is used for correcting the original image, i.e., converting the image in Fig. 1a to the one in Fig. 1b.

4. Details of RGB to HSV calculations can also be found, for example, at http://www.fact-archive.com/encyclopedia/HSV_color_space or at http://beesbuzz.biz/code/hsv_color_transforms.php, which also provides a computer program written in C or C++.

5. Some publications use color photographs converted to grayscale for estimating chlorophyll content. This is not suitable, because colors with different Hue parameters give the same gray, as illustrated in Fig. 3. Figure 3a shows an artificial color palette, with various shades of yellow and green, with some contribution of red, representing possible leaf colors. RGB and HSV parameters of colors in this palette are shown in Fig. 3b, c respectively. Grayscale conversion, which is a calculation of weighted sums of RGB values as: $0.2989 \times R + 0.5870 \times G + 0.114 \times B$ (17), is the same, (=204) for all colors resulting in an image of uniform gray (Fig. 3d), illustrating the inadequacy of using gray-converted leaf images for pigment analysis.

6. While Hue measurements offer a good tool for measuring chlorophyll content, it should be emphasized that they may only give an estimate of leaf photosynthesis. In our example using senescing tobacco leaves, Hue was in good correlation with chlorophyll content, but this may not be the case in other senescing species and in all conditions leading to chlorophyll loss. Leaf Hue measurements can be used as rough estimate in an initial screening of photosynthesis if a linear relationship (similar to the one shown in Fig. 1d) is established first with the plant species and leaf conditions studied, but for a precise study of leaf photochemistry, Φ_{PSII} should also be measured more accurately.

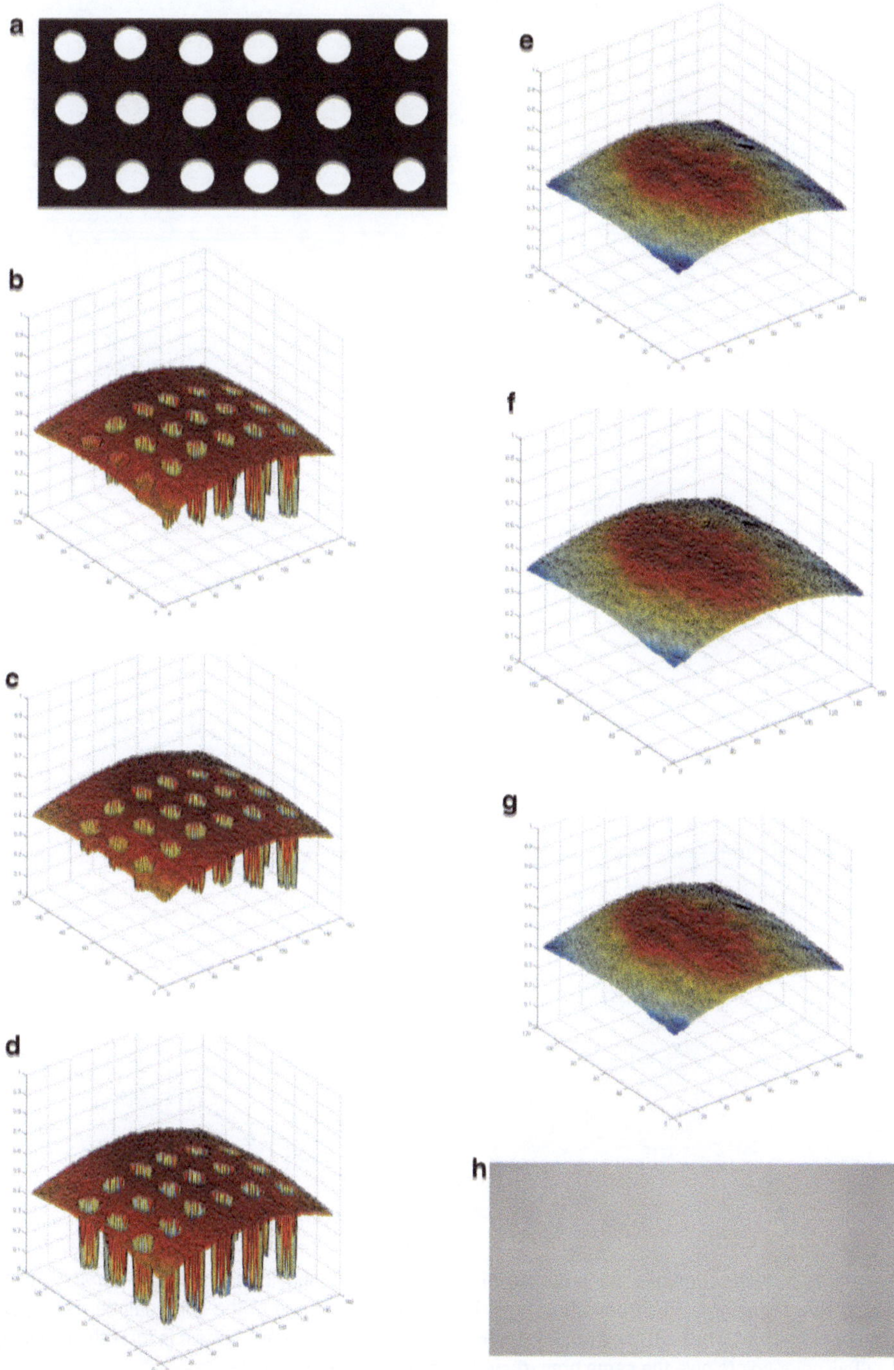

Fig. 2. Details of white compensation (*see description in* Note 3). (**a**) The mask image; areas of interest are *white*, the rest of the image is *black*. *Red* (**b**), *green* (**c**), and *blue* (**d**) pixel compositions were determined from the original image (in Fig. 1a), then original data in the areas of interest were replaced by interpolating the pixel values from the borders of the region. This gives the corresponding *red* (**e**), *green* (**f**), and *blue* (**g**) components of a background image. Note that the background image (**h**), which is sum of components (**e**), (**f**), and (**g**) is not of uniform color, mainly due to the unevenness of illumination. This background image was subtracted from the original image (Fig. 1a) in order to get the white-balanced image (Fig. 1b).

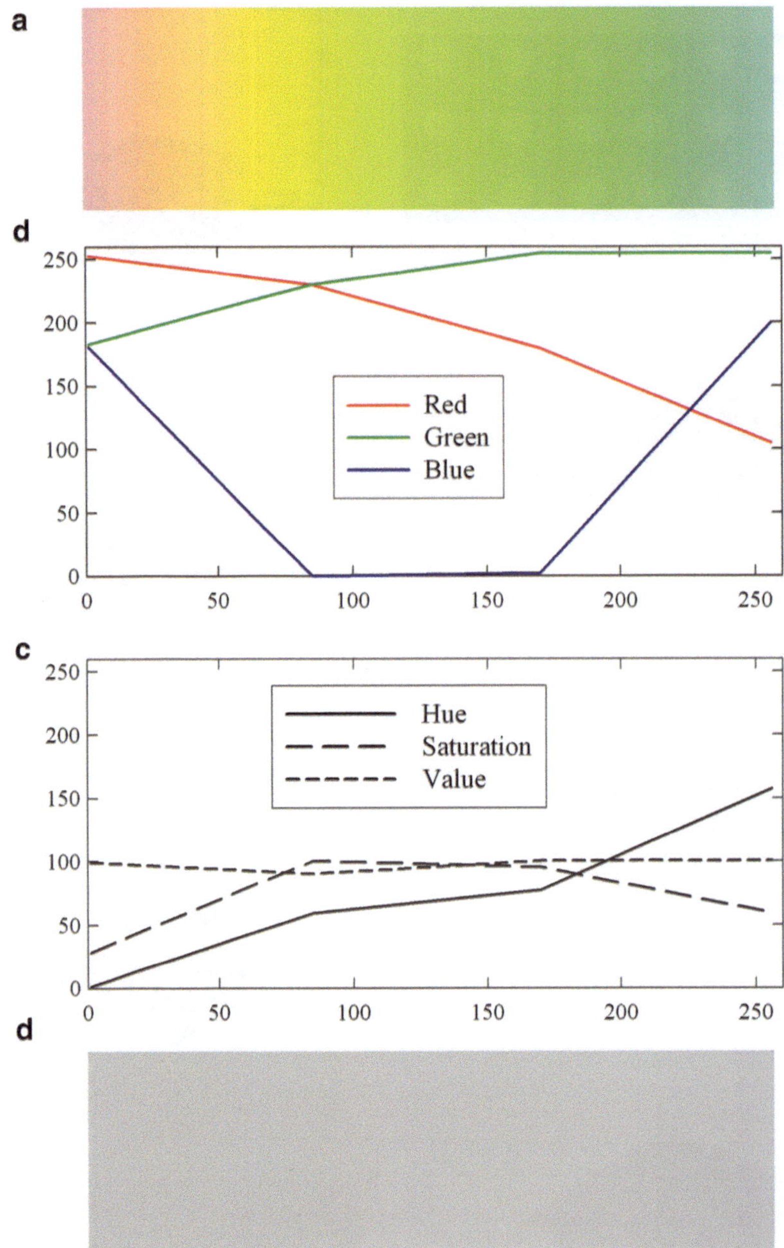

Fig. 3. Why are gray-converted images inappropriate for estimating chlorophyll content? (*see details in* Note 5) (**a**) An artificial color palette, changing colors from left to right. Each pixel in a column of the color palette was calculated from RGB parameters shown in (**b**). These RGB parameters correspond to HSV parameters shown in (**c**) with increasing Hue from left to right, but all color points give the same gray (**d**) when converted into grayscale.

References

1. Hörtensteiner S (2006) Chlorophyll degradation during senescence. Annu Rev Plant Biol 57: 55–77
2. Tang L, Kwon S-Y, Kim S-H et al (2006) Enhanced tolerance of transgenic potato plants expressing both superoxide dismutase and ascorbate peroxidase in chloroplasts against oxidative stress and high temperature. Plant Cell Rep 25:1380–1386
3. Hegedűs A, Janda T, Horváth VG et al (2008) Accumulation of overproduced ferritin in the chloroplast provides protection against photoinhibition induced by low temperature in tobacco plants. J Plant Physiol 165: 1647–1651
4. Singla-Pareek SL, Yadav SK, Pareek A et al (2008) Enhancing salt tolerance in a crop plant by overexpression of glyoxalase II. Transgenic Res 17:171–180
5. Porra RJ, Thompson WA, Kriedemann PE (1989) Determination of accurate extinction coefficients and simultaneous equations for assaying chlorophylls a and b extracted with four different solvents: verification of the concentration of chlorophyll standards by atomic absorption spectroscopy. Biochim Biophys Acta 975:384–394
6. Yang C-M, Chang K-W, Yin M-H et al (1998) Methods for determination of the chlorophylls and their derivatives. Taiwania 43:116–122
7. Chaerle L, Van der Straeten D (2000) Imaging techniques and the early detection of plant stress. Trends Plant Sci 5:495–501
8. Maxwell K, Johnson GN (2000) Chlorophyll fluorescence—a practical guide. J Exp Bot 51:659–668
9. Carter GA, Knapp AK (2001) Leaf optical properties in higher plants: linking spectral characteristics to stress and chlorophyll concentrations. Am J Bot 88:677–684
10. Richardson AD, Duigan SP, Berlyn GP (2002) An evaluation of non-invasive methods to estimate foliar chlorophyll content. New Phytol 153:185–194
11. Madeira AC, Ferreira AA, De Varennes A et al (2003) SPAD meter versus tristimulus colorimeter to estimate chlorophyll content and leaf color in sweet pepper. Commun Soil Sci Plant Anal 34:2461–2470
12. Lenk S, Chaerle L, Pfündel EE et al (2007) Multispectral fluorescence and reflectance imaging at the leaf level and its possible applications. J Exp Bot 58:807–814
13. Cassol D, Silva FSP, Falqueto AR et al (2008) An evaluation of nondestructive methods to estimate total chlorophyll content. Photosynthetica 46:634–636
14. Majer P, Sass L, Horváth VG et al (2010) Leaf hue measurements offer a fast, high-throughput initial screening of photosynthesis in leaves. J Plant Physiol 167:74–76
15. Schreiber U (1989) Detection of rapid induction kinetics with a new type of high-frequency modulated chlorophyll fluorometer. Photosynth Res 9:261–272
16. Genty B, Briantais JM, Baker NR (1989) The relationship between the quantum yield of photosynthetic electron transport and quenching of chlorophyll fluorescence. Biochim Biophys Acta 990:87–92
17. Pratt WK (1991) Digital image processing. Wiley, New York

Chapter 7

High-Resolution, Time-Lapse Imaging for Ecosystem-Scale Phenotyping in the Field

Tim Brown, Christopher Zimmermann, Whitney Panneton, Nina Noah, and Justin Borevitz

Abstract

The high spatial and temporal resolution of data required for high-throughput phenotyping has typically been all but impossible to obtain in field populations of plants. When studies of individual and population genetic variation and microclimate sensor data are combined with phenology data, a landscape-level view of how populations respond to changing environments can be obtained. This chapter will discuss the development of a multi-billion pixel ("gigapixel") camera system that enables the collection of phenology data at up to hourly intervals from in situ plant populations. Such gigapixel time-lapse imaging systems represent a key technological advancement for enabling high-throughput phenotyping in field settings. Gigapixel resolution image datasets allow researchers to record life-history (phenology) data across an entire landscape over multiple seasons. Image data can be wirelessly transmitted to a remote server where it can be accessed online within hours of capture. The time-lapse panoramic images are browsable through an interactive web tool that can be used to compare plant phenology with environmental sensor data collected simultaneously from the field. The high spatial and temporal resolution data can be used to identify individual plant phenology, which can in turn be used to generate complete population level phenotype data. The Gigavision platform is especially powerful when coupled with next-generation population genomic analysis. The Gigavision system permits the rapid identification of the phenotypes and genotypes responding to natural selection in wild populations.

Key words: Phenology, Near remote sensing, Gigapan, Time-lapse, Landscape ecology, Phenomics, High-throughput phenotyping

1. Introduction

Addressing challenging ecological questions, such as predicting climate change impacts and effective ecosystem management require improved quantitative models of how plants and climate interact across scales, from individual to ecosystem levels. At the

Jennifer Normanly (ed.), *High-Throughput Phenotyping in Plants: Methods and Protocols*, Methods in Molecular Biology, vol. 918,
DOI 10.1007/978-1-61779-995-2_7,

individual genetic level, we must better understand how genetic variation and structure within individuals and populations affect ecosystems at the landscape scale (1, 2). At the other end of the spatial scale there is need for a better understanding of how phenological changes at community scales correlate with changes observable in satellite products such as MODIS and Landsat (3). However, ground-based monitoring of phenological stages in individual plants over large spatial and temporal scales has traditionally been labor intensive, time consuming, and difficult. Understanding how phenology is controlled by the local environment and/or internal genetic variation among individuals in field populations is likewise arduous and limited in the numbers of plants that can be analyzed. Although studies of phenology using satellite imagery can provide coarse spatial scale observations at up to daily frequencies (3, 4), neither traditional ground-based sampling nor satellite imagery has sufficient spatial *and* temporal resolution to enable collection of the plant-level data needed for field phenomic studies (2, 3, 5). Consequently, there is a pressing need for new technologies to enhance data collection at the landscape scale. This type of data collection has been recently termed "near-surface" remote sensing (3, 6).

In addition to the need for better monitoring tools, efforts are needed to develop analysis tools to transform what have typically been descriptive, qualitative phenological observations into a quantitative framework. Furthermore, because high temporal and spatial resolution phenological datasets are rare, there is limited work exploring what time resolution is actually required for understanding key questions about the interrelation between environment and genetics in wild plant populations at the landscape scale. Additionally, for genomics applications, it would be helpful to develop experimental pipelines to expand lab-based genomic studies into quantifiable field experiments and to create field phenological protocols for identifying plants and plant populations of research interest for greenhouse studies.

The emergence of new, low-cost, high-resolution imaging systems, high-powered computers, and wireless and solar technologies provide the opportunity to revolutionize the scale and quality of phenological data that can be collected in the field. Since 2009, the Borevitz Lab at the University of Chicago has been working with TimeScience (Time-Science.com) to develop solutions to the above challenges. Using off-the-shelf components, we have developed a ruggedized camera system called "Gigavision" that can record hourly, multi-billion pixel ("gigapixel") resolution panoramic images, year-round in a field setting. The Gigavision camera system is solar powered and can upload images to a remote server over wireless 802.11 or cellular networks. Online image viewing and data collection are enabled through our interactive gigapixel image time-lapse player. Gigapixel scale images enable time-series recording of entire landscapes with a resolution sufficient to track phenological events at the scale of individual flowering plants. Once such a system is established in a fixed location, it completely changes the magnitude of phenological data that can be collected in that area.

There are currently three Gigavision systems deployed in the United States: two on Lake Michigan at a field site at the Indiana Dunes State Park and one in Utah at the University of Utah. Data streams from these cameras are available online at www.gigavision.org. Although there are numerous technical challenges associated with collecting and analyzing gigapixel imaging of data, we strongly believe that high-resolution imaging systems represent the future of phenological and observational field research.

2. Project Background: The Gigavision Concept

The basic idea of gigapixel imaging is that by capturing and merging a set of overlapping, highly zoomed-in images, it is possible to create a panoramic image that is hundreds or thousands of times the pixel resolution of a single image taken by the same camera (Fig. 1a). Such images can be taken with a robotic pan-tilt tripod mount or even by hand. Many existing pan-tilt imaging systems have been developed for a wide range of applications from real estate documentation to military surveillance (7). However, due to the technical complexity, cost and power requirements of such systems, gigapixel imaging has seen little application in biology. The GigaPan project at Carnegie Mellon University (in collaboration with NASA Ames Intelligent Robotics Group and Google) changed the situation by developing a sub-$300 automated pan-tilt head (the GigaPan Epic) that works with almost any consumer digital camera. An added innovation was the creation of a web site where any image greater than 50 megapixels can be uploaded, publicly viewed and annotated for free (http://www.gigapan.org). The GigaPan web site provides an excellent forum for the formation of a public community of users interested in gigapixel imaging. The result has been an explosion of gigapixel imaging applications with users uploading images with resolutions of more than 180 billion pixels (8). The GigaPan research group has also been active in encouraging scientific applications for gigapixel imaging (9) (http://gigapixelscience.gigapan.org/).

While the GigaPan Epic series and similar camera mounts are useful for occasional high-resolution documentation, there were no pan-tilt systems that we were aware of expressly designed for repeat time-series gigapixel photography in a field setting. Building a system that can reliably take images year-round in adverse weather conditions, and developing an automated data pipeline for image

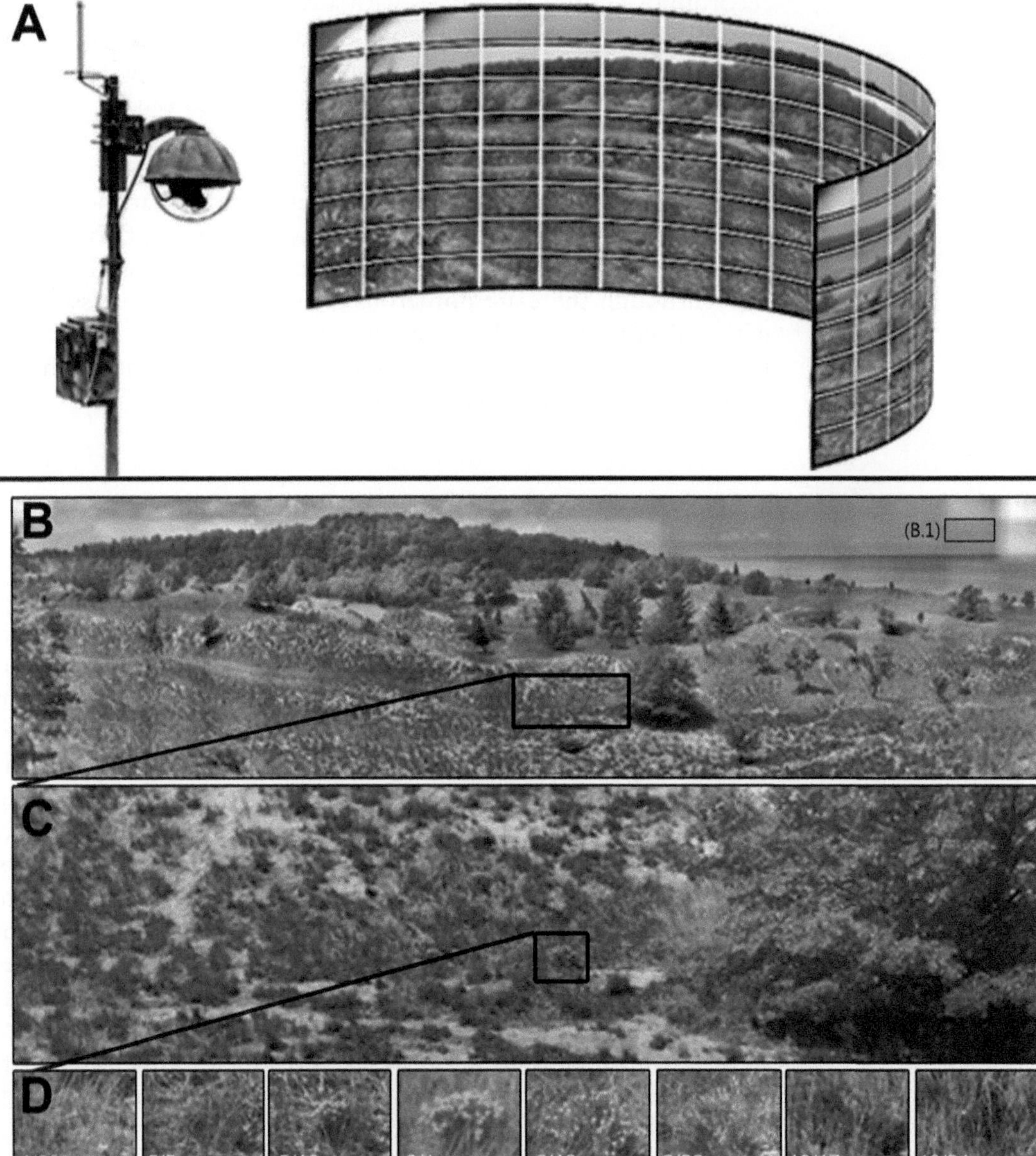

Fig. 1. The Gigavision process from data capture to data collection. (**a**) The Gigavision camera captures hundreds of highly zoomed, overlapping image in rows and columns. (**b**) Images are stitched into a single large panorama of 1–2 gigapixels resolution. (**b.1**) Approximate area captured by each 15 MP image. (**c**) High resolution images permit the user to zoom into any area and (**d**) collect time-series phenology data from individual plants.

stitching and online time-lapse visualization is a significant challenge. In addition, once the time-series gigapixel images are captured, the challenge of how to collect biologically meaningful numeric data from such images is not trivial.

The basic Gigavision camera system consists of a megapixel (MP) consumer grade DSLR camera or network-enabled (IP)

Table 1
The future of gigapixel time-lapse

The basic engineering requirement behind the Gigavision system is simple—create a camera and positioning system that can be programmatically controlled to capture a large number of high-resolution images in a panorama. Consequently, there are multiple routes to solving this problem and as new technologies develop, what constitutes the best method for generating panoramic time-series image sets may change as well. For example, many commercial surveillance cameras have pan-tilt-zoom (PTZ) capability, but until recently, they typically captured low quality 1/3 megapixel resolution images and were challenging to program. However, in the last year, surveillance camera manufacturers have begun releasing 2 MP resolution PTZ cameras. These cameras have integrated pan-tilt mechanisms that can be controlled with standardized protocols (e.g., Pelco D) enabling programmatic control of camera positioning, zoom and focus over a single Ethernet cable or remotely over the Internet. Although PTZ cameras are substantially lower resolution than a DSLR or fixed IP Camera, some cameras have a large enough zoom lens that one could potentially reach resolutions of 1.2 gigapixels or higher in a 360° field of view (FOV). In addition, the major camera manufactures have recently formed an international group to promote open standards (www.onvif.org) for interoperability in Internet enabled surveillance cameras. Such standards potentially allow the development of software controls that would work with any megapixel PTZ camera. Furthermore, some cameras (e.g., Axis Q series) have an on-board processor that can run software programs and capture images to an external drive or internal memory card, potentially eliminating the need for an external computer. This suggests a near future where plug-and-play gigapixel imaging systems with rugged, industry-tested hardware can be widely deployed at low cost. Such systems would require minimal technical skill to manage and could reliably provide gigapixel time-lapse imagery.

camera equipped with a high-resolution zoom lens. The camera is mounted on a set of pan-tilt robotic rotational motors (servos) controlled by a Phidgets-brand servo microcontroller, connected via USB to a low-power FitPC2 Windows-based PC. Each panoramic image is made up of a collection of overlapping rows and columns of full-resolution zoomed-in images (typically 150–600 images depending on the camera lens and the field of view required). After capture, images are flattened or "stitched" into a single panorama using the GigaPan Systems stitcher software (10). We estimate that a resolution of about 1 pixel/cm^2 is required to accurately monitor phenological change across most plant species in a study area including small ephemeral flowering plants. Our current gigapixel imaging systems provide a spatial resolution of approximately 0.5–1 pixel/cm^2 over a 10 ha area. These systems employ a 15 megapixel (MP) digital camera with a 150 mm zoom lens. With such a camera it takes only about 100 images to achieve gigapixel resolutions. The pixel resolution of cameras continues to increase as prices continue to go down, so it is reasonable to expect increasing gigapixel resolutions for the same or lower hardware costs every few years for the foreseeable future (Table 1). A sample parts list for a complete Gigavision system is included in Appendix A (Supplementary Materials) as a starting point for the reader.

3. Gigavision Technical Description

3.1. Camera and Lens Choice

In order to create gigapixel images, one needs the ability to programmatically tell the camera when to take a picture and then to download the pictures to a PC. Currently Canon is the only major camera manufacturer who provides a software development kit (SDK) for all their DSLR cameras (see Note 1). The SDK enables computer-based control of a DSLR via USB cable (see Note 2). The Gigavision software controls camera settings, shutter release, and image download using the Canon SDK.

The currently deployed Gigavision systems use either a 15 MP Canon T1i DSLR camera with a 150 mm lens (see Note 3) or a 5 MP network IP camera (StarDot NetCam SC5) with a 43 mm C-mount lens (35 mm lens equivalence = 250 mm). Maximum resolution for the NetCam-based systems is about 1.2 gigapixels. Ten MP network cameras should become more widely available in the next year, doubling this resolution. Maximum resolution for the DSLR-based systems is 3.5 gigapixels over a field of view of about 320° × 25° height (this could be easily increased to 5 gigapixels by using a Canon T2i, which has an 18 MP base resolution).

Although DSLR cameras typically capture higher resolution and higher quality images, network-enabled (IP) surveillance cameras are also a viable choice and are much more physically robust than DSLR cameras. IP cameras have a huge functional temperature range (−20 to +70°C vs. 0 to 20°C for DSLR cameras) and can be powered over an Ethernet cable, directly off the 12 V battery (the DSLR requires a voltage converter). This eliminates the need for USB camera control and power cabling for the camera. Although maximum resolution for IP cameras is currently only 5–10 MP, the cameras are smaller, allowing the use of a much larger zoom lens in the same sized housing (250 mm equivalent vs. 150 mm for the DSLR systems). By taking twice as many images at double the zoom, one can reach gigapixel resolutions with IP cameras.

3.2. Camera Control and Positioning

Camera positioning and image capture are enabled through custom-built software written in Microsoft Visual Basic (11). For DSLR camera control, a wrapper for the Canon SDK tools was created that provided camera access to the Gigavision control software in Visual Basic. The Canon SDK enables software control of image resolution and image capture and download. For IP camera-based systems, images are downloaded over Ethernet using the camera URL. Camera positioning is achieved through the use of two servos designed for moderate-duty robotics applications

Table 2
Reliability vs. cost

There are many factors that must be considered when choosing components for a Gigavision system because cost and reliability are directly correlated. For example, a netbook-based system that runs on AC power at a site with existing Ethernet infrastructure will cost much less than a winterized solar-powered wireless system with a redundant PC-system. Site accessibility and servicing costs for the camera site must also be considered in the total cost calculation as does the cost of data loss. If a field site is distant or difficult to access, the cost of repeat service visits can quickly outstrip the upfront costs of building a system with more expensive but reliable hardware. Thus, although the camera may work perfectly 99% of the time, a camera failure in the week in spring that happens to correspond with the flowering dates of plants of interest can render the remainder of the data from that entire year substantially less useful. A long time-series plant phenology dataset is only as good as the amount of data there is for all periods of phenologically significant plant activity for each year.

(Appendix A, Supplementary Materials). Servos are controlled with a Phidgets-brand 8-servo controller connected to the PC via USB. Phidgets controls were chosen because they provide a very user-friendly interface for programmatic control of system hardware. Phidgets supplies free software examples in a multitude of programming languages providing the user with quick software access to multiple hardware controls (see Note 4).

3.3. System Costs

Base cost for the Gigavision camera hardware (unassembled) ranges from $6,000 to $11,000 depending on the components chosen and the reliability required (Appendix A, Supplementary Materials). Currently wage costs for assembly, installation, and server configuration are the largest components of the overall system cost. We also recommend purchasing a mac-mini server and RAID hard drive system to act as a standalone web server and stitching platform. Total system cost for a Gigavision camera including assembly, installation, and training is in the $20,000–30,000 range (including the server). Infrastructure availability, system reliability, and maintenance requirements all directly correlate with the cost (Table 2). For a camera system that will be in a lab or easily accessible location, it may be cost effective to reduce initial hardware costs by purchasing lower quality components because the system can be easily monitored to assure reliable operation. An additional consideration is the visual resolution required for the project at hand. For example, if one is studying deciduous trees, significantly lower pixel resolutions would be required than for the study of small flowering plants. We expect that as off-the-shelf PTZ options become available (Table 1) and the data processing pipeline becomes streamlined, system prices will go below $10,000.

3.4. Control Software

The main Gigavision control software was written in Visual Basic and runs in Windows. The software has two components. (1) A "manager" program controls the off/on power sequence for the camera and servos, tracks system variables such as housing temperature and battery voltage, and allows the user to set the wake and sleep cycle for the computer. (2) The "recorder" program handles the actual servo movement, image capture, and processing. In the recorder program, the user configures the field of view of the panorama, thumbnail resolution, and capture frequency. These variables allow the software to calculate how many rows and columns of images at the current zoom level are required to capture the entire FOV of the panorama (Fig. 1). Using these data and the known range of the servos, the software calculates how many degrees to move the camera for each picture. During panorama capture, the camera is positioned for each photo, the image is captured and downloaded to the PC, then the camera is moved to the next position and the process is repeated. Image download can take a few seconds and is currently the longest part of this process. In general, it takes about 10 s to capture and process an individual image; thus it takes about 30 min to capture a 200-image panorama. Half an hour per image has little impact on general phenological measurement for plants, but it may be of concern for applications where the subject being photographed is potentially moving or changing on a faster time scale. For DSLR cameras, modifying the capture sequence to download images from the camera only after all the images in the panorama are captured would reduce time per image to closer to one second or less and permit full panoramas to be taken very quickly. This method would, however, require the camera to have a memory card (rather than only storing images on the computer), which introduces an additional potential failure point in the system.

3.5. Data Management

Gigavision cameras can upload images over a wired or wireless Internet connection or over a cellular data connection if fast Internet connectivity is not available at the field site. For systems with a fast Internet access, full-sized images are uploaded to a remote server where they can be automatically stitched into panoramas. Size of the raw image data for DSLR-based systems is about 500 MB to 1 GB per panorama. After stitching, the image tiles take up an additional 200–400 MB for a total of about 500 GB per year for a camera taking one image per day. If high bandwidth Internet is not available, the capture software can create a set of lower resolution "thumbnail" images that can be used to generate an 80 MP panorama in addition to the full-resolution images. The thumbnail images are small enough to be uploaded over a cellular network in less than an hour. Thumbnail image size is chosen to keep monthly data usage under the 5 GB cap typical to most cellular data plans. In systems without fast Internet, full-sized images can be stored on an external USB-powered hard drive and retrieved manually from the field at periodic intervals. "Next-generation" wireless technologies such as WiMax and 4G cellular systems typically do not have a monthly data cap and are now available in

Table 3
File syncing and data management

One significant challenge of working with gigapixel image sets is data management. If images are not removed from the capture machine in a timely manner, valuable phenological events can be missed due to the system hard drive being full. Images can be retrieved from the field through a number of routes. Ethernet or WiFi enabled Gigavision systems work under the simplest scenario, with images uploaded directly to the stitching server and then deleted from the capture machine.

For machines with only 3G connectivity, file management becomes more challenging. On these machines, the capture software stores both full-resolution images and smaller "thumbnail" images that can be stitched to form an 80 MP panorama. The thumbnail images can be uploaded over 3G and do not cumulatively use more than 5 GB of data transfer over the course of a month. However, since 3G connectivity is not always reliable, even the smaller images aren't always completely uploaded each time the computer is on and must be synced when connectivity is available. For non-Ethernet enabled systems, a technician must retrieve an external hard drive from the field at regular intervals. Once the drive is back in the lab, the full-resolution images can be synced with the server and stitched into panoramas. Finally, the capture machine must check with the stitching server and delete any images on the local drive that may have appeared on the server.

To address these issues we developed a multi-tiered approach for image management. A core requirement of the data management plan is that the sync processes be automated wherever possible and easy to use for anyone.

1. Image syncing is performed through batch files that script the linux FTP program rsync.
2. For low resolution "thumbnail" images on the capture machine, the script directs rsync to scan the local Gigavision images directory. For each image it finds, rsync checks if the image exists on the server. If the image exists it deletes the local copy, if the image does not exist, it uploads the image to the server and then deletes the local copy.
3. For the full-resolution images, rsync only checks if the image exists on the server and deletes the local copy if it does exist. If the image is not on the remote server, the local image is preserved. This allows the remote machine to preserve any data that hasn't yet been retrieved to the main server while deleting any images that may have been uploaded to the server via other mean (e.g., by hard drive retrieval).
4. At the lab, a modified version of the sync scripts is run on new hard drives brought back from the field. For these drives, all new images are uploaded to the server. Any images on the hard drive that are also on the server are deleted from the drive, so it will be empty and available to be brought out to the camera on the next trip to the field.

most major cities. Current 4G data rates (5–16 Mbps) are fast enough to enable daily upload of full-resolution gigapixel images over a cellular connection.

Image upload and remote syncing is enabled using "sync" scripts that control the open source file transfer program *rync* (12). Automated scripting assures that the most recent thumbnails and/or full-sized images have been uploaded and manages image deletion on the capture computer after the images are backed up on the server. Local image deletion after syncing is important to maintain sufficient disk space on the capture computer. Due to the volume of image data generated (up to 7,000 images and 10 GB/day), the sync scripts are a crucial system component for data management. See Table 3 for further details on image syncing.

3.6. Computer System

For solar-powered cameras, a system computer must be chosen that is low-power, can run off an unregulated 12 V nominal current, and is robust enough to withstand a wide temperature range. It is also essential that the PC can automatically restart after power loss and can be put into hibernation or a low-power sleep mode. The current Gigavision systems use a small form factor, Windows XP-based PC made by FitPC (Appendix A, Supplementary Materials), which consumes less than 7 Wh of power. There are many similar "car computers" and other mini-form factor computers available, but the FitPC2 had the lowest power usage and widest temperature operating range that we encountered. Because the FitPC is a small desktop computer (rather than a notebook), a mini trackball keyboard and 12 V compatible 10-in. monitor are required to complete the package. If greater hardware reliability is required, we recommend using the "Nano Biscuit" computer developed by Erdman (see www.video-monitoring.com for more details). However, the Erdman computer costs $3,500 more than the FitPC2. Using a solid state hard drive (SSID) in place of a conventional 2.5 in. notebook hard drive reduces power usage and extends the operating temperature range of the system. A potential alternative computing platform that we have not yet tested would be to use an industrial-grade mini-computer running the Windows Embedded operating system. These stripped-down PCs are typically found in specialized industrial control systems or outdoor-rated enterprise-grade wireless routers, which are very low power and can function in extreme environmental conditions. If power usage, operating temperature, and high reliability are less of a priority, a consumer grade "netbook" laptop computer can be used for about 1/3 the cost of the Fit PC. This approach is also more convenient in that it eliminates the need for the external monitor, keyboard, and mouse.

3.7. Power Usage and Power Management

The Gigavision system is optimized for solar power and draws less than 25 W. The average capture time is about 30 min per panorama. In our field systems that use a 3G cellular connection to upload images, the system remains on for a total of an hour at each capture interval to provide adequate time for image upload. In a typical field installation, system timing and power management is configured using the Windows task scheduler. System wake and hibernation times can be set by the user in the Gigavision Manager software. An external 12 V power timer (Flexcharge, DC timer) cycles power off during the night to further reduce power usage. External power cycling also assures that if the PC has locked up during the day it will be restarted the following day. Care should be taken when testing usability of potential PCs, as we found much variation in whether a system can be configured to wake from hibernation or sleep mode and whether the PC will automatically

Table 4
Description of the typical event sequence for the Gigavision system

System power management is enabled through a sleep/wake cycle managed by scheduled Windows tasks. An external 12 V power switch on the batteries forces the computer to reboot once a day allowing the computer to recover from any non-hardware system errors that might freeze the computer and cause the software-based system management sequence to fail. The drawback of using an external power timer is that it must be manually configured on-site. To reduce the need to change the external power timing while preserving a hard reset to improve system stability, the 12 V timer should be set to the far extent of normal daylight hours (e.g., 6 AM to 11 PM).

1. 7:50 AM: external 12 V power timer on the batteries restores power to the system. The PC is set to auto start when power is restored so the PC turns on and loads Windows.
2. 8:00 AM: the Windows task scheduler starts the Gigavision Manager program.
   (a) Note that if the system does not need to capture morning images, the task scheduler can be configured to run the sleep sequence just after system power is restored so the PC will sleep until the scheduler wakes it to take the next panorama.
3. 8:02 AM: Gigavision Manager switches on the power to the servos and the DSLR. The DSLR is power cycled for each image set; this improves camera reliability and reduces software lock-ups.
4. 8:05 AM: Windows Task Scheduler launches the image sync script process (Table 3).
   (a) The image syncing process runs independently of the capture process to improve reliability.
   (b) Syncing is started at the beginning of the capture process to give the system a full hour to upload the images over slow connections.
5. 8:05 AM: the Gigavision Manager launches the Gigavision Recorder software.
6. 8:06 AM: Gigavision Recorder software moves the servos to point the camera at the "top-left" of the camera field of view and starts the capture sequence. After each image is captured it is download from the camera, named with a timestamp and put into the folder for the current panorama. If thumbnail creation is enabled, the system also creates a smaller image set suitable for upload over low bandwidth cellular Internet connections.
7. 8:35 AM: Gigavision recorder finishes capturing the panorama.
8. 9:05 AM: a scheduled task runs a batch file that restarts the computer. A full restart after each capture sequence helps assure system stability and prevents driver issues that may arise due to USB hardware being in use during system hibernation.
9. 9:10 AM: a scheduled task runs a batch file that puts the computer to sleep until the next capture period.
10. For systems that are always on, the computer restarts and sleeps only at night. The Gigavision recorder software can be configured to automatically capture a new panorama every hour.

turn on when power has been cut and then restored (i.e., by a daily external timer or unplanned loss of power). See Table 4 for further details on the system event cycle.

Correct sizing of the solar array is crucial for system reliability, particularly if the system will be on during the winter. In addition, we have found that sealed Gel (AGM) batteries are more reliable than traditional lead-acid deep-cycle batteries. Although AGM batteries cost two to three times more than lead-acid, they more than make up for the cost in reduced maintenance hours and reliability. Even though a system may only need 70–80 W of power to capture three images a day, it is worth using a solar panel that doubles the anticipated power requirements (e.g., ~120 W) if the system is to remain working during winter months. The impact of temperature on batteries is a significant problem that needs to be accounted for at sites that experience extreme winter conditions and/or extended periods of limited sun. We recommend winterizing such systems by burying or insulating the batteries and reducing the system "on time" for winter (see Note 5).

4. Image Processing and Visualization

4.1. Basic Data Visualization Interface

The jpeg images captured by the Gigavision camera are automatically stitched by the GigaPan panorama stitching software (10). The GigaPan stitcher software takes a collection of images in rows and columns and merges them into a single large panoramic image. The GigaPan stitcher was chosen because it is low cost ($10 for current owners of GigaPan hardware) and it can be controlled via the command line. Panoramic images are typically stored, not as single images but as set of nested folders made by taking the large panorama at a particular resolution and cutting it into smaller 256 × 256 pixel tiles. When the panorama is loaded onscreen, the image tiles for the section of the image currently being viewed are reassembled into the image. As the user zooms and pans within the panorama, only the tiles in the current field of view are loaded (readers may be familiar with this type of interface from using Google Maps or Google Earth). As a user zooms further in, the tiles on the screen are expanded until a certain zoom level is reached and the image tiles at the next level down in the folder hierarchy are loaded.

The standardized folder structure created by the GigaPan software can be accessed with an open source software toolkit called Open Zoom (www.openzoom.org). Open Zoom provides a programming framework for building gigapixel image viewing software in Adobe Flash (see Note 6).

4.2. Data Processing and Panorama Creation

The image stitching server is a Mac Mini system running an automated script to detect newly uploaded images; when a new panorama is detected, the images are passed to the GigaPan stitcher program. The scripts also create an up-to-date database of all images and stitched panoramas on the system. These database files are used by the online gigapixel time-lapse player to track which images are available for playback.

4.3. File Naming and Folder Structure

To provide a predictable naming and folder structure for time-lapse playback, panoramas are stored in a time-stamped folder hierarchy and named using the camera name and a timestamp down to the hour. Thus by knowing the site name and date of the image capture, the playback software can accurately generate the full file path to all image files within a multi-year time-lapse panoramic image set.

4.4. Online Image Player and Data Collection

Online image visualization and data collection for Gigavision image is enabled through PHP scripts and a custom-built Adobe Flash application. The panoramic images are displayed in a web-based "player" with an adjustable timeline bar and tabs that provide quick access to common time periods such as "month" or "week" (Fig. 2). The image player loads a list of the available images for the camera being viewed. Users can zoom and pan within individual images and play a time-lapse from the recorded images of any location and zoom level within the panorama.

A data collection interface allows users to mark regions of interest called "bookmarks." The bookmark selection box is persistent over time enabling the user to create time-stamped metadata content about the region of interest. All bookmarks are auto-assigned a unique numeric identifier. Bookmarks are designed to be versatile and can be nested, permitting them to be used both for marking larger regions in an image and to mark individual items of interest such as a plant or tree. Each bookmark has a "Category" field that might correspond either to the name of a region of the field site or to a specific item that is repeated such as a species name. "Name" and "Notes" fields for each record provide additional flexibility for data collection. To enable collection of phenological data for a given plant, each bookmark contains a list of "Events" or phenophases. For each event, the user can click a button to mark the time and date when that phenophase was reached. For example, a researcher might bookmark the location of every individual of a small ephemeral flowering plant in their field site. Then for each plant, the researcher would record the date and time of the first appearance of leaves, first flowering, flower senescence, etc. This process can be repeated for every plant of all species in the camera field of view. Each bookmark has a "state" setting such as "In Review" or "Complete" to help multiple users manage data consistency.

Individual bookmarks can be auto-loaded via a unique URL enabling researchers to share links to a specific location, time period, or plant of interest in the image. Datasets created using the online interface are stored in standard comma separated (csv) format for easy offline data processing.

4.5. Technical Challenges to Time-Series Panorama Visualization

Smooth time-lapse playback and reliable data collection from gigapixel images requires that users be able to mark an area of interest such as an individual plant and then view the time-series images that correspond with that particular point on the landscape

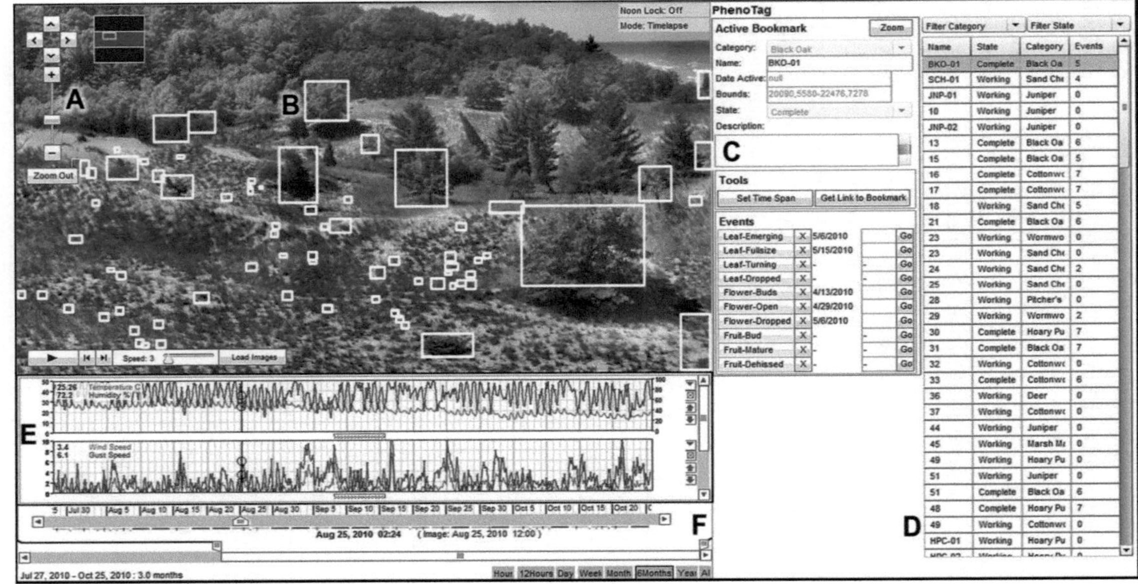

Fig. 2. Online data collection and time-lapse interface for Gigavision images. (**a**) Pan/zoom controls for navigating the images. (**b**) User-created bounding boxes called "bookmarks" for marking plants or areas of interest. Each bookmark has a unique ID, a category or species name from an editable drop-down box, notes and a status field for tracking whether observations in that bookmark are complete or require review, etc. (**c**) Each bookmark is associated with a list of "events" marked with a timestamp. Events are very flexible and can be used to mark anything of interest that may have occurred at a location. Typically events are used to mark each of the specific phenophases of interest for the particular species marked by the bookmark (e.g., "bud burst," "leaf drop," etc.). (**d**) A sortable list of bookmarks and marked areas provides a quick overview of the data collected thus far. Each entry in the bookmark list indicates how many events have been keyed out for a particular bookmark (e.g., plant). Clicking on a bookmark in the list zooms to that plant or location within the panorama. (**e**) Weather data or any other time-stamped data can be co-visualized with the time-series panoramic images. This makes it easier to visually correlate observed phenology with potential abiotic drivers of phenology events such as storms or extreme temperature events. (**f**) Interactive timeline permits quick navigation within multi-year data-sets. Time-period buttons allow users to zoom in to a particular time period (e.g., "year," "week," "day"). Timeline sliders allow more continuous, dynamic adjustment of time period being viewed. Color coded markings on the timeline indicate image availability within the time period being viewed.

over long periods of time. This presents a major challenge for processing the panoramic images because there are numerous interacting issues causing the panoramas not to line up cleanly between subsequent images. This variation in the field of view of the camera causes the location of the pixels corresponding to a particular region of interest to vary between photos. Consequently, any variation in the congruence of a given image pixel with a physical point on the landscape rapidly expands into a large error in the visualized position of objects in the image set. Problems that contribute to this issue range from lack of precision in the servos, to variation in the output of the GigaPan stitcher, to planned or unplanned camera changes that affect camera resolution or position. Although steps can be taken to minimize the above factors, issues with panorama alignment and data continuity are an unavoidable consequence of taking gigapixel resolution images over long periods of time and thus must be considered when developing a gigapixel imaging program.

On an image-by-image basis, servo precision is an issue. The servos we chose for the camera system are reliable in the 0.5–1° range. Although this is quite precise for a typical hobby servo, for an object 150 m from the camera, an error of less than 1° in camera positioning results in the object being photographed shifting by more than 2 m. Although high precision pan-tilt systems can be purchased, they typically cost $4,000 or more, which would significantly increase system hardware costs. In addition to the servo error for each panorama, there are larger more infrequent sources of error that cause one-time shifts in panorama position. These types of error can be caused by the lens zoom setting being changed or the need to replace the camera due to hardware failure or upgrade. Likewise, because the camera is moved by the servos multiple times a day, the camera itself can gradually shift on the pan-tilt mount. Finally, the whole camera housing may move slowly over time if, for example, the ground settles or the housing is buffeted by high winds.

An additional source of image variation stems from inconsistencies in output by the GigaPan stitcher. Because the GigaPan stitcher is designed to work with single, unrelated panoramas, it is not optimized for reliably stitching multiple time-series versions of the same panoramic view. As overall image quality varies with weather events and time of day, panorama output quality and consistency can also vary substantially between images.

4.6. Aligning Time-Series Panoramic Images

To address the above issues we implemented a multilayer process for improving alignment between panoramas. A key component of this is a manually created "camera configuration" file that stores essential information about each "configuration" that a camera has been in. The camera configuration file specifies a series of time ranges that correspond to any major event such as movement of the camera housing, replacement of the camera, etc. For each configuration, the file contains an x and y offset and a zoom factor. It also stores the total image resolution of the panorama for each camera configuration. In addition, each configuration specifies a "master panorama" that is used to automatically align the rest of the images in the time range covered by that configuration.

We also developed an automatic alignment system that processes each panorama after it has been stitched. Feature recognition algorithms are used to compare a subset of tiles in each panorama with the same tiles in the master panorama. The algorithms generate a list of control points in the tiles and then calculate a list of X/Y pixel offset values for control points in the image. This process generates a spatial "offsets" file for each panorama. The Gigavision player loads the "offsets" file and then aligns the tiles to the control points as they are displayed. This process results in much greater congruity between time-series images even in the face of substantial changes in camera configuration or image resolution.

Feature recognition algorithms cannot always make a match between the tiles of different panoramas. While the gross features of an image may remain the same, daily and seasonal environmental variation (morning freeze/thaw events, snowfall, leaf-out, etc.) can render the landscape unrecognizable to automated feature recognition software. Likewise, time of day has a major impact on control point detection. To help reduce these errors, the time-lapse player can be configured to only play images from a particular time of day (e.g., "noon" or 11 AM to 2 PM, etc.). In addition, images that are particularly bad for some reason can be automatically or manually flagged to be omitted from the time-lapse. GigaPan recently released an improved version of the stitcher that may help with many of the above issues by allowing users to supply the stitcher with a control panorama with which it will attempt to align new image sets.

Although the above solutions improve usability of the panorama data, the long-term goal for this type of image data should be to develop a system where one could determine how any individual pixel in a given image maps onto the actual physical location on the landscape being photographed. Satellite image data are currently processed in this manner with each pixel in a satellite image being mapped to a GPS coordinate on the earth. Generating a similar dataset from gigapixel time-lapse cameras would provide researchers with an important new data layer to augment conventional satellite data. Such a solution (see Note 7) could potentially be enabled by adding a high precision three-dimensional compass to the camera package. Farther in the future, low-cost LIDAR or similar imaging systems could be used to provide one-time or occasional datasets of extremely high precision landscape data that could be matched with image pixels from the panoramas.

4.7. Infrastructure Requirements

Although hard drive space, bandwidth, and computing costs continue to drop, the Gigavision system can push the limits of a departmental IT infrastructure. The first camera we developed generated approximately 55 million images in less than 9 months. Consequently, it is essential to have an upfront plan and the financial resources available for management and backup of this scale of data.

5. Conclusions

Although there are many technical challenges that must be overcome to generate gigapixel time-series image sets, the current pace of technological change assures us that gigapixel imaging will only become easier and more ubiquitous. We anticipate a day when time-lapse photography and gigapixel imaging systems are a standard feature at long-term field sites. In particular, for larger budget research sites such as those hosting fluxnet towers or NEON infrastructure, the availability of high-resolution long time-series imagery would greatly enhance existing research efforts.

As noted in the introduction there is a need for better tools to turn qualitative observational data on phenology into quantitative datasets suitable for rigorous numeric and genetic analysis. The Gigavision system helps this process by enabling temporally precise monitoring of every plant in a landscape. In addition to improving general phenological studies and measurement of climate change impacts on wild plant populations, an ideal application for the Gigavision system is enhancing the flow of information between lab and field studies. High-resolution imagery from field sites allows identification of behavioral outliers in wild populations; these plants can then be grown as cloned populations in precisely controlled greenhouse experiments to examine the relative effects of genes and environment. Likewise, the availability of high temporal and visual resolution data in field populations greatly enhances our ability to quantitatively describe plant growth in common gardens and field plantings of select genotypes of interest identified from greenhouse studies.

Gigapixel imaging also offers a unique opportunity for enabling citizen science, public education, and socially networked science applications. The Gigavision online interface provides an interactive tool that is engaging to non-scientists and can demonstrate to the general population how phenological changes on the landscape actually take place. An effective feature of the GigaPan Project web site is that users are encouraged to explore images and mark (via bookmarks) items of interest. Gigavision image sets add an additional level of interactivity via time-series data that can document change on the landscape across over time periods of decades.

As climate change, human impacts and similar pressures increasingly threaten ecosystem function, it is crucial that we develop effective tools for monitoring landscapes in high resolution. Gigapixel scale time-lapse imaging systems also have the potential to revolutionize landscape ecology. Such systems can exponentially increase the visual and temporal resolution of phenological data that is currently being collected. Gigapixel time-series imagery provides a quantitative tool both to measure population, group and individual scale variation in a field setting and to identify outliers of interest. The dropping cost of whole-genome sequencing suggests that it won't be long before it is cost effective to sequence all trees and annuals (if not all plants) in a field site. Taken together, these advances provide a significant new toolset for enabling next-generation phenomics (5) combining quantitative measures of phenological variation observed with Gigavision cameras with micro-scale climate data and individual and population level genomic data.

6. Notes

1. Potential developers should be forewarned that Canon requires users to register to receive their SDK and it often takes months or longer to receive access to the software. Canon also provides almost no technical support for their SDK and the sample programs are poorly documented. Consequently, potential developers must have a moderately high level of programming expertise to work with the camera control software. Web search: "Canon SDK" for a link to the Canon developers program.
2. Panasonic has an SDK but their cameras require the user to manually push a button to enable USB control of the camera; Nikon has an SDK but it works only with a limited selection of models.
3. Note that a DSLR camera with a zoom lens and attached USB and power cords is quite bulky and a significant challenge has been to find camera housings large enough to permit a 360°×25° range of motion. We are currently using a 16 in. dome housing which is the largest camera housing we could find on the market (Appendix A, Supplementary Materials),

but even with this huge dome, the camera lens can only be extended to about 150 mm.

4. Although the Phidgets controls work reliably when wired correctly, care must be taken to reduce electrical interference due to grounding issues when switching power on the servos and camera. To reduce electrical interference, all USB cables must use 24-gauge or heavier wiring wherever possible (look for "24/2 C" on the cable). In addition, all electrical wires should be 14- or 12-gauge wherever possible. All wiring and USB cables must be wrapped with ferrite cores (see for example: DigiKey, part number 445-2047-ND). We also recommend reading the documentation provided by Phidgets on reducing electromagnetic interference.
5. Adequate winterization is crucial if a system is to run year round in very cold climates. A significant problem is that if the system runs out of power due to lack of sun, one cannot just assume the system will start again because overnight extreme cold temperatures can drop the voltage of an already drained battery below the typical 9 V cutoff of the charge controller. At that point the charge controller no longer attempts to charge the battery and the system will be offline until either the batteries are directly wired into solar panels to recharge, or the batteries are removed from the field, warmed and recharged—no small feat with 120 lbs of batteries at a remote field site in winter conditions.
6. Microsoft's Photosynth project (http://photosynth.net) provides an alternative programming interface for online interaction with gigapixel images that we did not investigate. Photosynth supports visualization of High Dynamic Range panoramas.
7. A 3 gigapixel resolution image is approximately 650 million times the resolution of MODUS satellite imagery (Gigavision camera ~1 pixel per square centimeter, MODIS, 1 pixel = $(250\ \text{m})^2 = 650$ million cm^2).

Appendix A
Sample full parts list and sourcing information for a Gigavision System

Item	Cost (2011)	Part number	Company URL	Notes
Camera (DSLR)				
Canon Rebel T2i 18MP	$850	B0035FZJI0	Amazon.com	Body only
250 mm lens	$230	B0011NVMO8	Amazon.com	
Amazon SquareTrade 3 years warranty	$60	B001N82JN4	Amazon.com	Amazon 3 years warranty
Camera shipping	$30			
Canon DC adapter with battery insert	$45	HANKEN ACK-E8 AC	Amazon.com	
DSLR camera subtotal	$1,215			
Camera (IP)				
Stardot 5MP IP camera	$750		StarDot.com	
16–48 mm Lens, megapixel lens	$250			
IP camera subtotal	$1,000			
PC				
Diskless Fit-PC2, 2 GB RAM	$474		Fit-pc.com	6 weeks lead time on orders; Make sure to order PC with auto-start capabilities
Nano Biscuit PC (higher reliability PC)	$4,000		video-monitoring.com	*A higher reliability but substantively more expensive PC option is the Nano biscuit system made by Erdman*
PC shipping	$30			
Windows XP home	$100		Amazon.com	
80 GB solid state drive	$200	SSDSA2MH080G2R5 R	Amazon.com	
Super mini trackball keyboard	$56	KBSMOTB	Fentek-ind.com	

Keyboard cover	$13	KBSeal		
Keyboard shipping	$13			
10”, 12v monitor w/shipping	$260	CY201046	*Sunny Island Tech* lcd-touchscreen-monitor.com	Touch screen options available
(2) WD 200 GB USB powered HD for data transfer	$130, ~$65 each		Amazon.com	
PC subtotal	$1,276			
Pan Tilt System				
90° Aluminum 2”: mounting plate	$13	ALM-90-2.00	ServoCity.com	
3/8” bore clamping hub	$8	3172CH	ServoCity.com	
5-40 × 3/8” Pan Had Philips Machine Screws	$2	90272A126	ServoCity.com	
Gear Drive Pan System (5:1 ratio, metal gears)	$60	GDP785A-BM	ServoCity.com	
HS-785HB Servo (universal connector)	$60	33785S	ServoCity.com	
HS-785HB Servo (universal connector)	$40	33785S	ServoCity.com	
785 Servo Power Tube Gearbox-5.0 (5:1 ratio)	$50	SPG785A-5.0	ServoCity.com	
Production assembly	$30	AL200	ServoCity.com	
Shipping	$15			
PanTilt subtotal	$277			
Phidgets microcontrollers				

(continued)

Appendix A (continued)

Item	Cost (2011)	Part number	Company URL	Notes
Advanced Servo 8-motor	$90	C-100-SV-P1061	TrossenRobotics.com	
8/8/8 board	$80	C-200-P1018	TrossenRobotics.com	
0/0/4 Power switch	$60	C-200-P1014	TrossenRobotics.com	
Voltage sensor	$19	S-50-P1135	TrossenRobotics.com	
Current sensor	$31	S-50-P1122	TrossenRobotics.com	
Temperature sensor	$15	S-60-P1124	TrossenRobotics.com	
Phidgets subtotal	$295			
Power and power converters				
Solar components				
135w solar panel	$360		*N. AZ Wind & Sun* solar-electric.com	
10 Amp solar charge controller	$60		*N. AZ Wind & Sun* solar-electric.com	
60w Panel mount	$50			Usually easiest to build one on site rather than buy one
(2) 115Ahr Sealed AGM Deep cycle battery	~$400 ($200each)			We recommend finding a battery with an extended warranty, preferably from a local retailer in case returns or servicing are needed. In the US, Costco and Sears are good options

Flexcharge 12v timer	$80	Flextimer	solar-electric.com	The DCDC-USB by MiniBox may be a good alternative for an external 12v switch that can be reprogrammed from the PC. http://www.mini-box.com/DCDC-USB
Heavy duty tupperware or cooler for battery housing	$25			
Power regulation and distribution				
SWADJ—camera voltage regulator	$15	DE-SWADJ	DimensionEngineering.com	Not needed with Netcam
Breakout board—Camera	$10	Vreg Breakout	DimensionEngineering.com	
Breakout board—Hub	$10	Vreg Breakout	DimensionEngineering.com	
Btron DB8-ACF	$26	DB8-ACF	www.b-tron.com	8 port 12v fused power distribution board. *Note, these boards are a weak link in the system and easily shorted out. We haven't found a good alternative yet*
Btron Shipping	$15			
USB hub	$60	ST4200USBM		USB hub is optional but useful
Power tip and cord—camera	$10		Radio Shack	
12v case fan	$2		Radio Shack	
12v temp switch	$15		DigiKey.com	
Power subtotal	$958			

(continued)

Appendix A (continued)

Item	Cost (2011)	Part number	Company URL	Notes
Computer housing				
14 × 12 × 7" Vented Weatherproof NEMA Enclosure	$159.99	NB141207-00V	L-Com.com	
Universal Pole Mounting Kit-Pole Diameter 1-1/4 to 2"	$39.99	HGX-PMT16	L-Com.com	
USB cable, Waterproof Panel Mount A Female → A Male, 0.5 m	$17.50	WPUSBAX-05M	L-Com.com	Inside case connection for USB-dome wiring
USB Cable, Waterproof Panel Mount A Female → A Male, 0.5 m	$17.50	WPUSBAX-05M	L-Com.com	Inside case connection for USB-dome wiring
USB Cable, Waterproof Type A Male → B Male, 2.0 m	$19.00	WPUSBAB-2M	L-Com.com	USB cable to servo in dome
USB Cable, Waterproof Type A Male → B Male, 2.0 m	$19.00	WPUSBAB-2M	L-Com.com	USB cable to camera in dome
Premium USB Type A–B Cable, 0.5 m	$6.95	CSMUAB-05M	L-Com.com	USB cable for servo
Premium USB Type A–B Cable, 1.0 m	$7.45	CSMUAB-1M	L-Com.com	USB cable for sensors
Nylon Nut for Housing Size 13/16-28	$5.25	WPNUT-13/16-28	L-Com.com	
EPDM Rubber Gasket for Housing Size 13/16	$2.25	WPGASKET-13/16	L-Com.com	
Liquid Tight Cable Gland PG-16	$3.99	ASR-PG16	L-Com.com	
Shipping	$60			
PC housing subtotal	$359			

Camera housing				
16"-outdoor/indoor dome hsg w/ wall mount, clear	$340	VL-SDW16C	Video Alarm	
APM3—Aluminum pole mount bracket & female inserts	$36	VLM-APM3	Video Alarm	
APC2-Pole clamps for APM3, PM3, ACH13's: SS banding w/easy bolt together buckles	$25	VLM-APC2	Video Alarm	
Shipping	$50			
Mounting hardware, misc	$50			
Housing subtotal	$501			
3G wireless				
USB760 Card	$100		http://3gstore.com	
Antenna System	$100		http://3gstore.com	
Cell service (/year and $60/month)	$720		http://3gstore.com	
3G total	$920			
Misc wiring and hardware	$300			
System total	*Hardware total (with parts shipping): $6000–$11,000 Labor costs not included*			*Prices include estimated shipping*

See http://www.Gigavision.org for more suggestions and resources

References

1. Whitham TG, DiFazio SP, Schweitzer JA, Shuster SM, Allan GJ, Bailey JK, Woolbright SA (2008) Extending genomics to natural communities and ecosystems. Science 320: 492–495
2. Bailey JK, Schweitzer JA, Ubeda F, Koricheva J, LeRoy CJ, Madritch MD, Rehill BJ, Bangert RK, Fischer DG, Allan GJ, Whitham TG (2009) From genes to ecosystems: a synthesis of the effects of plant genetic factors across levels of organization. Philos Trans R Soc Lond B Biol Sci 364:1607–1616
3. Morisette JT, Richardson AD, Knapp AK, Fisher JI, Graham EA, Abatzoglou J, Wilson BE, Breshears DD, Henebry GM, Hanes JM, Liang L (2009) Tracking the rhythm of the seasons in the face of global change: phenological research in the 21st century. Front Ecol Environ 7:253–260
4. Nagler P, Scott R, Westenburg C, Cleverly J, Glenn E, Huete A (2005) Evapotranspiration on western U.S. rivers estimated using the Enhanced Vegetation Index from MODIS and data from eddy covariance and Bowen ratio flux towers. Remote Sens Environ 97:337–351
5. Houle D, Govindaraju DR, Omholt S (2010) Phenomics: the next challenge. Nat Rev Genet 11:855–866
6. Richardson AD, Braswell BH, Hollinger DY, Jenkins JP, Ollinger SV (2009) Near-surface remote sensing of spatial and temporal variation in canopy phenology. Ecol Appl 19:1417–1428
7. Leninger B, Edwards J, Antoniades J, Chester D, Haas D, Liu E, Stevens M, Gershfield C, Braun M, Targove JD, Wein S, Brewer P, Madden DG, Hassan Shafique K (2008) Autonomous real-time ground ubiquitous surveillance: imaging system (ARGUS-IS). In: Proceedings of SPIE, the International Society for Optical Engineering. Society of Photo-Optical Instrumentation Engineers, pp 69810 H.1–69810 H.11
8. Heckbert PS, Goldberg M, Donnell GO, Henderson R, Tew K, Sargent R (2010) How many pixels? Statistics from the GigaPan Web Site. In: Nourbakhsh IR, Sargent R (eds) Proceedings of the Fine International Conference on Gigapixel Imaging for Science, Nov 12–13, 2010. Carnegie Mellon University, Pittsburg, pp 1–8 Accessed on July, 2012 http://repository.cmu.edu/gigapixel/
9. Nourbakhsh IR, Sargent R (eds) (2010) The Fine International Conference on Gigapixel Imaging for Science. Carnegie Mellon University, Pittsburgh, PA. Accessed on July, 2012 http://www.cmu.edu/events/gigapixel-science/
10. http://gigapansystems.com/stitch
11. Microsoft Visual Basic 2003. Microsoft Corporation, Redmond
12. http://rsync.samba.org/

Chapter 8

High-Throughput Phenotyping of Plant Populations Using a Personal Digital Assistant

Raju Naik Vankudavath, Reddaiah Bodanapu, Yellamaraju Sreelakshmi, and Rameshwar Sharma

Abstract

During many biological experiments voluminous data is acquired, which can be best collected with portable data acquisition devices and later analyzed with a personal computer (PC). Public domain software catering to data acquisition and analysis is currently limited. The necessity of phenotyping large plant populations led to the development of the application "PHENOME" to manage the data. PHENOME allows acquisition of phenotypic data using a personal digital assistant (PDA) with a built-in barcode scanner. The acquired data can be exported to a customized database on a PC for further analysis and cataloging. PHENOME can be used for a variety of applications, for example high-throughput phenotyping of a mutagenized or mapping population, or phenotyping of several individuals in one or more ecological niches.

Key words: Computable phenotypic database, Computer-aided data acquisition, Functional genomics, High-throughput phenotyping, Personal digital assistant

1. Introduction

Since the domestication of crop plants, farmers have selected better yielding plant varieties based on phenotypes. Plant breeders have successfully combined desirable phenotypic characters from different cultivars/species of crop plants to achieve higher yields or disease resistance. Presently a great deal of emphasis is placed on genomics-assisted breeding, where the phenotype data of a large population is correlated with the genotype data of that species (1). Similarly ecologists determine plant phenotypes at different locations under different environmental conditions, such as availability

Jennifer Normanly (ed.), *High-Throughput Phenotyping in Plants: Methods and Protocols*, Methods in Molecular Biology, vol. 918, DOI 10.1007/978-1-61779-995-2_8,

and spectrum of ambient light, temperature, soil conditions, and nutrients. All these studies require collection of detailed records of different phenotype characters from a large number of plants (2). Additionally these records are required to be linked to other parameters such as genotyping/genomics data or environmental conditions. The collection, retrieval, and analysis of large amount of phenotype data requires availability of tools that expedite data collection and analysis (3). Often the data collected is to be correlated with other databases and correlation between different parameters that may involve genomic or environmental information. It is desirable that the available tools are high-throughput, allow researchers to accumulate, access, incorporate and organize phenotypic databases across populations, and also allow correlating the phenotypic information with other parameters such as genomic data (3, 4).

Development of innovative hardware devices in combination with advances in mobile computing have made available a range of gadgets that can be used for high-throughput phenotyping of plants (5). The personal digital assistants (PDAs) are approximately 1/4th of the size of notebook computers and offer mobility akin to cellphones. These PDAs accept input from a user either via a dual keyboard that is a touch screen and dedicated keyboard, or an incorporated handwriting recognition pad with a stylus. One of the important features of these PDAs is the integral barcode scanner that allows identification of the object using predetermined codes with various densities and dimensions of 1D, 2D, etc. (6, 7), and collection of information about the object using menu-driven software. Mobile computing is currently used in diverse sectors ranging from tracking the movement of airline baggage to keeping records of laboratory experiments (2, 8). It is expected that in future such devices may replace the conventional data notebooks used in research laboratories.

These devices are supported with a variety of open source or commercial software e.g., Wireless Internet Service provides Windows-CE users a stable link and contact between PDAs and the PCs for bidirectional data transfer (http://www.microsoft.com/windowsembedded/en-us/products/windowsce/default.mspx). In addition, a variety of programming languages are available to develop software applications for mobile computing. One popular programming language used for the PDAs is Compact Application Software Language (CASL) (http://www.caslsoft.com), which is based on the well-known computer language Beginner's All-purpose Symbolic Instruction Code (BASIC). The identification of objects with a barcode allows efficient database management and tracking of the data (8). In this chapter we describe the use of customizable phenotyping software named PHENOME developed using CASL. This PHENOME software application is loaded in a PDA with in-built barcode scanner for object identification. Using PHENOME software, a large number of tomato plants were phenotyped (9) belonging to an EMS-mutagenized population raised for TILLING (10, 11).

2. Materials

2.1. Hardware

Execution of the "PHENOME" software application depends on the hardware configuration of the PC and PDA (see Note 1 and Fig. 1). Advanced hardware configuration enhances the software capability for faster execution and larger storage of plant phenotype data.

1. PDA: We used the Symbol™ SPT1800 PDA. However any model of PDA enabled with a 1D-barcode Scanner, Touch Screen and USB connectivity to transfer data to the PC can be used.
2. PC: The minimum hardware configuration for the PC is 64 MB RAM, hard disk space of 20 GB with Windows XP.
3. Barcodes, Barcode Printers, and Barcode Scanners: We used 1D barcode labels for tagging the individual plants (see Notes 2 and 3). The barcodes were printed on polyester labels using a Zebra Technologies printer. The barcode scanner source codes for the SPT1800 model PDA device used for the study were downloaded from the following web link: http://www.caslsoft.com/links.htm#DevResources.

2.2. Software and Platform Technologies

1. The "PHENOME" software application was developed using seven important modules.
   (a) CASLide version 4.3 software (http://www.caslsoft.com).
   (b) PRCTools version 2.0 (http://prc-tools.sourceforge.net/).
   (c) CYGWIN version B.2.0 (http://cygwin.com/).
   (d) GCCTools (http://cygwin.com).
   (e) PalmOS Software Development Kit (Symbol PalmOne Desktop Software, included with Conduit and HotSync).
   (f) ODBC (object database connectivity).
   (g) Barcode labels (1D barcode).

   All of the above software modules are freely obtainable under GNU Public License.
2. The above-mentioned software used for the code development, software application execution, data-synchronization processes of the software application, and the web site link addresses for those software tools downloaded are cited in the PHENOME software application package that can be obtained from "Supplementary Material" on the publisher's web site (http://extras.springer.com).

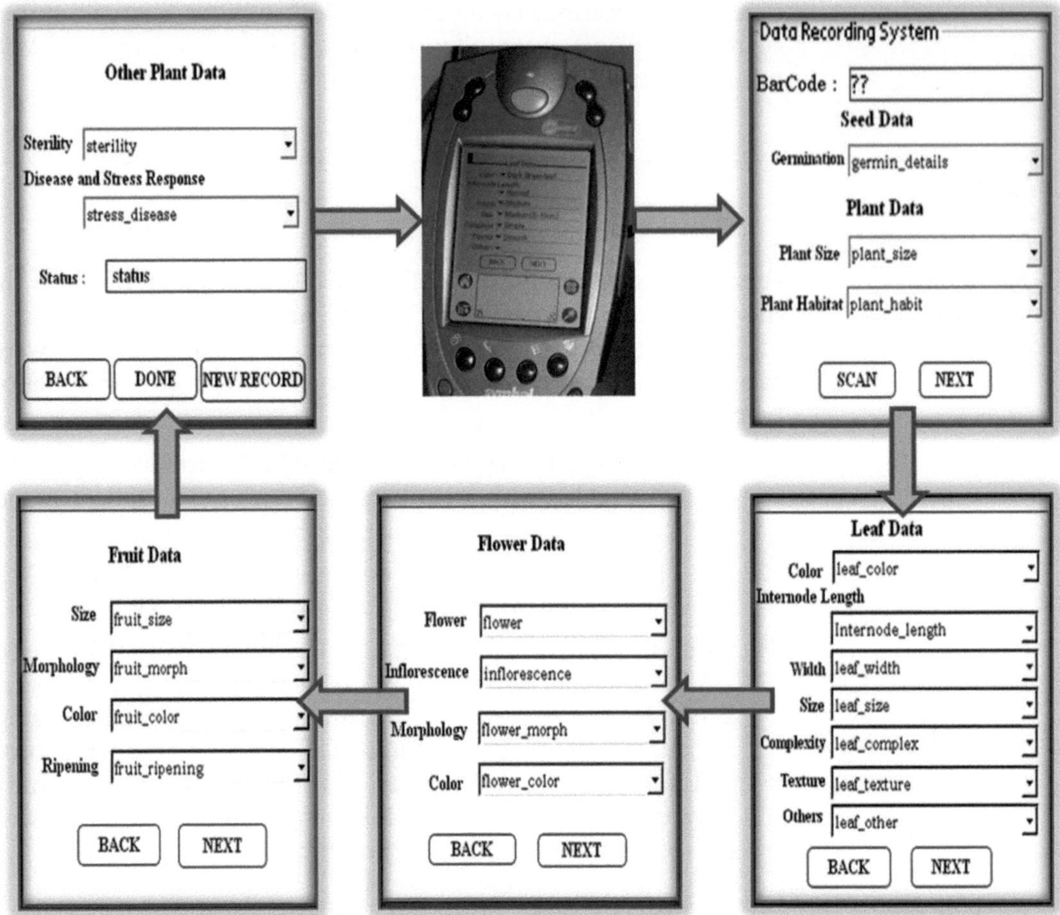

Fig. 1. The cycle of phenotype data collection using PDA with PHENOME software. Each plant is assigned a unique barcode and labels are tagged to plants. Using PDA phenotype data of plant is collected. Once the PDA is synchronized with PC, the data from PDA database is synchronized with MS-Access database. The process to collect and synchronize the plant phenotype data using PDA from the field is cyclic process and allows data collection at different developmental stages of plants.

3. The core software language CASL (see Note 4) was used to code and develop the PHENOME software application (see Note 5) for a PDA in the editor and compiler software CASLide v4.3.
4. In CASLide software, the integrated PRCTools module compiles the project file using GCCTools on Cygwin vB.2.0 Linux emulator software, on Windows XP OS to compile and produce the Palm OS (PDA Operating System) compatible p-code (pseudo-code).
5. An Object Data Base Connectivity (ODBC) System data source stores information about connecting the PC to the remote data device. A system data source is visible to all users on the

machine. The ODBC on Windows XP OS was configured for data-synchronization between the remote device PDA and the database (Microsoft Access) on a personal computer (PC).

3. Methods

3.1. PHENOME Software Development and Installation

3.1.1. CASLide Software Installation and CASL Programming Language

1. The CASLide was used for developing software for the PDA for designing the electronic forms and editing the executable programs code of the PHENOME software (Fig. 2).
2. To install "CASL v4.3" software browse the file "CASL43. exe" from "Supplementary Material" on the publisher's web site (http://extras.springer.com) and follow the default install settings to install "CASL43.exe." This will install base CASL 4.3 (includes: CASLPro, CASLWin, CASL PocketPC and supporting files) and CASL Sample Applications on PC with this path "C:\Program Files\CASLSoft\CASL43\."

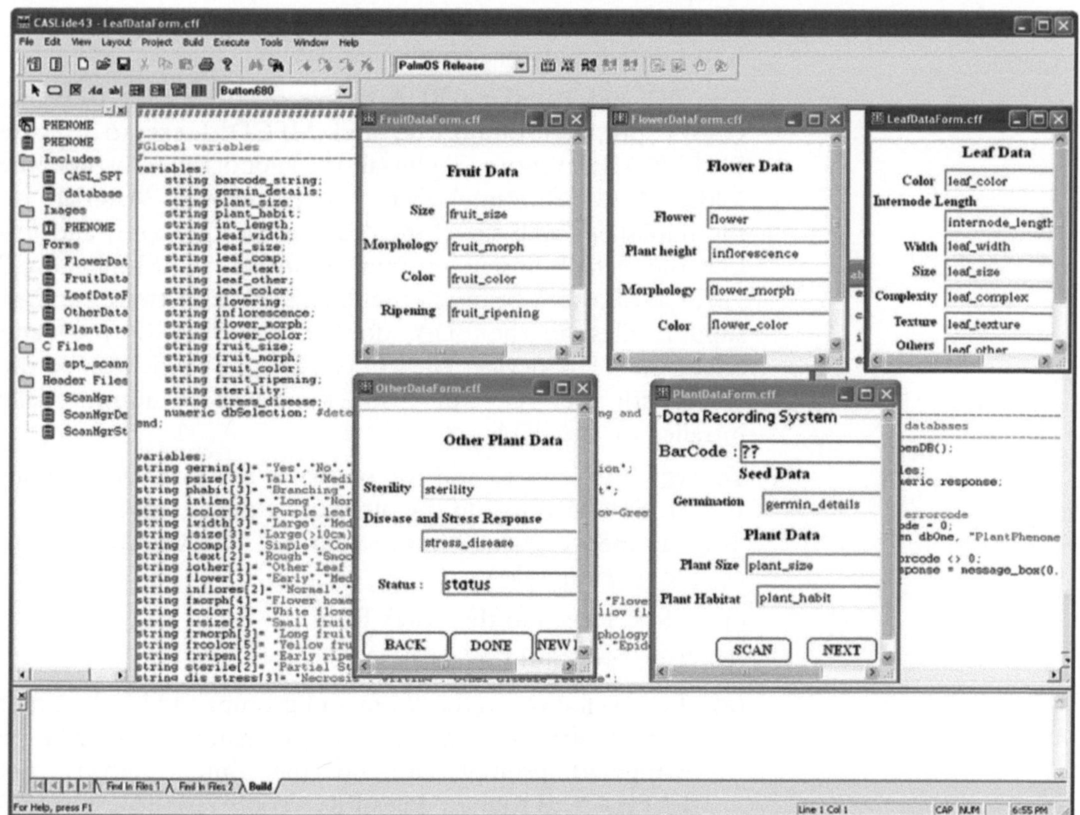

Fig. 2. CASL integrated development environment of PHENOME software application project file.

   (a) Now, to configure conduit software on your PC browse the CASL43 folder installed on the PC i.e., "C:\Program Files\CASLSoft\CASL43\," copy these files viz., "CASLcn20.dll" and "CondReg.exe" and paste it in the drive path "C:\."
   (b) Setting the HotSync Conduit (for integration of CASL database files on the PDA and PC): Open the "Command Prompt" (Start->Run->type "cmd" and click "OK" button) and follow the steps.
      (i) C:\Documents and Settings\Username>cd..
      (ii) C:\Documents and Settings >cd ..
      (iii) C:\ >CondReg "C:\."
      (iv) C:\ >exit.
3. To install "CASL Components" software, browse the file "Casl components.exe" from "Supplementary Material" on the publisher's web site (http://extras.springer.com) and follow the default install settings to install "Casl components.exe." This will install the following software: GCC Installer, PRC-Tools Setup (C and C++ Tool Chain, Documentation and Sample Programs), Palm SDK, CASL Productivity Pak, and Cygwin B20 Linux Emulator Setup.
4. The PC and PalmOne PDA device need support files viz., CASLrt.prc, CASLfonts.exe, and MFCDLL.exe to execute PHENOME software. To install the support files, browse the folder "CASL Runtime" from "Supplementary Material" on the publisher's web site (http://extras.springer.com) and follow the default install settings to install "CASLfonts.exe," "MFCDLL.exe." To install the PHENOME software runtime support files on the PDA, double-click "CASLrt.prc" and select the PDA device, click OK. Now, just synchronize the PDA with the PC, and CASLrt.prc will be automatically installed on the PDA.
5. You will also have to add the prc-tools bin directory to your PATH. You need to add the directory C:\Program Files\PRC-Tools\H-i586-cygwin32\bin to your AUTOEXEC.BAT. Here are the steps to follow:
   (a) Click Start, and then click Run.
   (b) Type cmd and then click OK.
   (c) Type as follows in the command prompt and hit the Enter button (alternatively, copy this command and paste in the command prompt and hit the Enter button). Set PATH="C:\Program Files\PRC-Tools\H-i586-cygwin32\bin;%PATH%."
   (d) Type exit (this closes the command prompt DOS window) and hit the Enter button.

Fig. 3. Photograph showing a large tomato population with bar-coded labels tied to each plant. The labels are weather resistant and last for the plant's lifetime. The *inset* shows the PDA used for phenotyping this population.

To modify the plant phenotypic catalog (see Note 6 and Fig. 3) in the PHENOME software (after installation), follow the steps described below.

6. To open the code editor, double-click on the CASLide v4.3 software icon on the PC.
7. To open the PHENOME software source code for editing in CASLide v4.3 software, click on the File menu, then click on Open Project. To browse PHENOME source code, click on the Browse button for a file D:\PHENOME\PHENOME PROJECT\PHENOME.CPJ, select and open the file in the editor.
8. PRCTools is an assembly of compilers that supports C and C++ programming languages (GCC and C++ compilers) for the PDA and compiles these project program codes and generates files with the extension .MAK. The assisting Linux emulator Cygwin software transforms the compiled project file from .MAK to Palm OS's version executable .PRC file, which is consequently accepted by the HotSync manager of the PalmOne Desktop software for installing the complete personalized software on the PDA.

The details of variable data types for different plant characters in project file and block diagram of PHENOME software application (Fig. 4) are presented in the following sections.

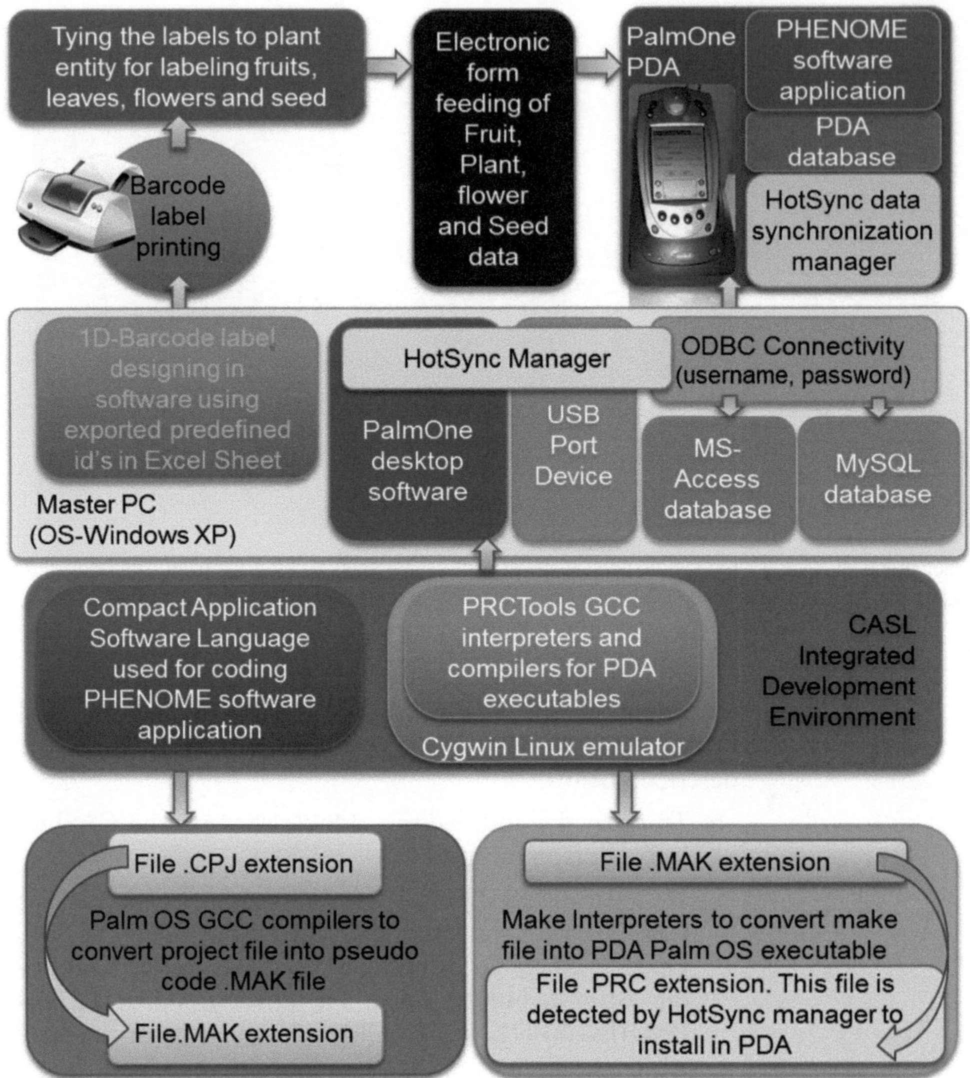

Fig. 4. The schematic representation of the PHENOME software application. Phenome is installed on the PC along with the required software viz. CASLide V4.3, PRCTools, Cygwin, PalmOne Desktop, and BarTender softwares. Barcode labels for tagging the plant can be printed using standard software. PHENOME software was developed on CASLide software. The internal making for PHENOME was processed through PHENOME.CPJ project file to PHENOME.MAK make file using PRCTools software. Thereafter PHENOME.MAK makes file to PHENOME.PRC executable for PDA using CYGWIN software. PHENOME.PRC is installed on PDA using PalmOne Desktop software with HotSync manager. HotSync manager is also used to synchronize the data from PDA to MS-Access database on PC.

3.1.2. Barcode Labels

The essential requirement for data acquisition is a unique barcode identity for every individual plant. The barcode labels were designed using BarTender v7.5.1 Enterprise software (Fig. 5).

1. Collate the plant identity numbers in MS-Excel. Retrieve the excel data sheet into Bartender software in design mode and print the custom-designed labels.

Fig. 5. Barcode designing and printing using bartender software on PC.

2. Print the labels on synthetic polyester labels using a Zebra Technologies printer. Alternately labels can also be printed on non-erasable, tear proof paper using a conventional laser printer.
3. The complete Barcode Labels design, configuration steps and procedure are provided in the BARCODE.PPT file which you can download from "Supplementary Material" on the publisher's web site (http://extras.springer.com) with images and configuration descriptions. The detailed configuration can be found in Bar-Code Label Printing Software Help Topics.
4. Tag each individual plant with these labels. Read these labels with the PDA barcode scanner.

3.1.3. Project Files and Data Types

1. The PHENOME personalized software application is assimilated with six sub-module program files (a) project files, (b) include files, (c) C files, (d) image files, (e) form files, and (f) header files.
2. The project file with file extension name .CPJ is the primary file of the project. It provides particulars of the software application prerequisites.

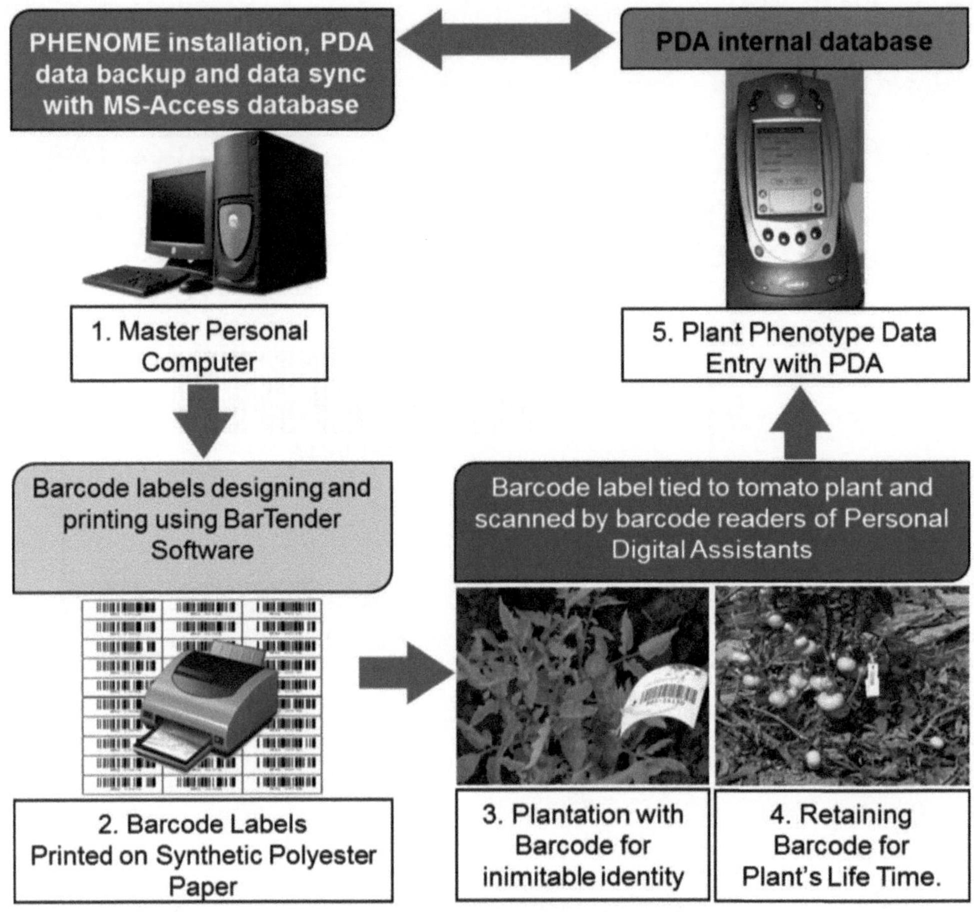

Fig. 6. The flow of interactive display of PHENOME software on PDA for user input. Data is fed into PDA after identifying plants by scanning the barcode label. Different electronic forms are displayed on by tapping on NEXT or BACK button. By tapping the DONE button on the last form, record is submitted in the PDA database.

3. In project file, the remotely referenced/imported sub-module "include" contains (a) "spt_scanner" files and (b) "database" files. The "spt_scanner" files are accountable for accessing the functionality of the barcode scanner operation, feature interface definition to "C file" functions. The "database" file assimilated with database connectivity definitions for saving phenotype data i.e., PDA data has a destination database name on the PC viz. MS-Access database "database_source_name: 'CaslExcel'."

4. The "form files" are multiple electronic forms (see Note 7 and Fig. 6) files that were used to design the customized look of the PHENOME software for recording phenotype characters of individual plants. The data are collected and stored using string data type, through electronic form that also assists to retrieve, view, and edit the data records on PDA.

5. The PHENOME software application's barcode scanner auxiliary files called "header files" or "library files" are (a) "ScanMgr," (b) "ScanMgrStruct," and (c) "ScanMgrDef."
6. The "Image files" are the files used for creating logo of the PHENOME software that is displayed on screen during loading of the PHENOME software.

3.1.4. Modifying Values in PHENOME Software

1. To add a new value to a plant catalog in the PHENOME source code, select the PHENOME file in the left hand panel of CASLide v4.3 software editor.
2. Navigate to the section "ADD/REMOVE A NEW VALUE TO/FROM EXISTING PLANT CHARACTERISTIC" and select the line that contains the phenotypic characteristic name that needs to be modified/added/removed.
3. For example for modifying phenotypic characteristic for "germination," in order to add a new value "*Fast Germination,*" the value can be added as follows:
   (a) Original: string germin [4] = "Yes," "No," "Seedling Lethality," "Slow Germination."
   (b) Modified : string germin [5] = "Yes," "No," "Seedling Lethality," "Slow Germination," "*Fast Germination.*"
      (i) Here, the declaration "germin" is phenotypic characteristics name, and [4] is the number of values for respective characteristics.
4. After adding all the required values, save the project, make the PRC file, and load it into the PDA. How to make the prc file is explained in Subheading 3.1.7.

3.1.5. Adding New Values in PHENOME Software

1. To add a new plant catalog in the PHENOME source code, select the PHENOME file in the left hand panel of the CASLide v4.3 software editor.
2. In the PHENOME file navigate to the section titled "ADD/REMOVE THE NAME OF THE NEW CHARACTERISTIC" and add the new line of phenotypic characteristic as follows:
3. string <characteristic name>[number of values] = "Value 1," "Value 2," "Value n."
4. Whatever modifications of phenotypic characteristic are made in the PHENOME file, the same modifications also have to be implemented in "database" file under the "include" category in the left panel of CASLide v4.3 software.
5. Navigate to the section titled "ADD/REMOVE THE NAME OF THE NEW CHARACTERISTIC" and add the new line of phenotypic characteristic name as follows:
6. field <characteristic name>.

7. Navigate to the section titled "ADD/REMOVE A LINE TO DISPLAY THE NEW CHARACTERISTIC" and add the line as follows:
8. <characteristic name>.display = <characteristic name>.
9. Navigate to the section titled "ADD/REMOVE A LINE TO INSERT DATA INTO THE FILE" and add the line as follows:
10. <characteristic name> = <characteristic name>.display.
11. Navigate to the section titled "ADD/REMOVE A LINE TO RESET THE CHARACTER" and add the line as follows:
12. <characteristic name>.display = "".
13. Then access the "Forms" category in the left panel of CASLide v4.3 software and click on the appropriate phenotypic characteristics form viz., FruitDataForm, LeafDataForm, FlowerDataForm, PlantDataForm, etc., where new phenotypic characteristics are to be displayed on the data recording system of the PHENOME software.
14. Add the list box in the same opened form using the tool box menu and write the label name of the phenotypic characteristic beside the list box.
15. To provide the new phenotypic characteristic values in the list box, double-click the list box to get the "Common Properties" window and enter the phenotypic characteristic value names in the field "list names" as provided in the PHENOME file.
16. Then save the project and make the prc file, then load it into the PDA (how to make the prc file is explained in Subheading 3.1.7).
17. In case if you wish to delete a phenotype character, you need to repeat the steps outlined in this section by deleting a character in the respective section and saving the project to make the .prc file to load in the PDA.

3.1.6. Data Recording System

1. The PHENOME data acquisition form displayed on the touch screen of the PDA enables one to enter the data using a drop-down menu (see Notes 7, 8 and Fig. 6).
2. Five electronic forms to obtain phenotype characters of each plant are as follows (a) "PlantDataForm," (b) "LeafDataForm," (c) "FlowerDataForm," (d) "FruitDataForm," and (e) "Other DataForm." The number of these electronic forms can be increased, reduced, or modified by users.
3. A new variable can be added to each of the above forms by altering the program code, respectively, in the primary file, extension name .CPJ (similar characteristic objects alteration should also be done in data acquisition forms).

3.1.7. Creating PDA Executables

The primary file extension name .CPJ of the project is compiled by PRCTools supported by CASLide v4.3 editing software using two different modes for three platforms, a total of six modes. This allows debugging the program code and releasing the software application executable file to test on Windows XP OS prior to installing on the PalmOS and PalmPilot.

PRCTools produces the pseudo-code (p-code)/object/executable files in two different format file extension names, .CSP (needs interpreter to execute) and .PRC (does not need interpreter and executes directly on the PDA) for Windows and PalmOS(PDA)/PalmPilot, respectively.

CASLWin interpreter executes the compiled p-code .CSP file i.e., Phenome software application, for testing on a Windows XP operating system PC prior to installing on the PDA.

To make .prc files follow the step outlined below.

1. Select the "PalmOS Release" from the list box that is present in the menu bar.
2. Click on the "Build" menu and click on "compile" to compile the project file.
3. Once the compilation is successful, then click on the "Build" menu and click on "Make PRC."
4. If the PRC is successfully created, then the build window will display "Successful PRC file creation." In case of failure the errors will be displayed.
5. Click on the "Build" menu and click on "PRC Install" to install the PHENOME.PRC in the PDA.
6. HotSync manager module installs p-code .PRC file i.e., Phenome software application, on PDA and synchronizes data between PDA and PC (Fig. 7).

3.1.8. ODBC Connectivity

The remote phenotype data transfer/synchronization from PDA to PC takes place upon configuring the database name as mentioned earlier in the "database" file i.e., "CaslExcel," in ODBC's "Data Source Name" parameter. With the same data source name, any one of the following database source files can be created for readily accepting the synchronized data from PDA viz., MySQL, Oracle, MS-Access, Excel, etc. For example, we used MS-Access database for plant phenotype data synchronization (Fig. 8). Alternatively password protected DBMS viz., Oracle/MySQL may be directly associated by saving the same password in the "database" program code.

1. Create a blank MS-Access database file in "D:\PHENOME\CASL_ODBC\" and save it as "CaslExcel.mdb" and table name as "PlantPhenome." This is the same name as mentioned in "database" file in PHENOME project. In the table provide first field id as REC_ID (Data Type: Number) and

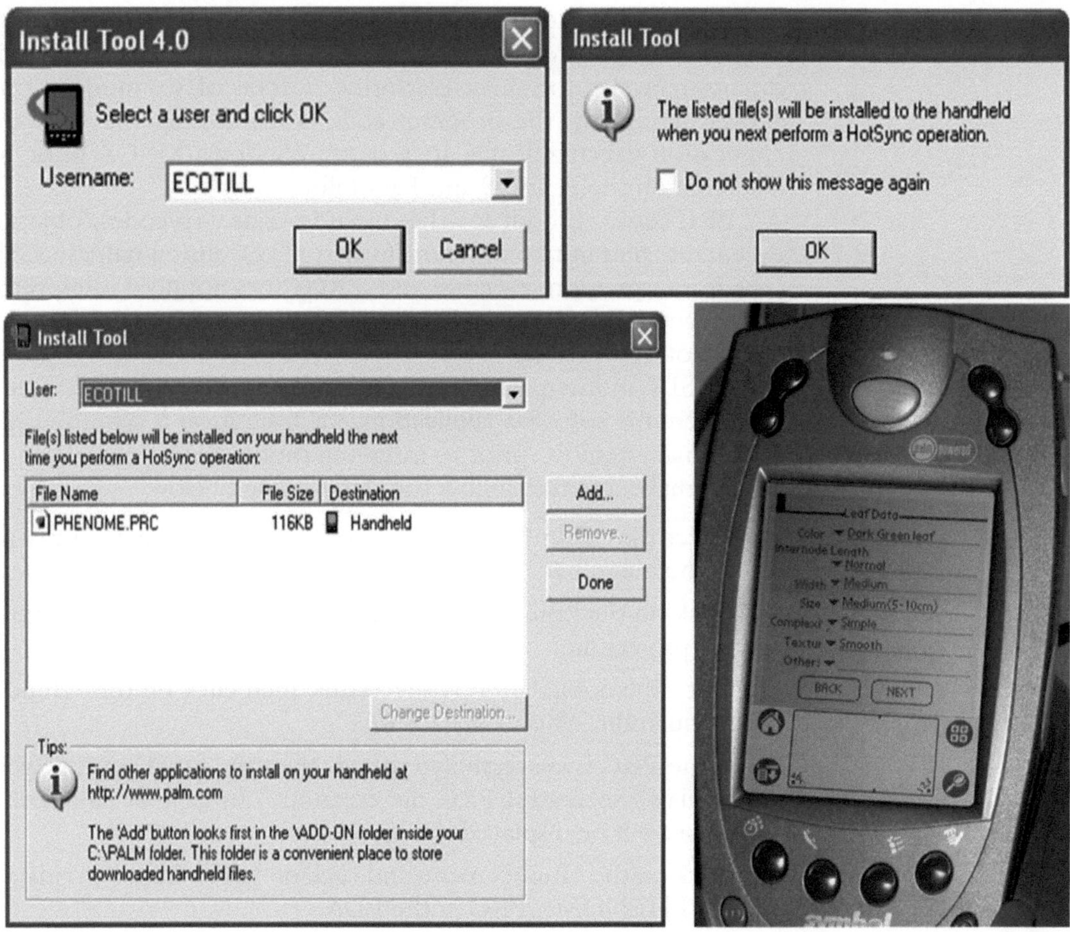

Fig. 7. Installing PHENOME software application using HotSync manager and conduit module of PalmOne desktop software.

make it as Primary Key, second field id as REC_STATUS (Data Type: Number). This step requires the Microsoft office package to be installed on the PC. Alternatively free office packages viz., Open Office, IBM Symphony's Office Package, etc. can be used. To open the Microsoft Access file click Start, click All Programs, click Microsoft Office, and then click on "Microsoft Access 2003" and save the file as mentioned. Alternatively you may check how to create the MS-Access database file at this web link www.officetutorials.com/Access%202003%20tut.DOC.

The next step describes how to configure the System DNS ODBC database source name (database source name "CaslExcel" which is same as "database source name" in the phenome project "database" file).

1	REC_ID	REC_STATUS	Barcode	Germination	Plant Size	Plant Habit	Leaf Colour	Internodal L	Width	Leaf Size	Complexcity	Texture
2	8233005	0	F82-0001B55	Yes	Medium	Aborted growth	Dark Green leaf	Normal	Medium	Medium(5-10cm)	Compound	Smooth
3	8233006	0	J82-0001B66	Slow Germination	Small	Aborted growth		Normal	Medium	Small(<5cm)		Smooth
4	8232974	0	M3206M3SED-0002Z	Yes	Tall	Aborted growth	Purple leaf	Long	Large	Large(>10cm)	Simple	Rough
5	8232994	0	M6206M3SED-0002H		Extremely small	Aborted growth	Purple leaf	Long	Large	Large(>10cm)	Simple	Rough
6	8233007	0	M82-0001B6H	No	Medium	Aborted growth			Short	Medium(5-10cm)	Compound	Smooth
7	8233011	0	M82-0001B5L	Seedling Lethality	Small	Aborted growth						Smooth
8	8233008	0	M82-0001B6U	No	Medium	Other plant habit		Short	Medium	Medium(5-10cm)	Compound	Smooth
9	8233010	0	M82-0001B6W	No								
10	8233009	0	M82-0001C5J	Slow Germination								
11	8232982	0	M8206F3SED-0002S		Extremely small	Aborted growth	Purple leaf	Long	Large	Large(>10cm)	Simple	Rough
12	8232980	0	M8206M3EED-0002V		Extremely small	Aborted growth	Purple leaf	Long	Large	Large(>10cm)	Simple	Smooth
13	8232967	0	M8206M3LED-0003F	No	Medium				Short			Rough
14	8233000	0	M8206M3SBD-0002B		Small plant	Aborted growth	Purple leaf	Long	Large	Large(>10cm)	Simple	Rough
15	8232997	0	M8206M3SEA-0002F		Extremely small		Purple leaf		Large			Rough
16	8232984	0	M8206M3SED-0002S		Extremely small	Aborted growth	Purple leaf	Long	Large	Large(>10cm)	Simple	Rough
17	8232968	0	M8206M3SED-0003S	Yes	Tall	Aborted growth	Purple leaf	Long	Large	Large(>10cm)	Simple	Rough
18	8232975	0	M8206M3SED-0003V	Yes	Tall	Aborted growth	Purple leaf	Long	Large	Large(>10cm)	Simple	Rough
19	8232983	0	M8206M3SED-0008U		Extremely small	Aborted growth	Purple leaf	Long	Large	Large(>10cm)	Simple	Rough
20	8232969	0	M8206M3SED-0031D	Seedling Lethality	Small	Other plant habit	Yellow leaf		Short	Small(<5cm)	Compound	Smooth

1	Others	Flowering	Inflorescence		Flower Color	Fruit Size	Morphology	Fruit color	Ripening	Sterility	Other Disease respons
2		Medium	Abnormal	Flower organ size	Pale Yellow flower	Large fruit					
3		Medium	Abnormal	Flower organ width	Pale Yellow flower	Large fruit	Rounded fruit		Late ripening		
4	Other Leaf Development	Early	Normal	Flower organ size	White flower	Small fruit	Long fruit	Yellow fruit	Early ripening	Partial Sterility	Necrosis
5	Other Leaf Development	Early	Normal	Flower homeotic mutation	White flower	Small fruit	Long fruit	Yellow fruit	Early ripening	Partial Sterility	Necrosis
6	Other Leaf Development	Medium	Abnormal	Flower organ width	Pale Yellow flower	Large fruit	Rounded fruit	Dark red fruit	Late ripening		
7					Pale Yellow flower				Late ripening		
8		Medium	Abnormal	Flower organ width	Pale Yellow flower	Large fruit	Rounded fruit	Yellow fruit	Early ripening		
9											
10		Late			Strong Yellow flower				Late ripening		
11	Other Leaf Development	Early	Normal	Flower homeotic mutation	White flower	Small fruit	Long fruit	Yellow fruit	Early ripening	Partial Sterility	Necrosis
12	Other Leaf Development	Early	Normal	Flower homeotic mutation	White flower	Large fruit	Long fruit	Yellow fruit	Early ripening	Partial Sterility	Necrosis
13			Abnormal		White flower			Orange fruit		Full Sterility	Wilting
14	Other Leaf Development	Early	Normal	Flower homeotic mutation	White flower	Small fruit	Long fruit	Yellow fruit	Early ripening	Full Sterility	Necrosis
15			Normal		White flower	Small fruit		Yellow fruit			Necrosis
16	Other Leaf Development	Early	Normal	Flower homeotic mutation	White flower	Small fruit	Long fruit	Yellow fruit	Early ripening	Partial Sterility	Necrosis
17	Other Leaf Development	Early	Abnormal	Flower homeotic mutation	White flower	Small fruit	Long fruit	Yellow fruit	Early ripening	Partial Sterility	Necrosis
18	Other Leaf Development	Early	Normal	Flower homeotic mutation	White flower	Small fruit	Long fruit	Yellow fruit	Early ripening	Partial Sterility	Necrosis
19	Other Leaf Development	Early	Normal	Flower homeotic mutation	White flower	Small fruit	Long fruit	Yellow fruit	Early ripening	Full Sterility	Necrosis
20	Other Leaf Development	Medium	Abnormal	Flower organ width	Pale Yellow flower	Large fruit	Rounded fruit	Orange fruit	Late ripening	Full Sterility	Other disease respose

Fig. 8. The output of PHENOME application in MS-Access file on PC after synchronizing the PDA with the PC.

2. To open ODBC Data Source Administrator, click Start, point to Settings, and then click Control Panel. Double-click Administrative Tools, and then double-click Data Sources (ODBC).

3. Click the System DSN tab, then click the Add button (a System DSN is available to all users of the computer, including Services. If you want to create a DSN that is only available to you, create User DSN) and from the "Create New Data Source" dialog box select "Microsoft Access Database (*.mdb)" click on the Finish button [alternative step: double-click on the Microsoft Access Database (*.mdb)].

4. In the dialog box of "ODBC Microsoft Access Setup" type the Database Source Name as "CaslExcel" and provide the description of the database type you have created (a description is optional).

5. Click the Select button, from the "Select Database" dialog box in the "Directories" section browse the folder for the database file "D:\PHENOME\CASL_ODBC\" (this was created in earlier steps) and select "CaslExcel.mdb" in the "Database Name" section then click OK to set the file path (alternatively double-click "CaslExcel.mdb"), click OK to set the database source and click OK to set the System DSN in ODBC. To associate and set up the PDA database file backup with MS-Excel spreadsheet on the PC, see Notes 9–12 for how to configure the ODBC-system DSN.

6. You will also have to confirm whether RPC and RPC Locator system services were started. To make sure these services are running follow these steps (if these system services were already started ignore the steps):
   (a) Click Start, and then click Run.
   (b) Type services.msc and then click OK.
   (c) In the list of services, right-click on "Remote Procedure Call (RPC)" and then click Properties.
   (d) In the Startup type list, select Automatic and click Apply.
   (e) Verify that the Service status is started, if the Service Status is stopped click on the Start Button.
   (f) In the list of services, right-click on "Remote Procedure Call (RPC) Locator" and then click Properties.
   (g) In the Startup type list, select Automatic and click Apply.
   (h) Verify that the Service status is started, if the Service Status is stopped click on the Start Button.

3.1.9. HotSync Technology

The data acquired using the PDA is synchronized to the ODBC's predefined dbms/MS-Access database file or MS-Excel 1997–2003 file by connecting PDA to PC. Whenever the PDA is connected/synchronized to the PC, an automatic application data-backup file is also generated by hotsync manager (5). USB-to-serial converter driver software may be needed if the PDA is connected to PC through a USB-to-serial converter cable. Install the converter cable model driver software from third party vendor cd on the PC. The data synchronization takes place between the PDA and the PC using the conduit module of CASLide. It is advisable to take the data backup in Excel Spreadsheet prior to every synchronization of the PDA database file after the first data backup population of the file. These backup files will assist in comparing and analyzing the data collected at different development stages of the plant. This also allows multiple users to use the same PDA to collect phenotypic plant data from field. However, users have to make the data-backup file from database file when the PDA is connected to the PC.

1. HotSync Manager has to be restarted for the first time connectivity with the PDA. Right click on the hotsync icon from the system tray ("system tray" is located on the right side of the "task bar" on the PC), select "exit" on the pop-up menu. To restart the HotSync Manger click Start, click All Programs, click SPT Desktop, and click HotSync Manager.
2. To set the connection medium on the HotSync Manager, left click on HotSync Manager Icon from system try, set the tick mark on "Local USB," "Local Serial," and "Modem." Now the connection between PDA to PC is ready to synchronize.

Sometimes PDA connectivity on the PC may need to be checked by connecting an alternative USB port for proper connectivity on the PC.

3.1.10. Installation of PHENOME Software

1. To install the PHENOME software, follow the outlined steps: Download/browse the PHENOME software folder from "Supplementary Material" on the publisher's web site (http://extras.springer.com), copy the folder PHENOME to your system in D:\PHENOME.
2. To install the Symbol PalmOne Desktop Software, double-click on SPTDSKEN-00-4.0.1.exe executable file (D:\PHENOME\SPTDSKEN-00-4.0.1.exe) and follow the instructions to install the software on the PC.
3. Provide the PDA name when prompted to recognize PDA portable device, when the PDA is connected to the PC.
4. Configure Mail Settings based on user's interest. It is recommended to select NO and click "Finish" button to exit the setup.
5. Double-click on PHENOME.PRC (D:\PHENOME\PHENOME.PRC), then click Done to load the PHENOME software in the ready-to-install module and Click OK to get set go. The HotSync Conduit in-built module i.e., "Install" icon present on PalmOne Desktop Software, "Install Tool" module temporarily stores the files to be installed on the two PDA. Genetill and ecotill were the names used for the PDA in the hotsync conduit for recognizing the PDA upon connecting to PC (see Note 13, and Fig. 7).
   PHENOME software is installed on PDA, once it is connected and synchronized to the PC.
6. The data files for unique barcodes can be generated using office application packages like MS-Excel and saved to the PC. To produce unique barcode labels for each plant, BarTender software version 7.5.1 (http://www.seagullscientific.com) or equivalent software can be used.

4. Notes

1. A PDA allows expansion of the computer stored database and permits an easy/quick field data synchronization/transfer to a PC that can be further categorized and investigated according to research project objectives (see Fig. 1).
2. Different types of barcodes available in the market. See the following web link: http://www.adams1.com/stack.html.
3. Though we have not tested them, 2D barcodes can also be used. The 2D barcode can be scanned using the current

generation of smart mobile phones where the integrated cameras can read 2D barcode. These phones can be customized to collect data and also pictures of the plant phenotypes.

4. CASL programming language is analogous to Visual Basics (VB) programming language; however unlike VB, it is a freeware and has provisions that permit development of program code for a PDA, which offers the benefit of the touch screen and integrated barcode reading.
5. The programs with file extension name PHENOME.CPJ are coded for personalized software "PHENOME" using the CASL programming language using editor tool/software CASLide.
6. In our study we used phenotype characters similar to those described by Menda et al. (see ref. 12) to record phenotypes of a mutagenized M2 population of tomato (Fig. 3).
7. The PHENOME software can be used to collect data at various stages of vegetative and reproductive growth in field. The touch screen and drop-down form allows easy navigation (see Fig. 6).
8. Presently obtainable typical data management system tools can also be used for collating phenotypic data for large populations (see ref. 13).
9. Alternatively in the place of MS-Access 1997–2003 database file one can create the data-backup file in MS-Excel 1997–2003 with the same name mentioned in the PHENOME project "database" file as "CaslExcel" (database source name: CaslExcel) and save it as "D:\PHENOME\CASL_ODBC\CaslExcel.xls."
10. For MS-Excel 1997–2003 files configure the System DSN in ODBC's "Database Source Name" as mentioned in the PHENOME project "database" file i.e., CaslExcel (database source name: CaslExcel; as provided in database file). To open the ODBC follow the steps as given in earlier section (see Step 2 of Subheading 3.1.8) then follow the steps as provided here:
    (a) Select the System DSN tab and click the Add button.
    (b) Select a driver for which you want to set up a data source [i.e., for an Excel sheet: Microsoft Excel Driver (*.xls)].
    (c) Sometimes it may be necessary to verify with another driver viz., "Microsoft Excel Driver (*.xls, *.xlsx, *.xlsm, *.xlsb)" available in the list, if synchronized data does not get updated/loaded/backup into the "D:\PHENOME\CASL_ODBC\CaslExcel.xls" file.
    (d) Type database source name "CaslExcel" and provide a description of the database type (for database reference description, this is optional).

(e) Select Version “Excel 97–2000” then click the “Select Workbook” button.

(f) In the “Select Database” dialog box from the “Directories” section browse the “D:\PHENOME\CASL_ODBC” folder and select “CaslExcel.xls” in the “Database Name” section then click OK to set the file path (alternatively, double-click “CaslExcel.xls”), click OK to set the database source and click OK to set the System DSN in ODBC. (Note: Do not forget to uncheck the check box for “Read Only” permission option on the dialog box by clicking “Options>>” button).

11. A reference copy of an MS-Excel file named “CaslExcel.xls” that was already populated by phenotypic data with the internal spreadsheet named “PlantPhenome” can be obtained from “Supplementary Material” at the publishers web site (http://extras.springer.com).
12. Now the PDA is ready to use and connect to the MS-Excel spreadsheet on the PC. The spreadsheet file will be populated once the data is fed into the PDA and synchronized with the PC.
13. The other way to load the PHENOME.PRC using PalmOne Desktop Software is as follows: click the “Install” icon present on the left panel on the PalmOne Desktop software, the “Install Tool” dialog box appears, click the “Add” button and browse the “D:\PHENOME\PHENOME.PRC” file, click the “Done” button to add the file in “Install Tool” for the selected PDA from the drop-down list, and press “OK” to load the PHENOME.PRC while synchronizing.

Acknowledgments

The development of PHENOME software was supported by the Department of Biotechnology, New Delhi, India grant entitled “Genome wide screen for tomato mutants by TILLING.”

Supplementary material

This chapter contains a supplementary material which can be found at the publisher’s website (http://extras.springer.com).

References

1. Bouch N, Bouchez D (2007) Arabidopsis gene knockout: phenotypes wanted. Curr Opin Plant Biol 4:222–227
2. Lussier Y, Liu Y (2007) Computational approaches to phenotyping high-throughput phenomics. Proc Am Thorac Soc 4:18–25
3. Exner V et al (2008) PlantDB—a versatile database for managing plant research. Plant Methods. doi:10.1186/1746-4811-4-1
4. Donofrio N et al (2005) 'PACLIMS': a component LIM system for high-throughput functional genomic analysis. BMC Bioinformatics 6:94. doi:10.1186/1471-2105-6-94
5. Rieger R, Gay G (1997) Using mobile computing to enhance field study. In: Proceedings of CSCL 1997. Ontario Institute for Studies in Education, Toronto
6. Monto H, Kumagai PM (2006) Development of electronic barcodes for use in plant pathology. Plant Mol Biol 61:515–523
7. Shellhammer SJ, Katz J, Goldman R (1996) Method and apparatus to scan randomly oriented two-dimensional bar code symbols. US Patent 5,523,552
8. Leakha H et al (2008) RGMIMS: a web-based Laboratory Information Management System for plant functional genomics research. Mol Breeding 22:151–157
9. Vankadavath RN et al (2009) Computer aided data acquisition tool for high-throughput phenotyping of plant populations. Plant Methods. doi:10.1186/1746-4811-5-18
10. McCallum CM et al (2000) Targeted screening for induced mutations. Nat Biotechnol 18:455–457
11. McCallum CM et al (2000) Targeting induced local lesions IN genomes (TILLING) for plant functional genomics. Plant Physiol 123:439–442
12. Menda N et al (2004) *In silico* screening of a saturated mutation library of tomato. Plant J 38:861–872
13. Köhl KI et al (2008) A plant resource and experiment management system based on the Golm Plant Database as a basic tool for omics research. Plant Methods. doi:10.1186/1746-4811-4-11

Chapter 9

High-Throughput Fractionation of Natural Products for Drug Discovery

Ying Tu and Bing Yan

Abstract

An automated high-throughput method applied to the production and analysis of libraries of natural products for high-throughput biological screening is described. The production of the library includes solid-phase extraction of crude extracts to remove polyphenols, followed by automated preparative high-performance liquid chromatography (HPLC) fractionation. Libraries of fractions are analyzed by an ultra-performance liquid chromatography–UV diode-array detection–evaporative light scattering detection–mass spectrometry system (UPLC/PDA/ELSD/MS) to provide information that facilitates characterization of compounds in active fractions. This system fractionates 2,600 unique natural product samples per year, providing fractions in 0.5–10 mg scale for creation of libraries that could be used for the screening of multiple targets to identify hits for various applications including drug discovery.

Key words: Natural products, Fractionation, High-throughput, Drug discovery, Automation, Preparative HPLC, LC/MS

1. Introduction

Natural products have been a major source of new drugs for medicines (1–3). Many successful drugs were originally synthesized to mimic the action of molecules found in nature, and the discovery of drugs from natural sources continues to provide important lead compounds (4–6). The slow and tedious isolation of highly diverse natural products and the labor-intensive nature of the manual operations hindered the success of natural product-based drug discovery and were partially responsible for the declining interest in natural products in the past decades.

Since then, several improved fractionation methods for natural products have been reported to fractionate crude extracts into

Jennifer Normanly (ed.), *High-Throughput Phenotyping in Plants: Methods and Protocols*, Methods in Molecular Biology, vol. 918, DOI 10.1007/978-1-61779-995-2_9, © Springer Science+Business Media, LLC 2012

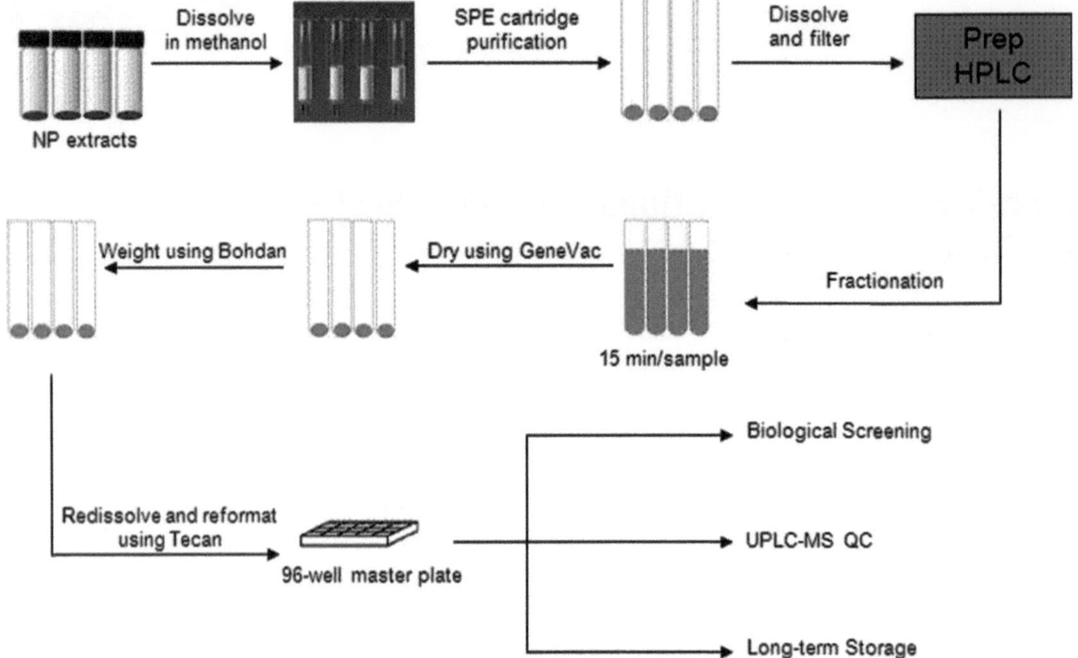

Fig. 1. Flow chart for the automated, high-throughput natural product fractionation system. Reproduced from (14) with permission from the American Chemical Society.

fractions that can be rapidly screened by high-throughput screening (HTS) (7–13). The active fractions can be further purified to determine the structures of the active compounds. A single- or multi-step solid-phase extraction (SPE) method was applied for natural products fractionation (8). The crude extracts were fractionated with solvent mixtures of increasing percentage of methanol and then concentrated via SPE resins. However, the whole process of fractionation and concentration took at least 4 days. Multichannel counter-current chromatography (CCC) and flash chromatography/preparative high-performance chromatography were also reported (9, 10). The fractionation used a large quantity of raw natural product materials with complicated protocols leading to relatively low throughput. Therefore, fast and automated natural products fractionation methods are urgently needed to match the throughput and the speed of modern HTS technology.

This chapter describes a high-throughput and high-resolution automated natural products fractionation method that applies the automation technology to natural product fractionation and reformatting processes that generate high-quality samples for HTS (Fig. 1). In our approach, natural product ethanol extracts are pre-fractionated with polyamide SPE cartridges and then are

	79547	79548	79549	79550	79551	79552	79553	79554	79555	79556	79557	79558
1	12	25.1	28	11.2	14.6	27.1	12	18.5	6.3	8.9	21.8	3.9
2	0.8	4.2	3.1	1	1.2	9.6	1.5	2.4	1.5	1.6	3.9	0.7
3	0	0.7	0.3	0.4	0.2	2.5	0.4	0.9	0.4	0.3	0.5	0.2
4	0.2	0.8	0.3	0.3	0.3	0.7	0.4	1.7	0.5	0.3	0.4	0.3
5	0.9	0.8	1.2	0.4	0.3	0.5	0.6	3.7	0.8	0.7	0.3	0.3
6	1.3	1.1	0.7	0.4	0.4	0.8	0.6	1.6	0.8	0.6	0.6	0.5
7	1.5	1.2	0.7	0.5	0.6	0.8	0.6	3.7	0.7	0.7	0.7	0.4
8	2.1	0.8	0.8	0.4	0.7	2.3	0.7	3.7	0.7	0.6	0.8	0.4
9	7.3	1.2	0.8	1.1	1	2.9	0.9	3.4	0.9	1.1	1.3	0.4
10	10.8	1.3	1	24.6	1.8	1.6	1.4	5.8	0.9	1.3	3.8	0.6
11	5.5	1.2	1	9	2.2	1.1	1.3	34.2	2.2	1.7	5.3	0.8
12	4	1.6	1.2	5	2.7	0.9	1.1	8.9	4	2.1	7.5	0.8
13	2.2	1.1	1.1	1.3	2.9	0.5	1	1.6	4.1	1.7	3.9	1.1
14	1.9	1.8	2.2	0.9	4.8	0.2	1	0.9	4.5	1.6	5.2	1.3
15	3.3	2.1	2.3	0.9	6.6	0.3	1.6	0.8	5.3	3.1	2.7	2.7
16	2.3	2	3.4	2.5	8.7	0.4	5.3	1.4	10.5	7	2.6	4.7
17	3.3	2.5	3.4	3.6	6.9	0.6	6.8	2.1	8.7	5.8	2	6.2
18	1.6	2	2	1.8	2.2	0.6	3.2	2	4.1	4	2	4.1
19	0.8	0.9	1.3	1	1.4	0.3	2.4	1.4	1.8	3.4	2.2	3.8
20	0.6	1.1	1.5	0.9	1.2	0.5	1.7	1.4	1.5	2.8	2	2.5
21	0.4	0.7	1	0.9	0.8	0.2	1.1	0.7	1.2	2.1	1.4	1.8
22	0.3	0.5	0.8	0.7	0.6	0.1	1.1	0.8	1.1	1.5	1.1	1.7
23	0.2	0.3	0.8	0.4	0.7	0.3	1.2	0.5	1.1	1.5	0.9	1.8
24	0.3	0.6	0.7	0.6	0.7	0.3	1.3	0.4	1.1	1.4	0.8	1.5

Legend: <0.5mg; 0.5-1mg; 1-5mg; > 5mg

Fig. 2. The weights of fractions from one batch (12 samples) from the natural product library (in milligrams).

fractionated with preparative HPLC. About 100 mg of natural product ethanol extract is fractionated into 24 fractions in this double fractionation process. Most fractions contain a mass of 0.5 mg or more, which can be used in the screening of multiple biological targets for drug discovery (Fig. 2). With this method, 2,600 natural product samples can be processed per year and more than 62,000 fractions can be obtained. The analysis of fractions with UPLC coupled with MS, PDA, and ELSD detectors can provide information for the identification and purity of compounds (Fig. 3) in the interesting fractions (14).

2. Materials

Plant materials were collected from various origins worldwide (National Center for Natural Products Research, NCNPR, at University of Mississippi).

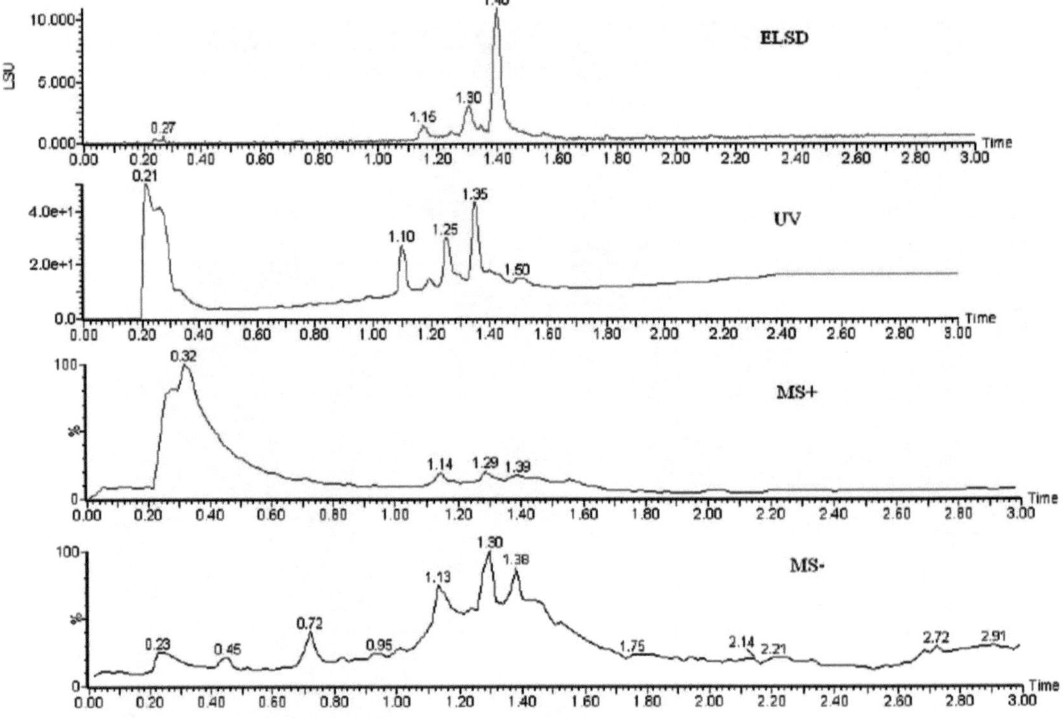

Fig. 3. UPLC/PDA/ELSD/MS chromatograms for fraction 11 of sample 77,771.

2.1. Prefractionation with Polyamide SPE Cartridges

1. SPE frit insertion tool for use with 6 mL PP SPE tubes (Sigma-Aldrich, St. Louis, MO).
2. Polyamide-filled SPE cartridge: 6 mL empty polypropylene SPE tubes filled with ~700 mg polyamide (Sigma-Aldrich, St. Louis, MO).
3. 48-Place positive-pressure SPE manifold (SPEware Corporation, Baldwin Park, CA).
4. Zymark TurboVap LV Concentration Workstation (Caliper Life Sciences, Hopkinton, MA).
5. Disposable glass tubes 16×100 mm (Fisher Scientific, Hampton, NH).
6. Ultrasonic bath FS60 (Fisher Scientific, Pittsburgh, PA).

2.2. Preparative HPLC Fractionation

1. Whatman GD/X syringe filters (Whatman Inc., Piscataway, NJ).
2. 96-Well HPLC plate: Waters 2 mL Square Collection Plate (Waters Corp., Milford, MA).
3. Shimadzu preparative HPLC system: a Shimadzu LC-8A binary preparative pump, Shimadzu SCL-10A VP system controller, Shimadzu SPD-M20A diode-array detector and

ELSD-LT II evaporative light scattering detector (Shimadzu Corp., Kyoto, Japan).

4. Gilson 215 auto sampler, fraction collector, and Gilson 207 test rack (Gilson, Inc., Middleton, WI).
5. Gemini C_{18} 110A column (30 mm × 50 mm, 5 μm, Phenomenex, Inc., Terrance, CA).

2.3. Solvent Evaporation and Weighing of Fractions

1. GeneVac HT series II high-performance solvent evaporation system and 24-well GeneVac racks (GeneVac Inc., Gardiner, NY).
2. Bohdan BA-200 Balance Automator (Mettler-Toledo AutoChem, Columbia, MD).

2.4. Quality Control with UPLC–MS

1. Waters Acquity UPLC–MS system: Waters Acquity UPLC system with PDA, ELSD detectors, and an SQ mass spectrometry detector (Waters Corp., Milford, MA).
2. Waters Acquity UPLC BEH C_{18} column (2.1 mm × 50 mm, 1.7 μm, Waters Corp., Milford, MA).
3. 384-Well QC plate (384PP, Corning Inc., Corning, NY).

2.5. Reformattting, Plating, and Storage of Fractions

1. Freedom Evo Tecan liquid handling system (Tecan Group Ltd., Mannedorf, Switzerland).
2. *REMP* automated sample processing and storage system (Tecan Group, Ltd., Zurich, Switzerland).
3. Biomek FXP laboratory automation workstation (Beckman Coulter, Inc., Brea, CA).
4. Robot reformatting system: the automated capper and decapper, central seal press, manual cap press, automated plate piercer are acquired from REMP. The Vprep conventional pipetting station, PlateLoc sealer, and 96LT 200 μL pipette tips are from Velocity 11 (now part of Agilent Technologies, Santa Clara, CA). LPR240 carousel-style plate hotels and STR240 automated incubators are from Liconic Instruments (Woburn, MA). The laser bar code scanner is from 3M (St. Paul, MN). TX40 robotic arms are built by Staubli (Duncan, SC). The automation system and its software (Cellario version 2.4) were developed by HighRes Biosolutions (Woburn, MA) (15).
5. 24-Well polypropylene rack (10 mL, Round Bottom, Whatman Inc., Florham Park, NJ).
6. 96-Well master plate (0.5 mL, V-bottom, Corning, NY).
7. 384-Well screening plate (384PP, Corning Inc., Corning, NY).
8. Tube-based STBR384 and STBR96 (REMP, Zurich, Switzerland).

2.6. Integrated Informatics

A custom informatics workflow Web application, Fractionation Workflow Application (FWA), was built using the Pipeline Pilot platform (version 7.5.2, Accelrys).

3. Methods

Plant materials are extracted with ethanol according to reported methods (16).

3.1. Prefractionation with Polyamide SPE Cartridges (see Note 1)

1. Crude extract (~100 mg) is dissolved into 6 mL methanol and is sonicated for 30 min.
2. The polyamide-filled SPE cartridge is rinsed with 6 mL methanol. Crude extract solution is brought onto the SPE cartridge and is then rinsed with 6 mL methanol.
3. The effluent is collected in a 16 mm × 100 mm disposable glass tube and then dried under a stream of nitrogen by using Zymark TurboVap LV Concentration Workstation. Fifty samples can be dried simultaneously in 5 h.

3.2. Preparative Reversed-Phase HPLC Fractionation

1. Samples after prefractionation purification are dissolved into 2,000 μL DMSO (see Note 2).
2. Sonicate the samples for 30 min.
3. Filter the samples through 0.45 μm filter membrane (see Note 3) into 96-well HPLC plate.
4. Samples are fractionated at a Shimadzu preparative HPLC system (see Note 4). The mobile phase consists of water (A) and methanol (B; 0 min, 98:2; 0.5 min, 98:2; 6.5 min, 0:100; 12.3 min, 0:100; 12.5 min, stop). The flow rate is 25 mL/min. Injection volume is 2,000 μL. Room temperature.
5. Fractions are collected into 16 mm × 100 mm disposable glass tubes. Each sample is fractionated into 24 fractions (collector: 30 s per fraction).

3.3. Solvent Evaporation and Weighing of Fractions

1. Fractions are transferred to 24-well GeneVac racks manually.
2. Fractions are dried with a GeneVac HT series II high-performance solvent evaporation system. The chamber is preheated to and maintained at 35°C. The SampleGuard Control temperature is set at 40°C, and the CoolHeat Enable pressure is set at 40 mbar. The running time is 18 h. Two hundred and eighty-eight fractions (from 12 samples) can be dried simultaneously.
3. Fraction collection tubes are pre-weighed using a Bohdan BA-200 Balance Automator (see Note 5). Tubes are held in a custom Gilson 207 test tube rack (75 tubes/rack).

4. The natural product fractions after drying are reweighed in GeneVac rack by the same Balance Automator (see Note 6). One hundred and fifty fractions can be weighed automatically in 1 h.
5. The net weight of compound is calculated from the difference between the two weights by using a FWA program developed on a Pipleline Pilot platform (Subheading 2.6).

3.4. Quality Control with UPLC–MS

1. Quality control (QC) of natural product fractions is performed on a Waters Acquity UPLC–MS system in 384-well QC plate format.
2. The column oven temperature is set at 50°C, and the flow rate is 0.6 mL/min.
3. The mobile phase consists of water containing 0.1% formic acid (A) and acetonitrile (B). The total run time is 3.0 min.
4. Both the positive and negative electrospray ionization modes (ESI) are applied. The capillary voltage is set at 3.4 kV. The extractor voltage is 2 V. Nitrogen is used as the nebulizing gas. Source temperature is set at 130°C. The scan range is m/z 130–1,400.

3.5. Reformatting, Plating and Storage of Fractions

1. Final plating of the natural product fractions is performed with a Freedom Evo Tecan.
2. Fractions in GeneVac racks are transferred to a 24-well polypropylene rack.
3. A Tecan worklist is generated with the FWA program by uploading the fractions information including sample ID, net weight, and source of raw material.
4. Fractions are dissolved in 500 μL of the appropriate solvent (see Notes 7 and 8) and sonicated for 10 s. Seventy microliters of sample is transferred to a 384-well QC plate, while the remaining sample is transferred to a 96-well master plate. The process is done automatically with a Freedom Evo Tecan liquid handling system (see Note 9).
5. Fractions in 96-well master plates are dried with the GeneVac HT series II high-performance solvent evaporation system. The chamber is preheated to and maintained at 35°C. The SampleGuard Control temperature is set at 40°C, and the CoolHeat Enable pressure is set at 40 mbar. Twelve 96-well master plates can be dried simultaneously in 5 h.
6. Fractions in the 96-well master plate are dissolved in 200 μL DMSO with a Biomek laboratory automation workstation (see Note 10).
7. Fractions in the 96-well master plate are reformatted into two 384-well screening plates (10 μL/well), three STBR384

(10 μL/tube), and one STBR96 (long term storage, 150 μL/tube) with the robot reformatting system (described in Subheading 2.5, Item 4).

8. Fractions are stored at −20°C in an automated storage archive, produced by REMP.

4. Notes

1. Polyphenols are undesirable in HTS and must be removed. Polyphenols act as promiscuous protein binders that may cause false-positive results in both enzymatic and cellular screening procedures due to nonselective enzyme inhibition and changes in cellular redox potential. Polyphenols are retained on the SPE cartridge, while non-polyphenolic compounds are eluted. Since compounds containing two or three phenolic hydroxyl groups do not bind to the cartridges, most flavonoids can be recovered (see refs. 14, 17).
2. Dissolve the samples into 2,000 μL DMSO the day before preparative HPLC fractionation. It takes some time for compounds to dissolve in DMSO.
3. Filtration of the samples before fractionation is important. There is always something insoluble in samples that may block the tubing and column of the preparative HPLC system if the sample is injected directly.
4. Run a standard compounds mixture before fractionation to make sure the good status of the preparative HPLC system. If the fractions cannot be collected properly, restarting the instrument can solve the problem in most cases.
5. Check the blank tubes carefully. Replace the tube if any crack or damage is found. Since the tubes are going to be gripped by robot arms several times at weighing and reformatting procedures, they will be easily broken if already damaged. There is a risk that the sample will be lost if the tube is broken during the robotic processes.
6. Fractions in GeneVac racks are warm right after solvent removal in the GeneVac system. It is necessary to wait until the fractions are cooled down to the room temperature before being weighed with Bohdan balance automator, otherwise the recorded values will be lower than the accurate values.
7. Two kinds of volatile solvent mixtures are used to dissolve the fractions. Early fractions are more polar than late fractions after elution. Therefore, fractions no. 1–12 are dissolved in solvent mixtures of methanol/water (50:50, v/v), while fractions no.

13–24 are dissolved in solvent mixtures of methanol/chloroform (50:50, v/v).

8. Solvent is added to the fraction tubes one by one instead of adding the solvent in groups. This step is followed by the immediate transfer of the samples solution to the appropriate wells/plates in order to avoid the evaporation of the solvent.
9. Make sure the blue covers are correctly placed onto tubes in the 24-well polypropylene rack to avoid the tubes being accidentally hit by the Tecan PNP arm.
10. Samples are dissolved in DMSO by using Biomek the day before robot reformatting in order to let the compounds dissolve more completely.

Acknowledgments

We thank Cynthia Jeffries, Hong Ruan, David Smithson, Cynthia Nelson, Jimmy Cui, Anang A. Shelat and Kip Guy at St. Jude Children's Research Hospital, and Xing-Cong Li, John P. Hester, Troy Smillie, Ikhlas A. Khan, and Larry Walker at the National Center for Natural Products Research, University of Mississippi for collaborations. This work was supported by the National Cancer Institute (P30 CA021765) and the American Lebanese Syrian Associated Charities (ALSAC).

References

1. Newman DJ, Cragg GM (2007) Natural products as sources of new drugs over the last 25 years. J Nat Prod 70:461–477
2. Newman DJ, Cragg GM, Snader KM (2003) Natural products as sources of new drugs over the period 1981–2002. J Nat Prod 66: 1022–1037
3. Newman DJ (2008) Natural products as leads to potential drugs: an old process or the new hope for drug discovery? J Med Chem 51: 2589–2599
4. Kingston DG (2011) Modern natural products drug discovery and its relevance to biodiversity conservation. J Nat Prod 74:496–511
5. Harvey AL (2008) Natural products in drug discovery. Drug Discov Today 13:894–901
6. de Sa Alves FR, Barreiro EJ, Fraga CA (2009) From nature to drug discovery: the indole scaffold as a 'privileged structure'. Mini Rev Med Chem 9:782–793
7. Schmid II, Sattler II, Grabley S, Thiericke R (1999) Natural products in high-throughput screening: automated high-quality sample preparation. J Biomol Screen 4:15–25
8. Thiericke R (2000) Drug discovery from nature: automated high-quality sample preparation. J Autom Methods Manag Chem 22:149–157
9. Eldridge GR, Vervoort HC, Lee CM, Cremin PA, Williams CT, Hart SM, Goering MG, O'Neil-Johnson M, Zeng L (2002) High-throughput method for the production and analysis of large natural product libraries for drug discovery. Anal Chem 74:3963–3971
10. Wu S, Yang L, Gao Y, Liu X, Liu F (2008) Multi-channel counter-current chromatography for high-throughput fractionation of natural products for drug discovery. J Chromatogr A 1180:99–107
11. Bugni TS, Harper MK, McCulloch MW, Reppart J, Ireland CM (2008) Fractionated

marine invertebrate extract libraries for drug discovery. Molecules 13:1372–1383

12. Bugni TS, Richards B, Bhoite L, Cimbora D, Harper MK, Ireland CM (2008) Marine natural product libraries for high-throughput screening and rapid drug discovery. J Nat Prod 71:1095–1098
13. Bindseil KU, Jakupovic J, Wolf D, Lavayre J, Leboul J, van der Pyl D (2001) Pure compound libraries; a new perspective for natural product based drug discovery. Drug Discov Today 6:840–847
14. Tu Y, Jeffries C, Ruan H, Nelson C, Smithson D, Shelat AA, Brown KM, Li XC, Hester JP, Smillie T, Khan IA, Walker L, Guy K, Yan B (2010) Automated high-throughput system to fractionate plant natural products for drug discovery. J Nat Prod 73:751–754
15. Cui J, Chai SC, Shelat AA, Guy RK, Chen T (2011) An automated approach to efficiently reformat a large collection of compounds. Curr Chem Genomics 5:42–47
16. Manly SP, Smillie T, Hester JP, Khan I, Coudurier L (eds) (2010) Unique discovery aspects of utilizing botanical sources. CRC Press/Taylor & Francis, Boca Raton
17. Hostettmann K, Marston AM, Hostettmann M (eds) (1998) Preparative chromatography techniques: applications in natural product isolation, 2nd edn. Springer, Heidelberg

Chapter 10

Conducting Molecular Biomarker Discovery Studies in Plants

Christian Schudoma, Matthias Steinfath, Heike Sprenger, Joost T. van Dongen, Dirk Hincha, Ellen Zuther, Peter Geigenberger, Joachim Kopka, Karin Köhl, and Dirk Walther

Abstract

Molecular biomarkers are molecules whose concentrations in a biological system inform about the current phenotypical state and, more importantly, may also be predictive of future phenotypic trait endpoints. The identification of biomarkers has gained much attention in targeted plant breeding since technologies have become available that measure many molecules across different levels of molecular organization and at decreasing costs. In this chapter, we outline the general strategy and workflow of conducting biomarker discovery studies. Critical aspects of study design as well as the statistical data analysis and model building will be highlighted.

Key words: Biomarker, OMICS technologies, Machine learning, Classification, Feature selection, Phenotype, Study design, Breeding, Plants

1. Introduction

Fueled by advances in genomics technologies, modern plant breeding has embraced genetic data as a source of information in breeding programs. So-called genomic markers allow the identification of candidate genes that impart a desired trait as well as an efficient selection in crossing experiments. Here, the genome sequence information and the presence of characteristic polymorphisms or other genome-level marker types constitute what can be considered molecular information. Terms such as "smart breeding" or "marker-assisted breeding" have been coined to capture this concept as a whole (1–5). Especially in the light of public resistance to genetically engineered crops, the application of these breeding

Jennifer Normanly (ed.), *High-Throughput Phenotyping in Plants: Methods and Protocols*, Methods in Molecular Biology, vol. 918, DOI 10.1007/978-1-61779-995-2_10,

strategies holds great promise to achieve a desired breeding objective without using genome modification techniques. However, marker information can also be used in a different way: in the form of molecular biomarkers as surrogates for complex phenotypes. Molecular biomarkers are measurable molecular compounds that are indicative of phenotypic properties, and often, even before these properties arise (6). Unlike genomic markers that are related to the DNA such as single nucleotide polymorphisms (SNPs), i.e., the essentially static genome of plant accessions, biomarkers used as phenotypic indicators fall into the molecular classes of transcripts, proteins, metabolites, or lipids, i.e., molecules that are expressed dynamically. For example, using a combination of different metabolites, it was possible to predict plant biomass (7) and freezing tolerance (8). In another study, it was shown that using metabolite level information significantly improved the performance of heterotic biomass predictions in offsprings compared to genetic marker information of the parental lines alone (9). A great advantage of molecular biomarkers is their independence from the availability of genetic information of the crop species. This is especially important in polyploid plants, where use and interpretation of genetic information (e.g., SNPs) is difficult.

The use of biomarkers may save much time and money during the breeding process, as only the targeted measurement of the marker molecules is necessary to select a plant, instead of performing cultivation experiments in the target environment. These cultivation experiments can be costly and time consuming especially when a complex, quantitative trait such as an abiotic stress tolerance (e.g., tolerance to cold, heat, drought) is the target of the breeding process.

Furthermore, complex phenotypes such as stress tolerance are often of multigenic origin and show considerable (gene × environment) interactions. Thus, complex phenotypes are difficult to associate with one or a few genetic markers. Molecular biomarkers offer the advantage of integration over many different individual genes and environmental effects. Thus, expression-based biomarkers are "closer" to the phenotype than genetic markers. In fact, they constitute a phenotype themselves: the molecular phenotype.

The surge in biomarker discovery programs not only in the plant and agricultural sciences (6), but also in diagnostic applications in the medical field (10–13) has been fueled by rapid technological advances substantially expanding our measuring capacity. Modern profiling technologies allow measuring molecules from all domains of molecular organization including transcripts, proteins, metabolites, and lipids at an unparalleled breadth even to the point that an entire complement of a particular molecular domain can be captured simultaneously. For example, modern transcript measurement technologies (e.g., gene expression microarrays or Next Generation Sequencing (NGS) technologies) permit monitoring essentially all expressed transcripts; i.e., the entire transcriptome. As for the other molecular levels—the proteome, metabolome, etc.—we are quickly making progress towards achieving the same goal. Thus, under the assumption that the molecular level of particular molecule class is informative of the phenotypic state, it will soon be possible to identify from the hundreds of thousands of candidate molecules occurring in living systems the subset of the best; i.e., most informative molecules for diagnostic and prognostic applications.

At the same time, technological advances have also made molecular profiling approaches a viable option from an economic perspective. While being able to measure the entirety of a particular molecular level is clearly desirable to identify the best possible biomarkers, it is not the goal to use whole-OME screens in actual biomarker applications. Instead, once selected, targeted measurement strategies can be employed to selectively measure specific compounds only. For example, NGS or gene expression microarray technologies used in the first phase of marker discovery may be replaced by PCR-based approaches for the targeted identification of only a few selected gene transcripts (12). Thus, not only can the economics of molecular marker applications be improved, but—and as importantly—the necessary quality assurance (QA) requirements can be reduced when in practice, as the number of variables and parameters can be reduced to a manageable number.

Biomarker discovery projects pose formidable challenges not only to the applied technologies, but also to the statistical interpretation of the acquired large datasets. Methods established in the field of machine learning are applied to identify most informative markers or a combination thereof and to translate marker level data into robust predictive, mathematical models. It is important to realize that there is a duality of molecules and mathematical models. In the minimal case, a single molecule may serve as a marker, e.g., prostate specific antigen for prostate cancer. At first glance, no statistical or mathematical model is required for a single biomarker to be used. However, even in the single-biomarker case, threshold values have to be defined based on distributions estimated by sampling, and thus, a mathematical model of the value distribution has to be obtained. Evidently, when several biomarkers are to be combined to improve the reliability of the phenotype prediction, the complexity of the involved mathematical model increases. Thus, when referring to biomarkers, one always also has a mathematical model in mind that combines different molecules into a single test. Accordingly, we will describe the process of biomarker discovery as a combination of molecular profiling and statistical/mathematical model building.

To ensure the highest possible validity and generalizability of the obtained predictive model, it is of utmost importance to know all influencing factors in the lab as well as in practice under field conditions to best control for confounding effects. Confounding factors correlate with both molecular markers and phenotypic traits, thereby leading to spurious direct associations between maker and trait. To safeguard against these pitfalls, great care needs to be exercised during the study design phase and during the statistical analysis. During a first rush to employ gene expression microarrays for clinical applications, many published study results claimed prognostic value of gene expression-based tests, only to be challenged later (14). The stability of these early biomarker tests proved insufficient and size and composition of test populations were revealed as the most critical factors, pointing to the importance of proper study design.

In this review, we will outline the general strategy, setup, and workflow of biomarker discovery projects in plant breeding programs. We will discuss the major precautionary steps during study design and will describe the main, general approaches to the statistical interpretation of the obtained data leading to predictive, mathematical models. It is not the intent to provide a detailed account of the employed statistical methods and mathematical models, nor an in-depth overview of the applied screening technologies and associated data acquisition issues.

2. Experimental Study Design and Experimental Data Acquisition

For an overview over the biomarker selection procedure as a whole, further described in the following paragraphs, we refer to Fig. 1.

2.1. First Steps: Problem Identification

Projects aiming at the discovery of novel biomarkers start with a series of simple, yet important questions in order to define the problem at hand. Which phenotypic trait *P* is to be assessed (e.g., drought tolerance in potatoes), and which quantitative parameter can be measured to capture the trait (which assay)? Biomarker selection requires a quantitative or ordinal parameter (e.g., starch content in potato tubers as a measure of performance; fitness in response to reduced water supply). The parameter associated with the trait is best measured as a numeric variable (e.g., fresh weight, dry weight). Ordinal (e.g., degree of flowering, degree of damage from drought stress), or binary (e.g., flowering/not flowering) values; i.e., quantifiable according to—ideally—objective standards can be used. Which molecules or molecule classes are likely to be informative with regard to the phenotypic trait, *P*, in question and thus may serve as biomarkers, *BM*? For example, in the case of potato drought tolerance, metabolite levels in potato tubers would be informative. Which measurement platform can be used in order to measure *P* and *BM*? Obviously, cost and practicality considerations are an essential component of this choice. Will the conditions

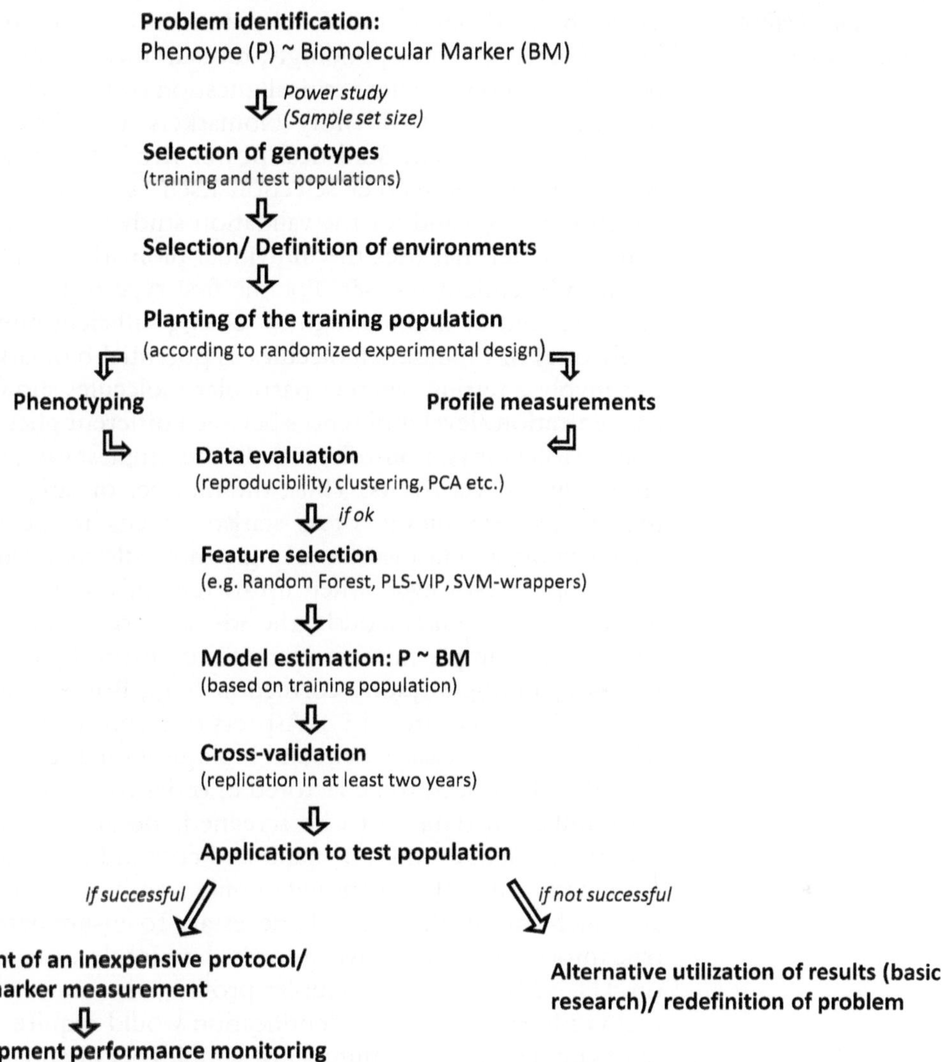

Fig. 1. Overview of process workflow in biomarker discovery studies. *P* phenotype, *BM* biomarker. *P* ~ *BM* denotes a specific model that links biomarkers to phenotypic endpoints.

under which the marker will be used allow the application of the selected technology? Will it be affordable? It should be noted that for a marker to be useful, it is not necessary to establish functional causality between marker level and trait. While evidently helpful, markers can be useful even without understanding the cause–effect relationship. That said, the selected marker type ought to be reasonable in a sense that the molecular marker domain could be informative and predictive, i.e., the marker domain (metabolites, proteins, transcripts) has some intrinsic relationship (either causative or informative) to the phenotype.

2.2. Statistical Power Considerations

Once the read-out and molecular domain in which to search for biomarkers and, correspondingly, measurement technologies have been decided on, the next critical question concerns the number of samples necessary to identify biomarkers at sufficient statistical confidence or power. Statistical power needs to be estimated on two occasions: biomarker selection itself (see section on feature selection below) and for the validation study in which the mathematical model that uses the individual biomarkers will be applied to an independent test set. For the first type of power study, the power should be high enough to select a sufficient number (minimally one) of candidate molecules as potential biomarkers. Often, this involves testing whether particular molecules show significant concentration/level differences between different phenotypic endpoints (tolerant vs. non-tolerant). In the simplest case, mean values are compared via *t*-tests. Thus, the number of samples necessary to establish significant *t*-test statistic needs to be established. Establishing significance for a single molecule may require only a few samples. However, when up to tens of thousands molecules will be tested simultaneously, the so-called *curse of dimensionality* strikes. The sample number increases dramatically because of the necessary multiple testing correction (e.g., Bonferroni correction or false discovery rate, (15)). Aspects of technical noise associated with the actual measurement and biological variance as well as cost considerations need to be factored in order to arrive at a meaningful number of samples to be screened. Because of the multiple testing problem on the one hand and cost and practicality considerations on the other, typically sample numbers in this first phase are much smaller than actually necessary to ensure detection of all true informative/predictive individual molecules. In a recent study on 1H-NMR human metabolic profiles, the authors concluded that confident biomarker identification would require a few thousand samples (16). The number of samples per phenotype category (e.g., high/ low) should at least be large enough to allow for a bare minimum of statistical rigor at the single variable level (>3 samples per phenotypic endpoint). However, at this typically low number meaningful cross-validation (see below) is hardly possible. Here, several dozens of samples per phenotype class are needed. Thus, the number of necessary training samples is notoriously difficult to decide on as the predictive performance of the model can only be assessed after a model was built; i.e., after using up all samples. If at all possible, pilot studies should be performed to obtain distribution characteristics of all candidate biomarker molecules.

The requirements for ensuring statistical power are substantially increased when the prediction model, in which all biomarkers are combined into a mathematical model, will be validated (see section on Model Validation below). Statistical power needs to be critically assessed before the validation study is conducted. The desired or minimally acceptable diagnostic/predictive signal needs to be determined. Often, the model will yield a numerical score indicative of phenotypic class. Thus, it needs to be determined what score difference between the different phenotypic classes is desirable and, furthermore, by which statistical test (parametric, such as *t*-test, non-parametric, such as Wilcoxon rank sum test; correlation measures, etc.) it will be tested. The biological and technical sources and magnitudes of variance associated with the score values need to be considered. This information will be obtained from testing the performance of the model in cross-validation (see below). The minimum effect strength must be determined when comparing the gain of a successful application of the method with its costs. The statistical power is the probability to detect such an effect and must be defined by the user. The probability of the failure to detect a true effect will be determined by this choice. Since such a failure means the failure of the whole procedure, a high statistical power should be chosen. These components—the statistical power, minimal effect strength, and variance estimates—and the significance level statistical software tools providing power analysis tools (e.g., Statistica by Statsoft Inc., or others) can be used to decide on the necessary number of validation samples. As the validation study is a "one-shot-only" exercise (see below), power analysis is of utmost importance to make sure that the validity of a test performance claim can be established. Note: Should more than one model be tested, multiple testing correction has to also be applied, scaling with the number of tested models.

2.3. Samples, Populations, and Environments: Making Design Choices

The previously described power analysis yields an initial number of samples size N required to detect biomarkers (training phase) and to validate the study at the desired effect strength—validation phase. The sample size N is initial in the sense that it may prove necessary to obtain additional samples for conducting the validation study because of weaker than expected model performance during training. The training and validation samples are selected according to this number, complying with the following rules.

1. The minimum sample size is N. The training and test populations should each consist of N different genotypes. Samples from the same genotype correspond to technical or biological repeats.
2. Ideally, the number of available samples is similar across the whole spectrum of phenotype outcome values to avoid biasing the marker selection to one particular expression of the phenotype.
3. Since a maximum genetic diversity is required in order to find generally applicable markers for the desired trait(s), population effects caused by genetic dependencies should be avoided as much as possible, e.g., cultivars should not descend from the same parent lines.

4. Samples of the training population must be genotypically distinct from those of the validation population. Otherwise, the test would essentially be a self-recognition test.
5. Very importantly, the different sample sets (training, testing/validation) should be well balanced with regard to all possible confounding factors such as location, treatment methods, environmental factors, even human personnel as this critically determines the generalizability of the test.
6. The selected training and test conditions should be as close to the foreseen use conditions as possible.

The environmental conditions in which the training and test populations are planted should be considered carefully and should cover the maximum possible variance regarding the phenotypes under study (e.g., locations with different soil or climate conditions). Often, it is desirable to capture the extremes of the phenotypic distributions as this allows for maximal contrast for marker selection.

It is of particular importance to control for all possible parameters and factors that may act as confounding factors; i.e., such factors that influence both the phenotype and the marker, but in a way that is inconsistent with objective of the study. For example, if drought-stress markers are to be identified, light may be a confounding factor. Under field conditions, long exposure to high light usually corresponds to dry climate conditions. At the same time, light induces the production of particular light-induced molecules. Thus, it may be concluded that those light-induced compounds are drought-stress markers. However, this association is spurious. Dry conditions can also exist at low light levels. The erroneously selected markers would then fail as they have been selected primarily as being high light-related, but not specifically drought-related. To avoid such problems, both case and control samples should be balanced with regard to as many parameters as possible except the one under study. Finally, when planting the training population and later the validation population, it is essential to employ a randomized experiment designs in order to mitigate biases possibly introduced by location (e.g., all sample plants of one genotype are located on a spot that is (unknowingly) less well irrigated than the rest of the field).

2.4. Harvest Time: Collecting Samples and Phenotype Information and Evaluating Data

To assess the phenotypic traits and biochemical quantities (e.g., metabolic profiles) of the training population, plant tissue samples and phenotype information have to be gathered. Depending on the study's aims, these data collections can be conducted once or on multiple occasions (e.g., in order to track reactions to stress conditions). To ensure reproducibility and comparability of samples and data gathered by independent field personnel, a detailed phenotyping and sampling protocol needs to be outlined before the field work is started.

For many crops, protocols for the determination of relevant performance parameters have been agreed upon and published by breeders associations and by the authorities that release new cultivars (e.g., the Bundessortenamt in Germany for potatoes). These protocols suggest classes for ordinal data such as developmental stages, stress, or disease symptoms or quality parameters of the harvest. Using these protocols saves times and allows comparison with older studies. For species without agricultural relevance, e.g., Arabidopsis, classification schemes of related crops, e.g., canola, can be adapted.

Sampling needs to take into account that many biochemical parameters such as metabolite levels can show strong diurnal variations. Plant tissue samples should thus be taken at specific timepoints during the day, ideally at a time when the measured levels reach a (temporary) steady state. There is also considerable variation between different parts of a single leaf and between different leaves of a single plant. Therefore, care should be taken to sample tissues of similar developmental stages and, if only parts of a tissue are sampled, to always sample the same part of the tissue. Recommendations for suitable organs may be obtained from sampling protocols, e.g., for mineral nutrient determination (17). Since the turnover rate of metabolic parameters can be very fast (~1sec for metabolites in leaf photosynthesis), rapid quenching by freezing in liquid nitrogen is essential after the harvest. During this quenching step, light conditions should be kept unaltered. The frozen plant material should be kept under liquid nitrogen or can be stored in a freezer at –80°C for up to 3 months until use. More detailed information on sampling and harvesting procedures to analyze metabolic parameters in plants can be found in (18, 19).

2.5. Profile Measurements

When assessing data from tissue samples, the capability of assigning profile information to known substances has to be taken into consideration. This capability depends on the nature of the profile data used in the study. Using gene expression data, for instance, allows the direct mapping of the profile data to gene sequences. By contrast, such an unambiguous assignment does not generally hold for metabolic profiles. In such a situation, the substances need to be validated by different experiments. Metabolite profiling operates on a large range of molecular species that have widely divergent physical and chemical properties. This requires various analytical methods, different equipments and separation technologies for analyzing the different families of chemical compounds found in plants (20). Because of the technical variability, it is always necessary to measure technical replicates. Likewise, biological replicates must be included to assess the biological variance. Importantly, markers should be robust against measurement noise and biological variation.

2.6. Data Management/Public Domain Data as Additional Sources of Information

It is advisable to think about data management and warehousing aspects early in the project. In particular, recording all meta-information—sample information, conditions, etc.—is of critical importance to later be able to assess the influence on individual factors and to possibly account for their confounding effects. Furthermore, when possible, public domain repositories may be considered as additional sources of information. With regard to gene expression data, much information is already available that can be brought into the study. For example, for *Arabidopsis thaliana*, thousands of gene expression profiles obtained from different conditions are available (www.arabidopsis.org) some of which may be relevant to current study. This knowledge may also help in the assessment of robustness of particular gene expression levels.

3. Statistical Analysis and Model Building

3.1. Data Evaluation/ Exploratory Data Analysis

Biomarker data obtained from profile measurements first have to be checked for quality. Of main interest are the reproducibility and technical robustness of the marker candidate substances. To ensure that these two conditions are fulfilled, at least one technical replication of the measurement results is necessary. A first computational analysis, using principal component analysis (PCA) and various clustering methods may give an impression of whether the data at hand contains enough information for the prediction task. If samples drawn from different phenotypic endpoints are separable in the first principal components that correspond to orthogonal directions in the space of all measured molecules of largest variance, it can be assumed that individual biomarkers or combinations thereof can be identified that are responsible for this separation. Furthermore, the characteristics of the marker level distributions need to be assessed, as this may influence the marker selection and also the type of machine learning method that can be employed in later steps. For example, many machine learning methods such as linear discriminant analysis (LDA) rely on the assumption of normally distributed values. However, if the distribution is non-normal, LDA cannot be applied, and decision trees (DTs) may be the better choice. Also, with regard to marker selection, the statistical test used to select them depends on their distribution—*t*-test for Gaussian data, non-parametric tests for other distributions (see below).

Oftentimes, quantities for molecular markers require a normalization step (e.g., quantile normalization for gene expression microarrays or normalization to selected housekeeping genes in PCR-based expression measurements). The influence of the chosen normalization method may be substantial, and thus should be chosen carefully and the effect of applying other methods be assessed during the model testing phase (see below). Also, issues of ongoing applicability need to be considered. If, for example, level data are compared against a reference sample mix, the availability of this reference mix needs to be ensured during the eventual application of the biomarker test.

The subsequent paragraphs describe the actual process of biomarker identification as based on and optimized after the successful metabolic biomarker selection approach described in (6). Simply put, the molecules to be identified are those whose levels/concentration are different in the different phenotypic endpoints. Thus, classical statistical testing may be sufficient for this task. However, testing differences of means or differences in value distributions do not allow an estimate of the prediction performance when confronted with novel samples (e.g., new crossings with new genotypes). Towards this end, a prediction model needs to be built that combines candidate biomarker(s) into a computerized/mathematical model. Only then, the practical value of the selected biomarkers can be estimated. This process falls into the domain of *machine learning methods*. More precisely, biomarker selection represents a task from the class of supervised machine learning problems, as the different classes (phenotypic endpoints) are known beforehand and are not obtained from clustering the data as in unsupervised learning methods. In the following section we provide an overview of the general concepts, strategies, and performance metrics used. For more information on this very large field in computer science, the interested reader may consult the many text books or review articles on machine learning methods (21–25).

3.2. Selection of the Model Class

Before the actual model building begins, suitable model classes for the problem must be chosen. The choice will depend on biochemical considerations and on the type of the phenotypic properties to be predicted. Numerical responses (e.g., fresh weight) will best be modeled by continuous functions. Very often linear models are the first choice, because their parameters can easily be estimated and interpreted. For nominal data (e.g., salt tolerant/non-tolerant plants), applying decision rules will be more appropriate. These models could be based on either linear or nonlinear decision boundaries. For ordinal data, proportional odds models, which are based on logistic or logit models, can be applied (26).

3.3. Feature Selection and Model Building: Machine Learning

The goal of this very central step is to actually identify marker molecules, also called features, that have a significant impact, or in other words, that bear information according to the classification model on the phenotypical property P under investigation across all environments and technical replications. How the features are combined into rules, more formally, into a mathematical or classification model, is the domain of model building. Both intimately related aspects—features selection and model building—are part of the machine learning field. Here, we are dealing with supervised machine learning problems; i.e., the class labels—the association of samples with distinct phenotypic states—is known. To select features means to deselect others. While this may sound like a trivial statement, it is important to realize that, typically, features are chosen from a very large set of candidate molecules. With high numbers of candidates and relatively few observations, the risks of selecting molecules based on spurious associations are substantial. Furthermore, even if several molecules may be identified as real, considerations such as assay robustness need to be factored in to select the most suitable biomarkers.

Features can be selected independently of the classification model or in a combined fashion ("embedded") such that the selection of features changes the model and becomes an integral part of model learning. In the first case, features can be selected before model building as done by so-called *filter methods*, or iteratively, such that the best possible subset of features is identified given the chosen prediction model as done by *wrapper methods.* While in filter methods the metric based upon which features are chosen may have no direct relevance to the classification model that is to be used later, in the wrapper approach, the actual model is used to evaluate the performance of the chosen feature set using cross-validation (see below). In embedded approaches, the selected features change the actual model.

In *filter methods*, conventional statistical tests such as t-test (in case of binary phenotypes) or ANOVA (in case of multi-level phenotypes = multiclass problems) are used to identify those markers that differ significantly between different phenotypic endpoints. Often, the Fisher's ratio, F_{ratio}, or its square root—a metric related to the t-test is used: $F_{ratio} = (m_1 - m_2)^2/(\sigma_1 + \sigma_2)^2$ with $m_{1/2}$ referring to the mean values in class 1 or 2 and $\sigma_{1/2}$ denoting the respective standard deviations. Unlike the t-test, this type of metric better identifies relevant markers as it compares the difference in the means to the standard deviations or variances and not to the standard error (stderr), which can be small simply because of many repeats ($stderr = \sigma / \sqrt{N}$). If the molecular features are not following a Gaussian distribution, non-parametric tests need to be used, such as Wilcoxon rank sum test. Correlation measures such as Pearson, Spearman correlation that correlate marker levels and phenotypic value are alternative filter methods when the phenotypic endpoint is continuous. Mutual information (MI) between predictor and outcome variable can also be employed as a filtering step (22). MI is particularly appropriate for nominal data. Otherwise, continuous data has to be transformed into discrete variables by binning. PCA allows determining sets of molecules associated with phenotypic differences. When samples from different phenotypic endpoints can be separated in their PCA-space associated with the principal components (PCs) explaining most of the observed variance, the loadings associated with each molecule and PC allow them to be ranked. Furthermore, PCA offers a multivariate approach; i.e., combinations of features can be selected at once (captured by the principal component vectors), while the above-mentioned statistical tests operate at the univariate level, thus not including any possible interactions.

In *wrapper methods*, given a chosen prediction model (for example, kNN, see below), the subset of features is sought that yields best prediction performance as typically judged by cross-validation. Thus, the feature selection is "wrapped" around the model. However, the combinatorics of this search problem is immense, and as we will see below, repeated cross-validation bears the risk of overfitting.

Embedded feature selection methods rank features relative to their predictive value when incorporated into a specific model (27). The selection of features is intimately tight into the model building itself. For example, in classification trees (also called decision trees), at every step growing the tree, a feature is selected based on the predictive value in the very model built so far. Similarly, stepwise forward/backward regression approaches choose features based on their merit when incorporated into a growing regression model. The partial least squares—variables importance in projection (PLS-VIP) method used by (6, 8, 9) is another example of an embedded method.

Evidently, combinations of filter, wrapper, and embedded methods are possible. For example, the wrapper method may be considered an outer loop around embedded or filter methods (28). However, it is important to note that typically biomarker selection proceeds in the context of a prediction model. For reviews on the topic of feature selection, see (20, 29, 30).

3.4. Classification Methods

As outlined above, the actual classification method not only represents how the selected biomarkers are combined to arrive at a prediction as to what class a particular sample belongs to, but also often is an integral part of the biomarker selection itself. Furthermore, the utility of the chosen biomarkers, their predictive value, is to large degree dependent upon the classification model. Which method to use depends on the problem at hand. While some methods potentially lead to interpretable results such that they link molecules with phenotypes, others remain to a large degree black boxes (such as artificial neural networks). Typically, several methods will be tested. A detailed discussion of all listed methods is beyond the scope of this chapter. The interested reader is referred to (21, 23–25, 31) and to the many available books on the subject.

3.4.1. Nominal Classification Problems

In Table 1, we list common classification methods for nominal classification tasks. Typically, nominal prediction tasks are discussed

Table 1
Selected supervised machine learning methods for binary classification problems

Algorithm	Description	Variable selection	Application
Fisher linear discriminant analysis (LDA)	Class separation via linear model and threshold	Prior or embedded stepwise filtering	Gaussian-distributed data
Logistic regression	Assumes nonlinear, logistic model	Prior or embedded stepwise filtering	Binary decision problems with class-associated numeric values
Support vector machines	Transformation into higher-dimensional space with subsequent linear separation	Wrapper method	General applicability
Decision trees	Introduces consecutive splits on individual features to increase class purity in the resulting partitionings	Embedded	Suitable also for combination of numeric and nominal features, leads to interpretable rules
Random forest (32)	Combines DTs and voting by learning many different trees with subsequent winner-takes-all voting	Embedded	General applicability
k-Nearest neighbors (kNN)	Prediction based on majority vote of neighboring samples	Filter and wrapper methods	Ideal for intermixed classes ("islands" of one class in the "sea" of another class) with no single linear or nonlinear separation possible between classes
Artificial neural networks	Modeled after neuronal processing	Filter or wrapper method	Nonlinear problems

as two-class problems. However, many classification approaches can be extended to multiclass problems as well. Most simply, they can be reduced to two-state problems—selected class against all alternative classes and subsequent "winner (class) takes all" (33).

3.4.2. Numeric (Continuous Endpoint) Prediction Problems

In cases of phenotypes expressed as a continuous numeric value (e.g., biomass), regression methods can be used. It may be reasonable to assume that P (the phenotype, here biomass) is predictable by a linear combination of features (Table 2). Thus, a linear model is assumed, and weights associated with all features (factors A_i in Table 2, MLR) need to be determined. This can be achieved, for example, by performing a greedy stepwise-forward multiple linear regression. Conventional pairwise correlation may yield many potential markers, but often they will be highly correlated amongst themselves, thus not providing additional information; the features are said to be redundant. Often, it may be advisable to combine such highly correlated features into a single meta-feature. This also increases the robustness of the classification as it protects against possible measurement failures for single molecules. For many of the listed methods, software packages are available either freely, for example R, or commercially as part of advanced statistics packages such as Statistica (Statsoft Inc.).

3.5. Model Performance Estimation: Self-Test and Cross-Validation

As a first test, a so-called *self-test* can be performed. Model training is done using all samples and the learned model is then applied back to all samples. This procedure provides a best-case performance estimate as all available information is used and then applied to predict itself. Thus, a self-test inherently leads to an overestimation of performance and does not provide a robust estimate of the so-called generalization error; i.e., the performance of the model when applied to new samples. To estimate the generalization error, the prediction performance of the model is tested using k-fold cross-validation (CV) methods (34). The training sample set is split into typically $k=10$ portions and all model parameters trained on all but one portion. The resulting model is then tested on the left out portion (the hold-out set). This is repeated $k=10$ times until all portions have served as test sets. At every iteration a suitable performance measure is computed (such as prediction accuracy, see below) and averaged over all k CV runs. If the available sample set size is large enough, the partitioning of samples may be random, otherwise care has to be taken with regard to balancing all influencing parameters such that maximal parameter balancing is enforced. The extreme case of cross-validation is referred to as leave-one-out test. Here, only one training sample is removed, $k=N$ (with N being the training sample number), and model training is conducted on all other samples, and the model is then applied to the one excluded sample. In any event, care needs to be exercised to ensure that the left out portions constitute non-redundant

Table 2
Selected supervised machine learning methods for numeric prediction problems

Algorithm	Description	Variable selection	Application
Multiple (non-) linear regression (MLR)	Combine F individual features into a linear model: $P = \sum_{i}^{F} A_i X_i + B$ where P is the phenotypic outcome variable, X_i are the features (molecular levels), A_i the associated weights, B the intercept	Prior or embedded stepwise filtering, e.g., greedy stepwise forward or backward until defined termination criterion is reached	Linear problems; nonlinear if nonlinear functional form is chosen
LASSO (35)	Similar to MLR, but with additional constraints on the number of variables via regularization	Prior or embedded stepwise filtering under condition that sum of weights is constrained	Linear problems

information relative to the respective portions used during the rounds of cross-validation. Otherwise, in effect, a self-test is performed and no valid generalization error is obtained. For example, in the case of plants, genotypes in the CV-test portion should be different compared to the training portion, as only then a proper estimate of prediction performance in truly "new" genotypes can be obtained. If, however, the robustness of the test with regard to biological or technical replicates is of interest, other sample selection schemes may be pursued that are less strict with regard to genotype.

If the set of available samples is small, rendering partitioning and subsequent cross-validation difficult (i.e., sampling without replacement), resampling via bootstrapping can be applied (36). Here, a new population of samples is created by randomly selecting *with* replacement from the given set of samples. Clearly, no new information is created by the bootstrapping approach, yet the repeated random sampling allows for a better estimation of the score distributions. Usually several rounds of bootstrapping are performed, each creating sets consisting of the same number, N, of sample that was available originally. As some samples are not chosen at all during the resampling process, the trained model using the N-bootstrap samples can then be applied to the non-used samples. On average, the portion of non-used samples converges to 36.8%; i.e., 63.2% of the samples are used in training. Thus, the bootstrap performance estimate tends to be pessimistic, as only a relatively small portion of samples is used during training.

3.6. Performance Metrics

For the assessment of the diagnostic or prognostic value of the chosen biomarkers and the associated classification model, a number of different measures have been established. Please note that the difference between diagnostic and prognostic applications only lies in the time of occurrence of the phenotypic endpoint, either currently present = diagnostic, or in the future = prognostic. Nonetheless, both situations can be viewed as either nominal classification problems (here we focus on binary decisions) or numeric predictions. First, we list a number of frequently used performance parameters for the nominal classification problem (Table 3), followed by suitable metrics for the numeric value prediction task (Table 4).

These measures are not applicable in the case of predicting actual numeric values for continuous parameters such as biomass. As much as it is meaningful, continuous parameters can also be transformed into nominal (high/low) values. Then the metrics listed in Table 3 can be used as well. However, typically, if numeric values are to be predicted, the specific measures listed in Table 4 allow assessing the concordance of predicted and actual values.

Table 3
Selected performance parameters for nominal (typically) binary classification tasks

<table>
<tr><th>Metric</th><th>Definition</th><th>Description/application</th></tr>
<tr><td>Mean prediction score comparison between different phenotypic endpoints</td><td>t-Test (assumption of normal distribution) or Wilcoxon rank sum test (non-parametric test) of prediction score values between classes</td><td>Answers whether test is statistically informative or not given the prior defined phenotypic endpoints</td></tr>
<tr><td>Confusion matrix</td><td>Class membership: Y(es), N(o):
<table><tr><td></td><td>predicted Y</td><td>predicted N</td></tr><tr><td>true Y</td><td>TP</td><td>FN</td></tr><tr><td>true N</td><td>FP</td><td>TN</td></tr></table></td><td>Overview of classification success. Ideally, FN = FP = 0. As absolute numbers are given, sample statistics becomes obvious</td></tr>
<tr><td>Accuracy, Acc</td><td>Acc = (TP + TN)/(TP + TN + FP + FN)</td><td>Correct predictions among all predictions. Broad performance estimate without any frequency bias correction</td></tr>
<tr><td>Precision or positive predictive value, PPV</td><td>PPV = TP/(TP + FP)</td><td>Portion of correct positive predictions among all positive predictions made</td></tr>
<tr><td>Negative predictive value, NPV</td><td>NPV = TN/(TN + FN)</td><td>Portion of correct negative predictions among all negative predictions made</td></tr>
<tr><td>Sensitivity or recall, Se</td><td>Se = TP/(TP + FN)</td><td>Portion of correctly identified positive cases from the whole positive set. Measures the ability to detect positive cases</td></tr>
<tr><td>Specificity, Sp</td><td>Sp = TN/(TN + FP)</td><td>Portion of correctly identified negative cases from the whole negative set. Measures the ability to detect negative cases</td></tr>
<tr><td>F-measure</td><td>F = 2 × (Precision × Recall)/(Precision + Recall); harmonic mean of precision and recall</td><td>Combines precision and recall into a single measure</td></tr>
</table>

Matthew's correlation coefficient, MCC	$MCC = \frac{(TP \times TN) - (FP \times FN)}{\sqrt{(TP + FP) \times (TP + FN) \times (TN + FP) \times (TN + FN)}}$; ranges between +1 (perfect prediction) and −1 (perfect anti-prediction)	Provides performance estimate for unbalanced set sizes in which random expectation in binary classification deviates from 50%
Receiver-operator-characteristic (ROC-curve)	Plot of TP-rate vs. FP-rate (=1 − specificity) as a function of score value sorted from positive to negative decision values	Allows deciding threshold score value at which TP-rate is sufficiently high at an acceptable FP call rate
Area under the ROC (AUC)	Numeric value of the area under the ROC; equivalent to probability that a randomly chosen score value associated with a positive case is higher than a randomly chosen negative case; ranges from 0.5 = random predictions to 1 = perfect predictions, less than 0.5 indicate anti-predictions	Commonly used global performance measure

Binary classes are designated positive and negative classes; e.g., drought tolerant vs. non-tolerant. Score values are assumed to be designed such that they are higher for positive than for negative cases

Abbreviations stand for number of *TP* true positives, *TN* true negatives, *FP* false positives, *FN* false negatives

Table 4
Selected performance parameters for numerical prediction tasks (e.g., biomass)

Metric	Definition/description	Application
Root mean square deviation, RMSD	$\mathrm{RMSD} = \sqrt{\sum_{i=1}^{N} (y_{i,p} - y_{i,a})^2 / N}$ over all cases $i = 1,\ldots,N$, p = predicted value, a = actual value	Provides estimate of absolute error of predicted value, $y_{i,p}$ relative to true value, $y_{i,a}$. Variation of this measure use the absolute difference $\lvert y_{i,p} - y_{i,a} \rvert$ of values
Pearson correlation coefficient, r	$r_{xy} = \frac{\sum_i (x_i - \bar{x})(y_i - \bar{y})}{(n-1) S_x S_y}$, where x are the prediction scores with associated phenotype values y and $s_{x/y}$ are the respective standard deviations	Estimate of how well the predicted values correlate in a linear fashion with the actual values; sensitive to outliers, i.e., correlation may appear high even though only caused be few extreme cases. r^2 measures "explained variance"
Spearman correlation coefficient	Pearson correlation of original values transformed into ranks	Ordinal data, robust against outliers

Which of the many performance parameters will be used depends on the question at hand. For example, if false negative predictions are more serious than false positive predictions, as for example patients wrongly classified as healthy when, in fact, they are not and urgently need treatment, it would seem that more weight needs to be put on low false negative rates and tests need to deliver high sensitivity. However, while it would obviously be desirable to have both, high specificity; i.e., a relatively high false positive rate may be acceptable. Similarly, economic considerations may place more emphasis on one or the other performance characteristic in plant breeding programs. The ROC (Table 3) provides a simple means to set the score thresholds accordingly, as it plots true positives vs. false positives in relation to the classification score value.

3.7. The Perils of Overfitting

Pushing the classification performance to highest possible success rates given the data bears the great risk of increasing the complexity of the model and fine-tuning it to the data at hand and *only* the data at hand. When applied to new samples, the performance will then generally be disappointingly low. In machine learning, this phenomenon is referred to as overfitting. Overfitting is one of the greatest risks in model learning as, because of the understandable desire to obtain best possible results, bias sneaks in at many different places and occasions, and the model developers have to be conscious of the risk of overfitting at all times. Cross-validation and applying the developed test to an independent hold-out test set are the two main means to safeguard against overfitting. In particular, the importance of the latter cannot be overemphasized as cross-validation can also be driven towards overfitting when repeated many times.

The number of markers used in the test is itself an important adjustable parameter to control for the complexity of the model. If too many markers are used, the chance of overfitting; i.e., tailoring the test to exactly those samples used in the model development is high. The optimal number of features can be assessed by monitoring the performance of the model in a cross-validation setting as a function of different numbers of markers. Ideally, an optimal number of markers becomes apparent as an optimum of cross-validation performance. With only a few markers, not all information is being used. With too many markers, performance will drop when applied to the left-out cross-validation set because of overfitting the model to the 90% set used in model learning, even though performance as measured in a self-test is high. Thus, a reduction of variables (i.e., markers) may be necessary. In classification trees, this variable reduction is called pruning. Penalties on high complexity prediction models can also be introduced by applying constraints, an approach called regularization, such as the LASSO method (35).

4. Validation Study

In this last step, the marker selection process and the model training is performed on the entire training data set. It is then validated against the data from an independent test population, the data of which are obtained with the same protocol as the training population, but have never been used during model training; i.e., not during cross-validation either! In many studies, performance estimates are only provided from cross-validation runs. However, cross-validation still bears the risk of overfitting and thus delivering overly optimistic performance estimates as well. It is certainly possible to conduct as many cross-validation runs—every time with tweaked parameters—such that eventually cross-validation results will be optimal. This can even be done as a systematic search. However, this is nothing but overfitting (see below)! Thus, the only true test is to apply the model to hold-out samples. Evidently, this hold-out set should follow the same principal characteristics with regard to all parameters as the samples used in model training. Furthermore, enough samples must be available to allow withholding an extra sample set.

It should be noted that the validation study is a "one-shot-only" test. Subsequent parameter optimization, though tempting, must not be done, as this again is nothing else but overfitting and no valid estimate of performance can be given. Also, the criteria by which the model is considered successful have to be established beforehand. If more than one model is being tested (for example, using different machine learning methods that in turn may or may not use different markers), the significance threshold for a successful model needs to be adjusted, as this amounts to a multiple testing question. Typically, Bonferroni correction can be used ($p_{pass} = 0.05/\text{number\_of\_models}$).

5. After the Experiment

If the selected putative markers worked well as parameters for the phenotype prediction, the next logical step would be to devise a test (optimally in the form of a small measuring device or test strips) for specific measuring of the biomarker substances in practice. Otherwise, the knowledge gained during the study can be applied to another round of basic research, possibly searching for additional substances that could be important for the desired trait(s). Because marker selection procedures are blind to the underlying true cause–effect relationships, and therefore, a substance selected as a marker may or may not be directly involved in the phenotype expression process, new lines of research may result from the set of selected markers to gain a deeper understanding in molecular basis of the phenotype under study.

5.1. Post-Release Monitoring

Because of cost and time considerations, biomarker identification studies typically use only a small number of test samples. Once applied in practice, the number of actual tests and associated phenotypic outcomes will quickly increase. It is important to monitor the performance of the test, particularly if the study aim is long term. Furthermore, frequent quality assurance tests may be necessary to check the reproducibility of the test. Primarily, this concerns the applied instrumentation and sample handling practices that may change over time.

Acknowledgments

Support for this work was provided by the BMELV-funded TROST and the BMBF-funded SEPSAPE projects.

References

1. McCouch S (2004) Diversifying selection in plant breeding. PLoS Biol 2:e347
2. Vale G, Francia E, Tacconi G, Crosatti C, Barabaschi D, Bulgarelli D, Dall'Aglio E (2005) Marker assisted selection in crop plants. Plant Cell Tissue Organ Cult 82:317–342
3. Oliveira MM, Negrao S, Jena KK, Mackill D (2008) Integration of genomic tools to assist breeding in the japonica subspecies of rice. Mol Breed 22:159–168
4. Moose SP, Mumm RH (2008) Molecular plant breeding as the foundation for 21st century crop improvement. Plant Physiol 147:969–977
5. Mackill DJ, Collard BCY (2008) Marker-assisted selection: an approach for precision plant breeding in the twenty-first century. Philos Trans R Soc Lond B Biol Sci 363: 557–572
6. Steinfath M, Strehmel N, Peters R, Schauer N, Groth D, Hummel J, Steup M, Selbig J, Kopka J, Geigenberger P et al (2010) Discovering plant metabolic biomarkers for phenotype prediction using an untargeted approach. Plant Biotechnol J 8:900–911
7. Meyer RC, Steinfath M, Lisec J, Becher M, Witucka-Wall H, Torjek O, Fiehn O, Eckardt A, Willmitzer L, Selbig J et al (2007) The metabolic signature related to high plant growth rate in *Arabidopsis thaliana*. Proc Natl Acad Sci USA 104:4759–4764
8. Korn M, Gartner T, Erban A, Kopka J, Selbig J, Hincha DK (2010) Predicting Arabidopsis freezing tolerance and heterosis in freezing tolerance from metabolite composition. Mol Plant 3:224–235
9. Gartner T, Steinfath M, Andorf S, Lisec J, Meyer RC, Altmann T, Willmitzer L, Selbig J (2009) Improved heterosis prediction by combining information on DNA- and metabolic markers. PLoS One 4:e5220
10. Paik S, Tang G, Shak S, Kim C, Baker J, Kim W, Cronin M, Baehner FL, Watson D, Bryant J et al (2006) Gene expression and benefit of chemotherapy in women with node-negative, estrogen receptor-positive breast cancer. J Clin Oncol 24:3726–3734
11. Paik S (2006) Methods for gene expression profiling in clinical trials of adjuvant breast cancer therapy. Clin Cancer Res 12:1019s–1023s
12. Deng MC, Eisen HJ, Mehra MR, Billingham M, Marboe CC, Berry G, Kobashigawa J, Johnson FL, Starling RC, Murali S et al (2006) Noninvasive discrimination of rejection in cardiac allograft recipients using gene expression profiling. Am J Transplant 6:150–160
13. Fan Y, Wang J, Yang Y, Liu Q, Fan Y, Yu J, Zheng S, Li M, Wang J (2010) Detection and identification of potential biomarkers of breast cancer. J Cancer Res Clin Oncol 136: 1243–1254

14. Michiels S, Koscielny S, Hill C (2005) Prediction of cancer outcome with microarrays: a multiple random validation strategy. Lancet 365:488–492
15. Hochberg Y, Benjamini Y (1990) More powerful procedures for multiple significance testing. Stat Med 9:811–818
16. Nicholson G, Rantalainen M, Maher AD, Li JV, Malmodin D, Ahmadi KR, Faber JH, Hallgrimsdottir IB, Barrett A, Toft H et al (2011) Human metabolic profiles are stably controlled by genetic and environmental variation. Mol Syst Biol 7:525
17. Bergmann W (1992) Colour atlas nutritional disorders of plants: visual and analytical diagnosis. Gustav Fisher Verlag, Jena. Germany
18. Geigenberger P, Tiessen A, Meurer J (2011) Use of non-aqueous fractionation and metabolomics to study chloroplast function in Arabidopsis. Methods Mol Biol 775:135–160
19. Fernie AR, Aharoni A, Willmitzer L, Stitt M, Tohge T, Kopka J, Carroll AJ, Saito K, Fraser PD, Deluca V (2011) Recommendations for reporting metabolite data. Plant Cell 23:2477–2482
20. Sumner LW, Mendes P, Dixon RA (2003) Plant metabolomics: large-scale phytochemistry in the functional genomics era. Phytochemistry 62:817–836
21. Hastie T, Tibshirani R, Friedman J (2001) The elements of statistical learning: data mining, inference, and prediction, 2nd edn. Springer, New York
22. Peng HC, Long FH, Ding C (2005) Feature selection based on mutual information: criteria of max-dependency, max-relevance, and min-redundancy. IEEE Trans Pattern Anal Mach Intell 27:1226–1238
23. Bishop CM (2006) Pattern recognition and machine learning. Springer, New York
24. Larranaga P, Calvo B, Santana R, Bielza C, Galdiano J, Inza I, Lozano JA, Armananzas R, Santafe G, Perez A et al (2006) Machine learning in bioinformatics. Brief Bioinform 7: 86–112
25. Kotsiantis SB, Zaharakis ID, Pintelas PE (2006) Machine learning: a review of classification and combining techniques. Artif Intell Rev 26:159–190
26. Mccullagh P (1980) Regression-models for ordinal data. J R Stat Soc Series B Methodol 42:109–142
27. Lal TN, Chapelle O, Weston J, Elisseeff A (2006) Embedded methods. In: Guyon G, Nikravesh, Zadeh (eds) Feature extraction: foundation and applications. Springer, New York, pp 137–162
28. Huda S, Yearwood J, Strainieri A (2010) Hybrid wrapper-filter approaches for input feature selection using maximum relevance and artificial neural network input gain measurement approximation (ANNIGMA). NSS '10 Proceedings of the 2010 Fourth International Conference on Network and Systems Security
29. Saeys Y, Inza I, Larranaga P (2007) A review of feature selection techniques in bioinformatics. Bioinformatics 23:2507–2517
30. Guyon I, Gunn S, Nikravesh M, Zadeh LA (2006) Feature extraction: foundations and applications (studies in fuzziness and soft computing). Springer, New York
31. Kantardzic M (2002) Data mining: concepts, models, methods, and algorithms. Wiley Hoboken, New Jersey, USA
32. Breiman L (2001) Random forests. Mach Lear 45:5–32
33. Lorena AC, de Carvalho ACPLF, Gama JMP (2008) A review on the combination of binary classifiers in multiclass problems. Artif Intell Rev 30:19–37
34. Kohavi R (1995) A study of cross-validation and bootstrap for accuracy estimation and model selection. Proc Int Conf Artific Intelli
35. Tibshirani R (1996) Regression shrinkage and selection via the Lasso. J R Stat Soc Series B Methodol 58:267–288
36. Efron B, Tibshirani RJ (1994) An introduction to the bootstrap. Chapman & Hall

Chapter 11

Highly Sensitive High-Throughput Profiling of Six Phytohormones Using MS-Probe Modification and Liquid Chromatography–Tandem Mass Spectrometry

Mikiko Kojima and Hitoshi Sakakibara

Abstract

We describe a method for highly sensitive high-throughput analysis of six major phytohormones; cytokinins (23 species), auxins (7 species), abscisic acid (ABA), gibberellins (GAs, 11 species), salicylic acid (SA), and jasmonic acid (JA). The method consists of solid-phase extraction using 96-well column plates and liquid chromatography coupled with a tandem quadrupole mass spectrometer. In order to improve the quantification limit of negatively charged compounds and especially GAs, we use a chemical modification with an "MS-probe" that contains a quaternary amine moiety. The method requires plant tissue of <100 mg fresh weight for quantification and enables us to analyze large number of plant samples at a time.

Key words: Abscisic acid, Auxins, Cytokinins, Gibberellins, Jasmonic acid, Mass spectrometry, MS-probes, Plant hormones, Salicylic acid, Solid-phase extraction

1. Introduction

Plant hormones are involved in the regulation of all phases of plant development, from embryogenesis to senescence. They regulate expression of genes via various signal transduction systems, and in many cases they regulate signaling and metabolic systems cooperatively. For a comprehensive understanding of the regulatory networks, analyses of the correlations between the concentrations of multiple hormones and between hormone concentrations and transcriptome data are required.

Jennifer Normanly (ed.), *High-Throughput Phenotyping in Plants: Methods and Protocols*, Methods in Molecular Biology, vol. 918, DOI 10.1007/978-1-61779-995-2_11, © Springer Science+Business Media, LLC 2012

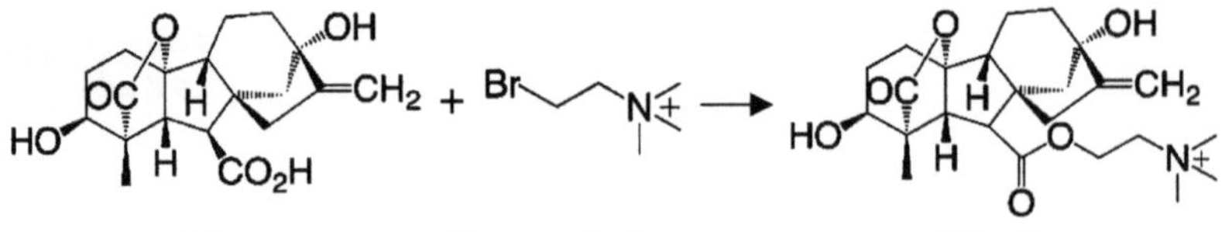

Fig. 1. Derivatization of GA_1 with bromocholine, an MS-probe. GA_1-P, probed GA_1.

Developments of mass spectrometric technologies in the past decade have enabled us to identify and quantify trace amounts of various phytohormones in plant tissue. For instance, some abundant hormones including abscisic acid (ABA), salicylic acid (SA), and jasmonic acid (JA) can be analyzed without purification (1). However, the simultaneous quantification of multiple phytohormones is still difficult because hormones have different chemical properties, and because their concentrations are much lower than those of common metabolites. To analyze phytohormones, it is therefore necessary to partially purify the compounds by column chromatographic techniques such as solid-phase extraction. To quantify hormones, liquid chromatography coupled with a tandem quadrupole mass spectrometer equipped with an electrospray interface (LC–ESI-qMS/MS) is now generally used, because of the wide dynamic range for quantification (10^4–10^5). In general, ESI-qMS/MS is less sensitive for negatively charged compounds, for example gibberellins (GAs), than for positively charged ones. In addition, since GA concentrations in plant tissues usually are quite low, a large amount of plant sample (on the order of grams) is needed.

To overcome this problem, we introduced the "MS-probe modification" methodology into plant hormone analysis (2). MS-probes possess a positively charged quaternary amine moiety (3), and the conjugation transforms natural compounds having a carboxyl group into positively charged compounds (Fig. 1). Thus, the peak intensities of negatively charged hormones, especially GAs, are greatly increased (Fig. 2).

In this chapter, we describe a method for the high-throughput profiling of six phytohormones (cytokinins, auxins, GAs, ABA, SA, and JA) using MS-probe modification and LC–ESI-qMS/MS. This method targets 44 compounds in total: 23 cytokinins including the conjugates, 7 auxins including the amino acid conjugates, 11 GAs including the precursors and metabolites, ABA, SA, and JA.

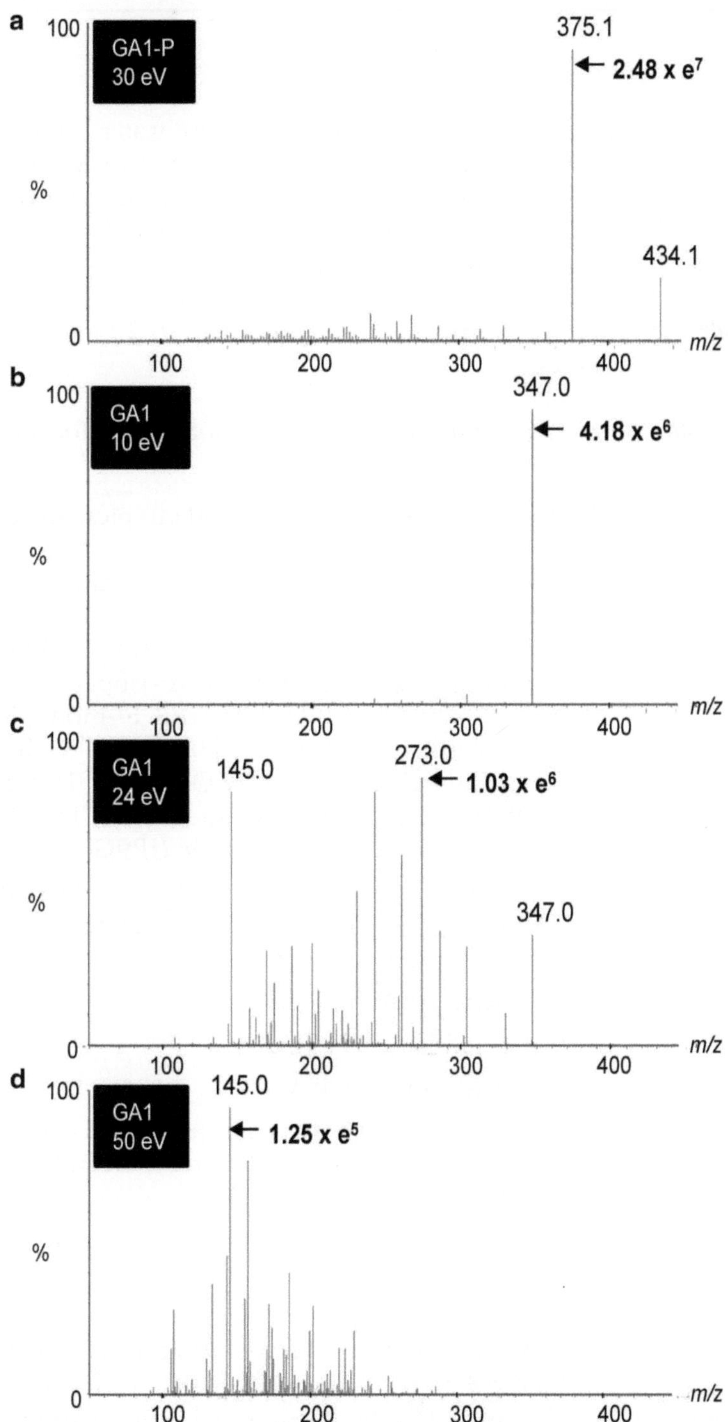

Fig. 2. Fragmentation pattern of GA_1 and GA_1-P in ESI-qMS/MS. Two nmol of GA_1 was derivatized with bromocholine, and a 1/2,000 aliquot was injected into LC–ESI(+)-qMS/MS. As a reference, one pmol of non-derivatized GA_1 was injected to LC–ESI(−)-qMS/MS. Precursor ions were set at 347.0 (*m/z*) for of GA_1 and 434.4 (*m/z*) for GA_1-P, and their daughter ions were scanned from 50 to 450 (*m/z*). (**a**) Daughter ion scan of GA_1-P with a collision energy at 30 eV. (**b–d**) Daughter ion scan of GA_1 with collision energies of 10 eV (**b**), 24 eV (**c**), and 50 eV (**d**). Values with *arrows* indicate the actual signal intensity.

2. Materials

Prepare all solutions using ultrapure water (prepared by purifying deionized water to attain a resistance of 18 MΩ/cm at 25°C; e.g., MilliQ water) and special grade reagents (purity >99.8%) unless indicated otherwise. Prepare and store all reagents at room temperature unless indicated otherwise.

2.1. Plant Materials

Ten to 100 mg fresh weight, or up to 20 mg dry weight of plant tissues.

2.2. Solvent Extraction

1. Extraction solvent: methanol/water/formic acid = 15/4/1 (v/v/v).

Use 150 mL of this solvent for 100 samples. Store at −30°C.

2.3. Internal Standards (see Note 1)

1. Stable isotope (SI)-labeled cytokinins: [$^2H_5$]*trans*-zeatin (tZ), [$^2H_5$]tZ riboside (tZR), [$^2H_5$]tZR 5′-monophosphate, [$^2H_3$]dihydrozeatin (DZ) riboside (DZR), [$^2H_3$]DZR 5′-monophosphate, [$^2H_6$]N^6-Δ^2-isopentenyl)adenine (iP), [$^2H_6$]iP riboside (iPR), [$^2H_6$]iPR 5′-monophosphate, [$^2H_5$] tZ-7-*N*-glucoside (tZ7G), [$^2H_5$]tZ-9-*N*-glucoside (tZ9G), [$^2H_5$]tZ-*O*-glucoside (tZOG), [$^2H_5$]tZR-*O*-glucoside (tZROG), [$^2H_3$]DZ-9-*N*-glucoside, [$^2H_6$]iP-7-*N*-glucoside (iP7G), [$^2H_6$]iP-9-*N*-glucoside (iP9G).
2. SI-labeled GAs: [$^2H_2$]GA_1, [$^2H_2$]GA_3, [$^2H_2$]GA_4, [$^2H_2$]GA_7, [$^2H_2$]GA_8, [$^2H_2$]GA_9, [$^2H_2$]GA_{12}, [$^2H_2$]GA_{19}, [$^2H_2$]GA_{20}, [$^2H_2$] GA_{24}, [$^2H_2$]GA_{44}, [$^2H_2$]GA_{53}.
3. SI-labeled ABA: [$^2H_6$]ABA.
4. SI-labeled IAA: [$^2H_5$]IAA.
5. SI-labeled SA: [$^2H_4$]SA.

   The compounds above can be purchased e.g., from OlChemim Ltd. (http://www.olchemim.cz/Default.aspx: Olomouc, Czech Republic).
6. SI-labeled JA: [$^2H_2$]JA; can be purchased e.g., from Tokyo Chemical Industry Co. Ltd. (http://www.tciamerica.com/: Tokyo, Japan).
7. SI-labeled indole-3-acetyl-L-amino acid conjugates: [$^2H_2$] indole-3-acetyl-L-Ala (IA-Ala), [$^2H_2$]indole-3-acetyl-L-Asp (IA-Asp), [$^2H_2$]indole-3-acetyl-L-Ile (IA-Ile), [$^2H_2$]indole-3-acetyl-L-Leu (IA-Leu) and [$^2H_2$]indole-3-acetyl-L-Phe (IA-Phe) were a gift from Dr. J. Hiratake (Kyoto University, Japan).

Dissolve the compounds with dimethyl sulfoxide to a final concentration of 500 μM (original solutions) and store at −30°C. Prepare the SI-labeled internal standard mix from the original solutions and water to obtain 0.01 μM of cytokinins, 0.1 μM of GAs, ABA, and IA-amino acid conjugates, and 1 μM of IAA, SA, and JA before use.

2.4. Solid-Phase Extraction Solvents

The volumes in parenthesis are for 100 samples.

1. Methanol (600 mL).
2. 1 M formic acid (300 mL): add 11.3 mL formic acid (26.5 M) to 288.7 mL water.
3. 0.35 M aqueous ammonia (100 mL): add 2.4 mL concentrated aqueous ammonia (28%) to 197.6 mL water.
4. 0.35 M aqueous ammonia in 60% (v/v) methanol (100 mL): mix 60 mL methanol with 137.6 mL water, then add 2.4 mL concentrated aqueous ammonia (28%).
5. 10% (v/v) formic acid (5 mL).
6. 0.5% (v/v) formic acid (100 mL).
7. 0.1 M aqueous ammonia (100 mL): add 676 μL of concentrated aqueous ammonia (28%) to 99.3 mL water.
8. 0.1 M aqueous ammonia in 60% (v/v) methanol (100 mL): mix 60 mL methanol with 39.3 mL water, then add 676 μL concentrated aqueous ammonia (28%).
9. 1.25 M formic acid in 70% (v/v) methanol (100 mL): mix 70 mL methanol with 25.3 mL water, then add 4.7 mL of formic acid (26.5 M).
10. 5% (v/v) formic acid (100 μL).
11. 0.1% (v/v) acetic acid (10 mL).
12. 0.1% (v/v) formic acid (5 mL).

2.5. Solid-Phase Extraction Columns

1. Oasis HLB 96-well plate 30 mg (Waters, Milford, MA, USA).
2. Oasis MCX 96-well plate 30 mg (Waters).
3. Oasis MAX 96-well plate 30 mg (Waters).

2.6. Dephosphorylation Reaction

1. *N*-cyclohexyl-2-aminoethansulfonic acid (CHES)–NaOH buffer, pH 9.8 (84.3 mL): prepare 0.5 M CHES–NaOH (pH 9.8) stock solution, and dilute with water to 0.1 M before use.
2. Alkaline phosphatase solution (1.7 mL): calf intestine alkaline phosphatase (Wako Pure Chemical, Tokyo, Japan). Dilute the alkaline phosphatase with dilution buffer [10 mM Tris–HCl (pH 7.0), 1 mM $MgCl_2$, and 40% (v/v) glycerol] to 10 U/μL and store at −80°C. Dilute with water to 1 U/μL before use.
3. Ten times Tris-buffered saline (9.3 mL): 1 M Tris–Cl, 1.5 M NaCl, pH 7.4.
4. 1 M HCl.

2.7. MS-Probe Modification

1. MS-probe (400 μL): bromocholine bromide (Tokyo Chemical Industry, Tokyo, Japan). Prepare 500 mM bromocholine bromide in 70% (v/v) acetonitrile.
2. 1-Propanol (7.5 mL).
3. Triethylamine (80 μL).
4. Prepare the MS-probe reaction solution by mixing the above three reagents with 2 mL water just before use (see Note 2).

2.8. LC–ESI-qMS/MS Analysis

Analytical grade methanol and acetonitrile (>99.9%) are used.

1. 0.06% (v/v) acetic acid (1 L).
2. 0.06% (v/v) acetic acid in methanol (1 L).
3. 0.05% (v/v) formic acid (1 L).
4. 0.05% (v/v) formic acid in acetonitrile (1 L).
5. Octadecyl silica C_{18} column: AQUITY UPLC BEH 2.1 mm × 50 mm, 1.7 μm (Waters).
6. Standard compounds of SI-labeled and non-labeled phytohormones.

2.9. Apparatus

1. Pulverizer: TissuesLyser (Qiagen, Hilden, Germany).
2. Automatic solid-phase extraction system: SPE215 (Gilson, Middleton, WI, USA).
3. Vacuum concentrator: SPD131DDA/RVT4104/OFP400 (Thermo Fisher Scientific, MA, USA).
4. Heat block for reactions at 80°C (e.g., Thermal cycler).
5. LC–ESI-qMS/MS: AQUITY UPLC/XEVO-TQS (Waters).

2.10. Other Components

1. Two-mL microcentrifuge tubes.
2. Zirconia beads (diameter, 5 mm).
3. 96-Well deep plates: nine plates are used for 96 samples.
4. 96-Well PCR plates.
5. Tape seals for 96-well PCR plates.

3. Methods

Carry out all procedures at room temperature unless otherwise specified. A flowchart of extraction and fractionation is shown in Fig. 3.

3.1. Sample Preparation

Harvest plant tissues (<100 mg fresh weight) into a 2-mL microcentrifuge tube with a zirconia bead, and freeze immediately in liquid N (see Note 3). Lyophilize the samples, if necessary.

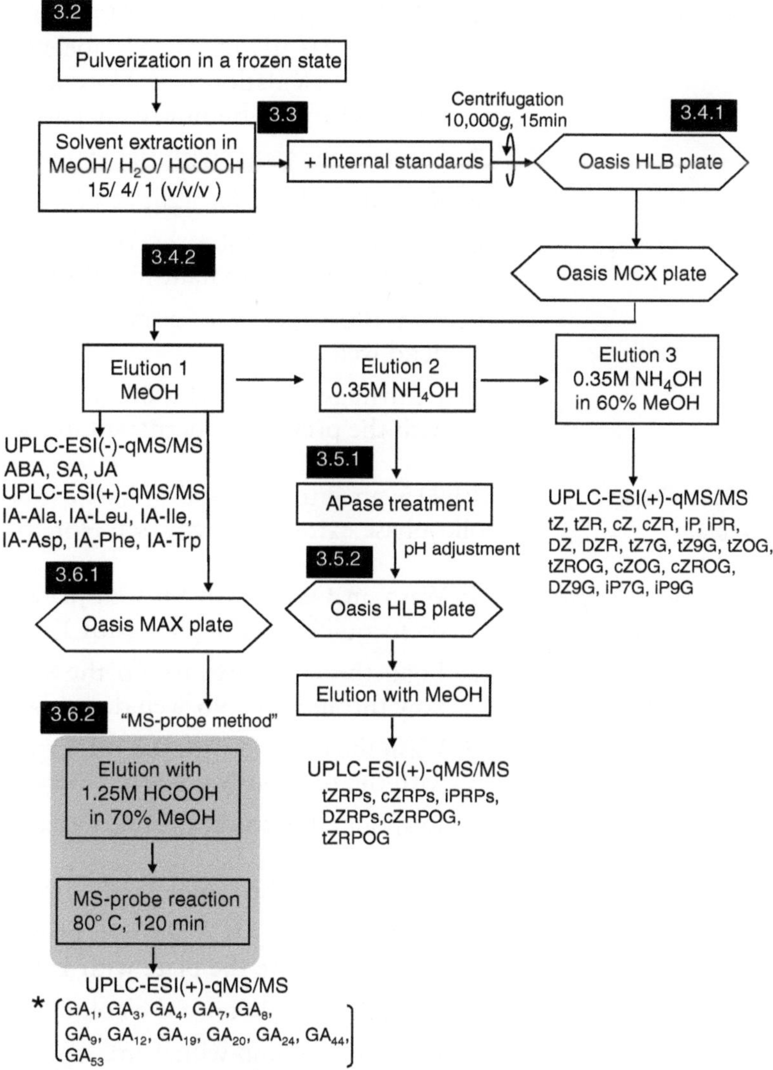

Fig. 3. Flowchart of the extraction and purification protocol for the hormone profiling. *Hexagonal boxes* represent separation columns, rectangular boxes represent other handling processes. Numbers in *white* on black background correspond to the protocol number in the text. *Compounds in parenthesis are detected in the MS-probe-derivatized form. *APase* alkaline phosphatase, *MeOH* methanol.

3.2. Pulverization

1. Set the tubes in the prechilled cassette of the TissueLyser (see Note 4), and crush the tissues (frequency: 20/s, for 30 s) to obtain a fine powder.
2. Repeat pulverization four times. The cassette is occasionally chilled with liquid N to keep the sample frozen.

3.3. Solvent Extraction

1. Add 1 mL prechilled (−30°C) extraction solvent to the tubes, and add 50 μL of the SI-labeled internal standard mix for recovery calculation (see Note 5). The tubes should be kept under −20°C to minimize enzymatic reactions.
2. Suspend the tissue with the extraction solvent by vortexing.
3. Keep the tubes overnight at −30°C to extract the hormones.
4. Centrifuge the tubes at 10,000 × *g* for 15 min at 4°C, and transfer the supernatant to a 96-well deep plate.
5. Add 200 μL of new extraction solvent to the pellet and suspend well.
6. Centrifuge again at 10,000 × *g* for 15 min at 4°C, and combine with the previous supernatant in the 96-well plate.

3.4. Solid-Phase Extraction

SPE215, an automatic solid-phase extraction system, is used for all solid-phase extraction steps (see Note 6).

3.4.1. Removal of Lipophilic Interfering Compounds by a Hydrophilic–Lipophilic Balanced Reversed Phase Column

1. Wash an Oasis HLB 96-well plate with 1 mL methanol, and then equilibrate the plate with 1 mL of 1 M formic acid.
2. Load the crude extract on the column plate and collect the flow through in a 96-well deep plate.
3. Wash the column with 0.3 mL extraction solvent and combine with the previously collected extract.
4. Evaporate the solvent using a vacuum concentrator.

3.4.2. Fractionation by a Mixed-Mode Cation Exchange Column

1. Reconstitute the sample with 1 mL of 1 M formic acid.
2. Wash an Oasis MCX 96-well plate with 1 mL methanol, and then equilibrate the plate with 1 mL of 1 M formic acid.
3. Load the sample onto the column plate.
4. Wash the column with 1 mL of 1 M formic acid.
5. Elute ABA, auxins, GAs, SA, and JA with 1 mL methanol to a 96-well deep plate (elution 1).
6. Elute cytokinin nucleotides with 1 mL of 0.35 M aqueous ammonia to a 96-well deep plate (elution 2).
7. Elute cytokinin nucleobases, nucleosides, and glucosides with 1 mL of 0.35 M aqueous ammonia in 60% methanol to a 96-well deep plate (elution 3).
8. Evaporate the solvent using a vacuum concentrator.
9. Reconstitute elution 1 with 1 mL of water, and transfer 50 μL to a new 96-well deep plate.
10. Add 1 μL of 5% formic acid to the 96-well deep plate and mix well before LC–ESI-qMS/MS analysis for ABA, auxins, SA, and JA.

11. Reconstitute elution 3 with 50 μL of 0.1% acetic acid before LC–ESI-qMS/MS analysis for cytokinin nucleobases, nucleosides, and glucosides.

3.5. Dephosphorylation of Cytokinin Nucleotides

Cytokinin nucleotides are detected as the corresponding nucleosides after dephosphorylation with alkaline phosphatase, because nucleotides are difficult to separate by reverse-phase column chromatography.

3.5.1. Dephosphorylation

1. Reconstitute elution 2 with 843 μL of 0.1 M CHES–NaOH, pH 9.8.
2. Add 17 μL of calf intestine alkaline phosphatase (1 U/μL).
3. Incubate for 1 h at 37°C.
4. Add 93 μL 10× Tris-buffered saline and 46 μL of 1 M HCl to adjust around pH 7.3 (see Note 7).

3.5.2. Desalting by a Hydrophilic–Lipophilic Balanced Reversed Phase Column

1. Wash an Oasis HLB 96-well plate with 1 mL methanol, and then equilibrate the plate with 1 mL water.
2. Load the sample onto the plate column.
3. Wash the column with 1 mL water.
4. Elute cytokinin nucleosides with 1 mL methanol to a 96-well deep plate.
5. Evaporate the solvent using a vacuum concentrator.
6. Reconstitute with 50 μL of 0.1% acetic acid before LC–ESI-qMS/MS analysis.

3.6. MS-Probe Derivatization

3.6.1. Semi-Purification by a Mixed-Mode Anion Exchange Column

1. Add 50 μL of 10% formic acid to the rest elution 1 plate at step 9 of Subheading 3.4.2 (i.e., total volume is 1 mL), and mix well.
2. Wash an Oasis MAX 96-well plate with 1 mL methanol, and then equilibrate the plate with 1 mL water.
3. Load the sample onto the plate column.
4. Wash the column with 1 mL of 0.5% formic acid, 0.1 M aqueous ammonia, and then with 0.1 M aqueous ammonia in 60% methanol.
5. Elute GAs with 1 mL of 1.25 M formic acid in 70% methanol to a 96-well deep plate (see Note 8).
6. Evaporate the solvent using a vacuum concentrator.

3.6.2. Reaction with MS-Probe (see Note 9)

1. Reconstitute with 99.8 μL of the MS-probe reaction solution.
2. Transfer the solution to a 96-well PCR plate and seal the top.
3. Incubate at 80°C for 120 min (see Note 10).
4. Put the tubes on ice to stop the reaction.

5. Evaporate the solvent using a vacuum concentrator.
6. Reconstitute with 50 μL of 0.1% formic acid and transfer to a 96-well deep plate before LC–ESI-qMS/MS analysis.

3.7. Quantification by Liquid Chromatography–Tandem Mass Spectrometry

The conditions for LC–ESI-qMS/MS depend on the apparatus used. General information is given below. Standard SI-labeled and non-labeled phytohormones are used to produce calibration curves.

3.7.1. Liquid Chromatography

Separation of phytohormones is achieved by liquid chromatography (Waters, AQUITY UPLC) with an octadecyl silica C_{18} column (Waters, AQUITY UPLC BEH 2.1 mm × 50 mm, 1.7 μm). Cytokinins are separated at a flow rate of 0.25 mL/min with linear gradients of solvent A (0.06% acetic acid) and solvent B (0.06% acetic acid in methanol) according to the following program: 0 min, 99.0% A + 1.0% B; 5 min, 70.0% A + 30.0% B; 8 min, 30.0% A + 70.0% B; and then with isocratic conditions: 9 min, 1.0% A + 99.0% B; 12 min, 99.0% A + 1.0% B. Other hormones are separated at a flow rate of 0.25 mL/min with linear gradients of solvent A (0.05% formic acid) and solvent B (0.05% formic acid in acetonitrile) according to the following program: 0 min, 99.0% A + 1.0% B; 2.0 min, 86.0% A + 14.0% B; 6 min, 84.0% A + 16.0% B; 13 min, 40.0% A + 60.0% B; and then with isocratic conditions: 14 min, 1.0% A + 99.0% B; 15 min, 99.0% A + 1.0% B. Retention times of each compound under these conditions are summarized in Table 1.

Table 1
Precursor-to-product ion transitions used for the quantification by LC–ESI-qMS/MS

Analytes	RT[a] (min)	Transition	CV[b] (kV)	CE[c] (eV)	IS[d]	RT (min)	Transition
tZ	3.49	220.1 > 136.1	2	16	d5-tZ	3.44	225.1 > 136.7
tZR	4.76	352.2 > 220.1	48	20	d5-tZR	4.73	357.1 > 136.0
cZ	3.84	220.1 > 136.1	2	16	d5-tZ	3.44	225.1 > 136.7
cZR	5.06	352.1 > 220.1	38	20	d5-tZR	4.73	357.1 > 136.0
DZ	3.65	222.2 > 136.1	56	20	d5-tZ	3.44	225.1 > 136.7
DZR	4.98	354.1 > 222.1	38	36	d5-DHZR	4.95	357.1 > 136.0
iP	6.49	204.2 > 136.1	2	12	d6-iP	6.44	210.1 > 137.1
iPR	7.11	336.2 > 204.1	42	16	d6-iPR	7.08	342.2 > 137.0

(continued)

Table 1 (continued)

Analytes	RT[a] (min)	Transition	CV[b] (kV)	CE[c] (eV)	IS[d]	RT (min)	Transition
tZ7G	3.03	382.2 > 220.1	52	22	d5-tZ7G	3.00	387.2 > 136.0
tZ9G	3.56	382.2 > 220.1	52	22	d5-tZ9G	3.53	387.2 > 136.0
tZOG	3.4	382.2 > 220.1	52	22	d5-tZOG	3.36	387.2 > 136.0
cZOG	3.65	382.2 > 220.1	52	22	d5-tZOG	3.36	387.2 > 136.0
tZROG	4.44	514.2 > 382.2	40	18	d5-tZROG	4.41	519.2 > 225.1
cZROG	4.71	514.2 > 382.2	40	18	d5-tZROG	4.41	519.2 > 225.1
DZ9G	3.72	384.1 > 222.1	6	22	d3-DZ9G	3.7	387.2 > 136.0
iP7G	4.75	366.1 > 204.1	46	20	d6-iP7G	4.71	372.1 > 137.0
iP9G	6.23	366.2 > 204.1	46	20	d6-iP9G	6.19	372.1 > 137.0
IAA	5.40	176.0 > 130.0	20	16	d5-IAA	5.28	181.0 > 134.0
IA-Ala	4.71	247.1 > 130.1	28	16	d2-IA-Ala	4.69	249.1 > 132.1
IA-Ile	9.51	289.2 > 130.1	32	26	d2-IA-Ile	9.50	291.2 > 132.1
IA-Leu	9.61	289.2 > 130.1	32	26	d2-IA-Leu	9.60	291.2 > 132.1
IA-Trp	9.54	362.2 > 130.1	26	28	d2-IA-Phe	9.77	325.2 > 132.1
IA-Phe	9.78	323.2 > 130.1	28	32	d2-IA-Phe	9.77	325.2 > 132.1
ABA	8.17	263.1 > 153.1	30	6	d6-ABA	8.12	269.1 > 159.1
SA	4.98	137.0 > 93.0	50	16	d4-SA	4.88	141.0 > 97.0
JA	9.23	209.2 > 59.0	34	12	d2-JA	9.21	211.2 > 59.0
GA_1-P	3.2	434.3 > 375.0	50	30	d2-GA_1-P	3.18	436.3 > 377.0
GA_3-P	3.1	432.3 > 373.2	50	30	d2-GA_3-P	3.08	434.3 > 375.2
GA_4-P	8.75	418.4 > 359.3	50	20	d2-GA_4-P	8.73	420.4 > 361.3
GA_7-P	8.59	416.2 > 356.9	50	20	d2-GA_7-P	8.58	418.4 > 359.3
GA_8-P	2.21	450.4 > 391.2	50	30	d2-GA_8-P	2.2	452.4 > 393.2
GA_9-P	10.73	402.3 > 343.2	50	30	d2-GA_9-P	10.72	404.4 > 345.2
GA_{12}-P	12.08	418.4 > 359.3	50	30	d2-GA_{12}-P	12.08	420.4 > 361.3
GA_{19}-P	7.92	448.3 > 389.2	50	30	d2-GA_{19}-P	7.91	450.3 > 391.2
GA_{20}-P	7.33	418.2 > 358.9	50	30	d2-GA_{20}-P	7.31	420.4 > 361.3
GA_{24}-P	9.52	432.3 > 373.2	50	22	d2-GA_{24}-P	9.5	434.3 > 375.2
GA_{44}-P	8.02	432.3 > 373.2	50	22	d2-GA_{44}-P	8.01	434.3 > 374.8
GA_{53}-P	9.05	434.3 > 375.0	50	30	d2-GA_{53}-P	9.04	436.5 > 377.0
IAA-P	3.23	261.0 > 201.9	50	20	d5-IAA-P	3.19	266.0 > 206.9

(continued)

Table 1 (continued)

Analytes	RT[a] (min)	Transition	CV[b] (kV)	CE[c] (eV)	IS[d]	RT (min)	Transition
IA-Ala-P	3.04	332.1 > 273.2	50	20	d2-IA-Ala-P	3.02	334.1 > 275.0
IA-Ile-P	7.99	374.2 > 315.2	50	20	d2-IA-Ile-P	7.96	376.2 > 317.2
IA-Leu-P	8.16	374.2 > 315.2	50	20	d2-IA-Leu-P	8.14	376.2 > 317.2
IA-Asp-P	2.71	376.3 > 317.0	50	20	d2-IA-Asp-P	2.7	378.3 > 319.0
IA-Trp-P	8.39	447.4 > 271.2	50	20	d2-IA-Phe-P	8.43	410.1 > 351.2
IA-Phe-P	8.45	408.1 > 349.2	50	20	d2-IA-Phe-P	8.43	410.1 > 351.2
ABA-P	4.64	350.2 > 291.2	50	17	d6-ABA-P	4.58	356.2 > 297.2

Internal standards used for recovery calculation are shown on the right

[a]Retention time

[b]Cone voltage

[c]Collision energy

[d]Internal standard

3.7.2. Mass Spectrometry

Analysis of phytohormones is achieved by tandem quadruple mass spectrometry (Waters, XEVO-TQS). Multiple reaction monitoring (MRM) is used for the identification and quantification of target phytohormones. Selection of precursor and product ions is carried out using unlabeled and deuterium-labeled standard compounds as summarized in Table 1. Cone voltage and collision energy to achieve maximum sensitivity under these conditions are summarized in Table 1.

3.7.3. Data Processing

Data are processed by the MassLynx software with TargetLynx (version 4.1, Waters).

4. Notes

1. Selection of SI-labeled internal standards depends on the aim of the research. Not all standards have to be used.
2. Water inhibits MS-probe modification. In this case, standard compounds for calibration curves are dissolved in aqueous solution (see Note 9). In order to unify the water concentration between the standard and sample reaction solutions, we add water to sample reaction solutions.
3. Remove water, soil, and other debris quickly but thoroughly from the sample surface. If the apparent sample weight includes weight from other materials, accurate concentration

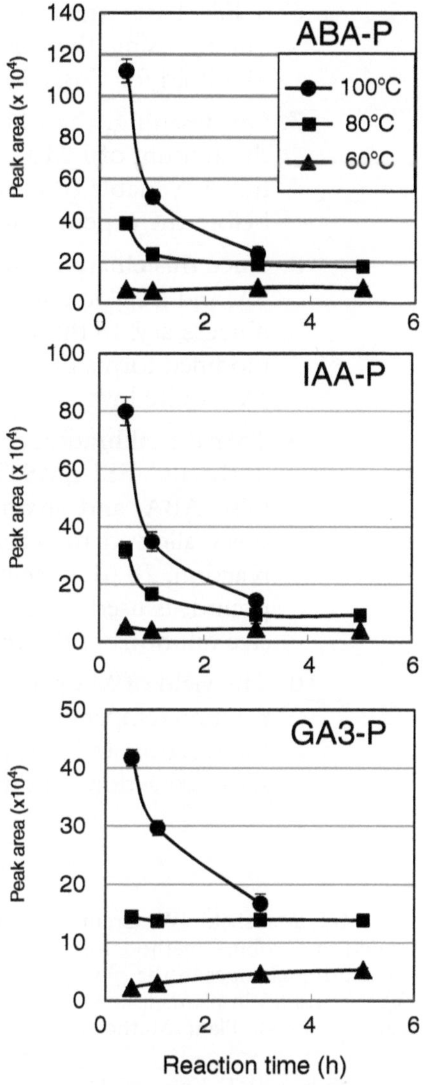

Fig. 4. Relationship between reaction period and yield of derivatives. ABA, IAA, GA_3 reacted with bromocholine at 60°C, 80°C, and 100°C for the indicated times. The modified ABA (ABA-P), IAA (IAA-P), and GA_3 (GA_3-P) were quantified with LC–ESI(+)-qMS/MS.

of phytohormones cannot be calculated. When the sample is over 2 cm long or solid (e.g., stem or fruit), chop it into small pieces (2–3 mm) with a razor blade for efficient solvent extraction.

4. Prechilling of the cassette serves to keep the samples frozen during pulverization. To prevent the cassette from cracking in the liquid N, precool it in a deep freezer (−80°C).

5. The amount of internal standards to be added depends on the nature of the sample. If large amounts of hormones are expected, more of the internal standards should be used.

6. Solid-phase extraction with Oasis 96-well plates can be carried out by using the corresponding Manifold (Extraction Plate Manifold for Oasis 96-well plates).
7. On desalting, the pH of samples must not be alkaline. Since the amount of 1 M HCl (about 46 μL) required for neutralization is variable, the necessary amount should be determined before the experiment using a pH meter.
8. Since this eluate contains not only GAs but also ABA, auxins, SA, and JA, you could quantify these compounds in this eluate if necessary. In this case, SA and JA should be detected as non-modified forms because the derivatization efficiency of SA and JA is quite low.
9. To make calibration curves for MS-probe-modified compounds in the LC–ESI-qMS/MS analysis, 100 pmol of GA standards (also ABA, and auxin standards, if necessary) are simultaneously allowed to react with the MS-probe in a well. In this reaction, 20 μL of standard solution (5 pmol/μL of each compound) is used. The calibration curves are drawn with successive dilutions of the reacted solutions.
10. The yield of MS-probe-modified compounds is affected by the reaction temperature and period. Higher temperature promotes the reaction and also the decomposition (see Fig. 4). Thus, we set the reaction conditions so that stable yields are obtained.

References

1. Forcat S, Bennett MH, Mansfield JW, Grant MR (2008) A rapid and robust method for simultaneously measuring changes in the phytohormones ABA, JA and SA in plants following biotic and abiotic stress. Plant Methods 4:16
2. Kojima M, Kamada-Nobusada T, Komatsu H, Takei K, Kuroha T, Mizutani M, Ashikari M, Ueguchi-Tanaka M, Matsuoka M, Suzuki K, Sakakibara H (2009) Highly sensitive and high-throughput analysis of plant hormones using MS-probe modification and liquid chromatography-tandem mass spectrometry: an application for hormone profiling in *Oryza sativa*. Plant Cell Physiol 50:1201–1214
3. Honda A, Hayashi S, Hifumi H, Honma Y, Tanji N, Iwasawa N, Suzuki Y, Suzuki K (2007) MPAI (mass probes aided ionization) method for total analysis of biomolecules by mass spectrometry. Anal Sci 23:11–15

Chapter 12

Qualitative and Quantitative Screening of Amino Acids in Plant Tissues

Will I. Menzel, Wen-Ping Chen, Adrian D. Hegeman, and Jerry D. Cohen

Abstract

The comprehensive analysis of metabolites (metabolomics) and expressed proteins (proteomics) in any given biological system forms the center of modern efforts to define the critical functions of biological systems. Because amino acids play important roles in primary and secondary metabolic pathways as well as serving as the building blocks of proteins, they have been important targets for efforts at metabolic profiling. Amino acids have been analyzed using a number of procedures, including separation by high performance liquid chromatography (HPLC), gas chromatography (GC), liquid chromatography (LC), and capillary electrophoresis (CE). Mass spectrometry (MS) remains the primary analytical and detection system for metabolic profiling, including amino acid analysis, due to its accuracy and the information content obtained by such analyses, which facilitates the identification and measurement of large numbers of biomolecules. MS methods also add the capability of monitoring isotope distributions of molecules for metabolic flux analysis. Here we describe a GC–MS method that is suitable for analysis of amino acids in sub-milligram quantities of fresh plant material and that is easily adapted to high-throughput screening approaches.

Key words: Amino acids, Gas chromatography, Mass spectrometry, Methyl chloroformate, Stable isotopes

1. Introduction

Gas chromatography coupled with mass spectrometry (GC–MS) provides a low cost bench-top platform for metabolite profiling that is easy to use and highly sensitive. However, because the functional groups on the amino acids make them unsuitable for direct GC analysis, fast and efficient derivatization techniques are needed to obtain the full potential of GC–MS for high-throughput analyses.

Jennifer Normanly (ed.), *High-Throughput Phenotyping in Plants: Methods and Protocols*, Methods in Molecular Biology, vol. 918, DOI 10.1007/978-1-61779-995-2_12, © Springer Science+Business Media, LLC 2012

Derivatization for profiling amino and organic acids in biological fluids using alkyl chloroformates is a particularly useful technique. Husek first introduced the basic procedure, which has since been investigated by many laboratories for its potential for quantitative and reproducible derivatization of amino acids (1). Using alkyl chloroformate derivatization has several advantages: (1) a rapid one-step reaction can be carried out directly in aqueous solution without the need for sample heating; (2) the reagent costs are very low; and (3) it is easy to separate the resulting derivatives from the reaction mixture by organic solvent partitioning, resulting in less potential contamination at the MS step. These methods can be automated, and the derivatization adds only a relatively low molecular weight modification, which proves advantageous for the calculation of isotopic abundance (2).

To conduct studies of stable isotope enrichment in growing plants, we developed a rapid microscale method based on methyl chloroformate for determination of amino acid profiles from minute amounts of plant tissues (3). The method involves solid-phase ion exchange followed by derivatization and analysis by GC–MS. The procedure has a number of advantages for several variations on high-throughput analyses. First, the procedure allows direct derivatization of the amino acids after elution from the solid-phase ion exchange medium without sample evaporations. Sample extraction and derivatization are quick and suitably efficient to provide highly sensitive analyses. For example, the quantification of 19 amino acids eluted from the cation exchange solid-phase extraction step from a single cotyledon (0.4 mg fresh weight) or three etiolated 7-day-old *Arabidopsis* seedlings (0.1 mg fresh weight) was easily accomplished on a standard GC–MS system running in the selected ion monitoring (SIM) mode. This method was especially useful for monitoring mass isotopic distribution of amino acids as illustrated by *Arabidopsis* seedlings that had been labeled with deuterium oxide and $^{13}CO_2$ (4, 5). Following are three examples of the application of amino acid analysis to plant phenotyping.

1.1. Determining Stable Isotopic Enrichment of Metabolically Labeled Plant Tissues via Mass Isotopomer Distribution Analysis

When plants are grown in the presence of isotopically enriched substrates such as $^{13}CO_2$ or $^{18}O_2$ gas, heavy water ($^{2}H_2O$), or hydroponic media prepared with ^{15}N-labeled nutrients, the labeled atoms are incorporated into the plant's metabolites and proteins in place of the natural abundance atoms that normally comprise these compounds. This process is often referred to as metabolic labeling, because it uses the organism's native metabolism to incorporate the labeled atom type (6). The rate at which the plant's metabolites and proteins become labeled, or "enriched," during exposure to labeled substrates is determined, in part, by the amount of these compounds moving through metabolic pathways (flux), and the rate at which the compounds are recycled (turnover).

Mass isotopomer distribution analysis (MIDA) enables investigators to determine the enrichment of metabolites and proteins

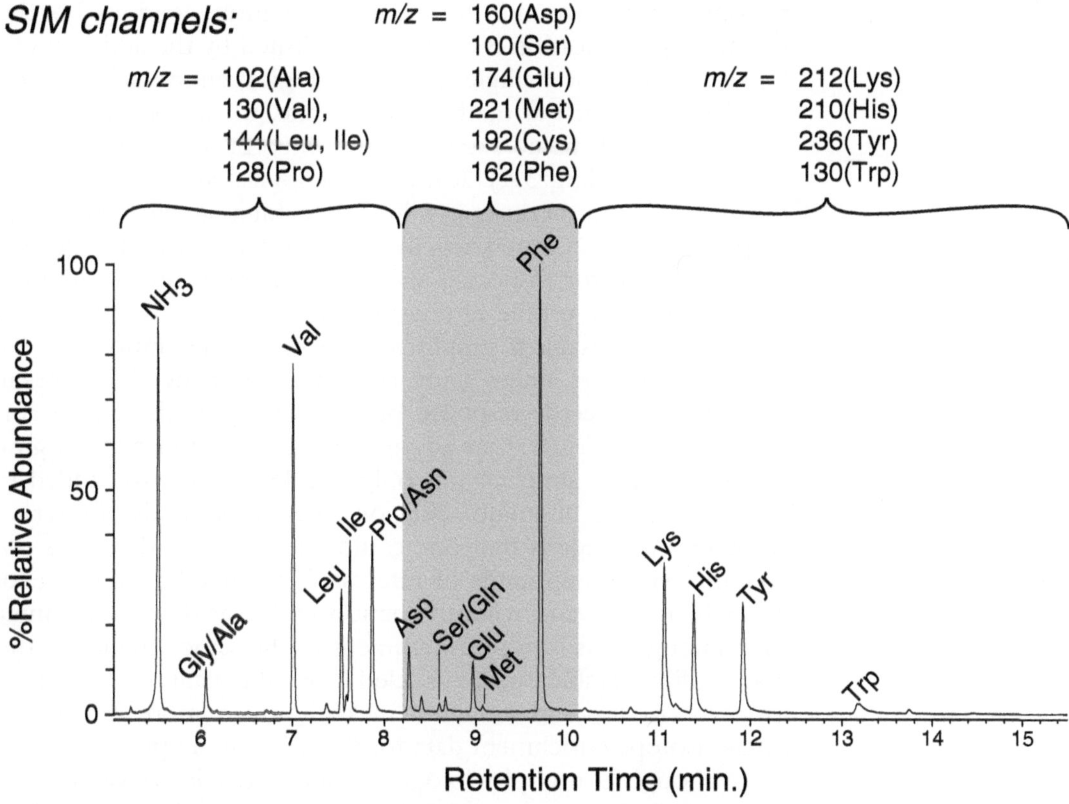

Fig. 1. Derivatized amino acids extracted from soy were analyzed by GC–MS using selected ion monitoring (SIM). SIM is a scanning mode in which only fragments of specified mass-to-charge ratios (*m/z*), or channels, are transmitted to and detected by the mass spectrometer detector during a specified time window. This specificity results in increased sensitivity compared to full-scan mode. Product ions (four to six channels) were monitored over three discrete elution time windows based on the known elution properties of the amino acid derivatives.

throughout the plant (7, 8). Therefore, in MIDA, the stable isotope acts as a marker for flux and turnover analysis. The method described above facilitates high sensitivity MIDA of isotopically labeled and unlabeled tissue, even on a single *Arabidopsis* cotyledon (~0.4 mg, (3)). This has enabled researchers, for example, to assess the efficiency of labeling protocols and to measure metabolic flux and metabolite and protein turnover in tissue of plants grown in a variety of stable isotope labeling conditions, including deuterium oxide ($^{2}H_2O$, (3, 4)), $^{13}CO_2$ (5), and ^{15}N-enriched hydroponic media (4).

1.2. Screening Plants for Variations in Amino Acid Abundance: Amino Acid Profiling with a Non-Physiological Amino Acid Internal Standard

This method facilitates amino acid profiling in minute amounts of tissue, on the scale of a single *Arabidopsis* cotyledon (4), when the GC–MS system acquires data in SIM mode (see Fig. 1 for details on SIM). To accurately compare chromatographic peak heights and spectral abundance values across samples, it is necessary to account for variations in amino acid recovery and in analyte detector response (see Note 1). By including internal standard compounds from the earliest possible steps in the analysis, it is possible to control for variation in sample recovery during workup. For this method, quantification is easily accomplished by the addition of a known amount of a single non-physiological amino acid internal standard (IS), such as methionine sulfone, to each sample before proceeding with tissue extraction (see Step 2, Subheading 3.1 or Step 2, Subheading 3.4). A non-physiological amino acid is used for quantification because it will likely exhibit similar behavior to its physiological counterparts in terms of derivatization, chromatography, detector response, and fragmentation, but its addition will not alter the profile of physiological amino acids in the plant extract. It is advisable to empirically derive response ratios between the IS and analyte amino acids, as subtle systematic differences in recovery or detector response between compounds can be corrected for by their use. One advantage of using a non-physiological amino acid IS is simplicity, as only one IS needs to be added to quantify a range of amino acids, or other compounds of interest. Another advantage is that one could choose an IS that does not co-elute with compounds of interest. This may be valuable; for example, if one wants to both quantify amino acids and determine their endogenous isotope enrichment in the same sample. In this case, adding a stable isotope-labeled IS would result in overlapping isotopic envelopes that would make it much more difficult to obtain isotope enrichment data for the original sample.

Although a non-physiological amino acid is convenient for routine screening, often for other types of experiments the best choice of IS for quantification is to use a mixture of heavy atom (^{13}C, ^{15}N, or ^{18}O) stable isotope-labeled amino acids. As the labeled and unlabeled amino acids are virtually chemically identical (except for negligible isotope effects), there should be little to no difference in behavior between these compounds during extraction, fractionation, or MS detection, and response ratios do not need to be computed. While heavy atom isotopes are accompanied by small isotope effects, the use of lighter isotope labels such as deuterium (^{2}H)-labeled compounds may result in significant kinetic or equilibrium isotope effects. These isotope effects may result in differential compound stability, variation in solvent extraction partition behavior, and can dramatically shift chromatographic retention times. These effects may cause serious problems in some types of analyses, but deuterated compounds still provide excellent internal control and are often the only commercially available option for stable isotopically labeled standard compounds. Heavy atom-labeled reagents are increasingly available and can be purchased as prepared mixed standards such as the [^{13}C]-labeled algal amino acid mixture offered by Cambridge Isotope Laboratories, Andover, MA, USA. (Note: this mixture does not contain tryptophan, which can be added separately.)

1.3. Targeted Amino Acid Analysis: High-Precision Quantification Using Stable Isotope Dilution Analysis

Where one or more amino acids are of particular interest, stable isotopes are excellent ISs for chromatographic and mass spectrometric analyses, enabling high-precision quantification with only a few extra steps. One advantage to the analysis of a small targeted set of amino acids is that the selection of isotopically labeled amino acids is often greater than what is available as complete amino acid mixtures. Because stable isotope analogues, or "isotopologues," are chemically identical to a compound of interest, the isotopologue will have identical extraction characteristics. Similarly, isotopologues will have identical, or nearly identical performance in chromatography columns and mass spectral detectors. For amino acids, standards labeled with ^{2}H (deuterium), ^{13}C, ^{15}N, and ^{18}O may be used, depending on application, availability, and cost considerations. Note that deuterated standards, although often the most economical, may exhibit some isotopic effects, such as slightly different retention times and back-exchange with natural abundance ^{1}H. In contrast, ^{13}C-, ^{15}N-, and, in many cases, ^{18}O-labeled standards exhibit no significant isotope effects. This means that [$^{13}C_6$]-phenylalanine, in which all six carbons in the benzene ring have an extra neutron in their nuclei, will elute at essentially the same time in a chromatographic column as endogenous [^{12}C]-phenylalanine from a plant extract. In addition, both the enriched and natural abundance amino acids will exhibit the same fragmentation patterns in the mass spectrometer, facilitating easy comparison of enriched and natural abundance isotopomers.

Isotope dilution is a method for quantification in which the amount (y) of an amino acid, or other compound of interest present in a sample can be determined by adding a known amount (x) of a stable isotope-labeled standard to a sample and using the isotope dilution equation (Eq. 1) to calculate the extent of dilution of the labeled compound by its natural abundance isotopologue in the sample. The enrichment of the standard before addition to the plant sample (C_0) and after addition (C_f) is determined via mass spectrometric analysis, as is the correction factor for actual concentration including minor isotopomers (R). For a more extensive discussion of this method, see Barkawi et al. (9).

$$y = \frac{(C_0 / C_f - 1)x}{R} \quad (1)$$

2. Materials

Prepare all solutions using deionized distilled water that is free of phthalate ester contamination (distilled in a glass still equipped with only glass and polytetrafluoroethylene (PTFE; "Teflon®")

tubing and stored in glass containers). Analytical grade reagents should be prepared and stored at room temperature, unless otherwise indicated. Follow appropriate regulations for disposal of waste materials.

2.1. Tissue Extraction

1. 0.01 M hydrochloric acid (HCl).
2. Deionized, glass-distilled water (ddH_2O).
3. Internal standard: methionine sulfone or stable isotope-labeled amino acid(s) in concentration range from 20 to 200 μg/mL (see Note 2) (store at 4 °C).
4. Ball bearings for tissue pulverization: 3 mm (1/8 in.) 316 stainless steel bearing balls (ToolSupply, Orlando, FL, USA; http://stores.ebay.com/ToolSupply).
5. Disposable 1.5 mL microcentrifuge tube pestles (Item Number: 749521-1500, Kimble Chase Kontes, Vineland, NJ, USA).
6. Mixer-Mill bead mill (Model MM 300, Retsch GmbH & Co. KG, Haan, Germany).
7. Microcentrifuge tubes: 1.5 mL Seal-Rite Natural microcentrifuge tubes (Catalog Number: 1615-5500, USA Scientific, Ocala, FL, USA).
8. Microcentrifuge: Marathon 16KM (Fisher Scientific, Hampton, NH, USA).
9. Vortex mixer: Vortex-Genie 2 (Model G-650, Scientific Industries, Bohemia, NY, USA) equipped with TurboMix attachment (see Note 3, Model Z511439, Sigma-Aldrich).

2.2. Solid-Phase Extraction

1. Vacuum manifold: Baker-10 Extraction System (Catalog Number: 7016-0, JT Baker Chemical Company, Phillipsburg, NJ, USA).
2. Waste tray (to fit in vacuum manifold).
3. Solid-phase extraction (SPE) columns: Alltech Extract-Clean strong cation exchange SPE columns; bed weight: 100 mg; column size: 1.5 mL; Manufacturer Number: AT209800 (Grace Davison Discovery Sciences, Deerfield, IL, USA).
4. SPE tips: TopTip packed mini-spin columns packed with Dowex 50-type strong cation exchange resin (Item Number: TT2-TWSCX.96, Glygen Corporation, Columbia, MD, USA; see Note 4).
5. Disposable glass Pasteur pipettes: 14.6 cm size.
6. Methanol/water wash: 80 % methanol (HPLC grade)/20 % ddH_2O.
7. Elution solution: A 1:1 (volume/volume) solution of 8 M ammonium hydroxide and methanol (see Note 5).
8. Glass culture tubes: 12 mm × 75 mm size.

2.3. Derivatization

1. GC–MS vial insert: 50 μL (Item Number: CTI-2405, Chrom Tech, Inc., Apple Valley, MN, USA).
2. GC–MS vials: Crimp-top wide-mouth vials, 2 mL volume, 12 mm × 32 mm (Catalog Number: CTV-1104, Chrom Tech, Inc., Apple Valley, MN, USA).
3. Crimp cap: with PTFE/rubber septum (Item Number: CTC-1108, Chrom Tech, Inc., Apple Valley, MN, USA).
4. Cap crimper: Wheaton EZ Crimper for 11 mm standard seals (Item Number: W225301, Wheaton Science Products, Millville, NJ, USA).
5. Mini glass culture tube: 6 mm × 50 mm size.
6. Methyl chloroformate: repackaged from manufacturer's bottle into 25 mL glass bottles using a glove bag purged with nitrogen gas. The 25 mL bottles should be stored in a −20 °C freezer. Allow to equilibrate to room temperature prior to use (see Note 6, Item Number: M35304, Sigma-Aldrich).
7. Pyridine: should be stored in a −20 °C freezer. Allow to equilibrate to room temperature prior to use (see Note 6).
8. Vortex Mixer for test tubes: Maxi Mix II (Type 37600, Thermo Scientific).
9. 50 mM sodium bicarbonate solution (see Note 5).
10. Positive displacement pipettes: 25, 50, and 250 μL sizes (Microman Models M25, M50, M250, respectively, Gilson, Middleton, WI, USA).
11. Aluminum foil.
12. GC–MS analyses are performed using a single quadrupole GC–MS system equipped with a fused silica capillary column (HP-5MS, 30 m × 25 mm ID, 0.25 μm film thickness; Agilent J&W Scientific, Folsom, CA). The procedures were developed using a Hewlett-Packard 5890 (GC)/5970 mass selective detector (MSD) with a 70 eV electron impact (EI) source.

3. Methods

For analysis of 20–50 mg of tissue.

3.1. Tissue Extraction

Procedures can be carried out at room temperature.

1. Place 20–50 mg (see Note 7) fresh weight (FW, see Note 8) plant tissue in a microcentrifuge tube.
2. Add 1 mL 0.01 M HCl, 10 μL internal standard, and two 3 mm stainless steel balls to sample.

3. Grind samples using bead mill for 3–5 min at 25 s^{-1} frequency.
4. Shake in vortex mixer with TurboMix attachment for 10 min.
5. Centrifuge samples at 14,000 × *g* for 3 min (see Note 9).

3.2. Solid-Phase Extraction

This procedure should be carried out in a fume hood. Be careful to avoid breathing fumes from the ammonium hydroxide solution.

1. Condition SPE column by pipetting 1 mL of 0.01 M HCl onto column (see Note 10) and drawing the liquid through the column at 0.5 mL/min using a vacuum manifold, being careful to avoid drying the column bed (see Note 11). Collect liquid in the vacuum manifold's internal waste tray.
2. Rinse column by drawing 1 mL of ddH_2O through column. Repeat twice, for a total of three rinse steps. Collect rinse liquid in waste tray.
3. Load amino acids by using a 14.6 cm Pasteur pipette to transfer supernatant from samples centrifuged in Subheading 3.1 to the SPE column, and then draw the supernatant through SPE column. Collect liquid in waste tray.
4. Wash SPE column by drawing two aliquots of 1 mL of methanol/water wash through column. Collect liquid in waste tray. Replace waste tray in vacuum manifold bed with glass culture tubes in tube holder, in preparation for elution step.
5. Elute amino acids by drawing 250 μL of elution solution through column slowly, over the course of 2 min, and continue to elute by vacuum until the column bed is nearly or completely dry (see Note 12).
6. Sample is now ready for derivatization or storage (see Note 13).

3.3. Derivatization

This procedure should be conducted in a fume hood. Be careful to avoid breathing fumes from the derivatization reagents and chloroform. Use positive displacement pipettes for this section (see Note 14).

1. Transfer 50 μL of the amino acid-enriched eluent (obtained from the SPE column, as described in Subheading 3.2) into a 6 mm × 50 mm mini glass culture tube.
2. Derivatize amino acids by adding 5 μL of pyridine and 5 μL of methyl chloroformate (MCF) to sample (see Note 15). While adding derivatization reagents, use pipette tip to stir the reaction mixture.
3. Let stand for 1 min.
4. Add 90 μL of chloroform to sample.
5. Add 90 μL of 50 mM sodium bicarbonate to sample.
6. Vortex well using Maxi Mix, being careful to avoid loss of sample (see Note 16).

7. Gently tap until two phases (layers) form. If the bottom phase is still turbid or there is not a clean interface between the two liquid phases, vortex again.
8. Transfer 50 μL of the chloroform (bottom) phase to a new mini glass culture tube containing a few crystals of anhydrous sodium sulfate. There will still be some residual chloroform left in the mini glass culture tube in which the derivatization reaction took place.
9. Cover with aluminum foil to prevent evaporation.
10. Let stand for 30 min.
11. Transfer sample to a 50 μL GC–MS vial insert in a GC–MS autosampler vial, being careful to avoid transfer of the sodium sulfate crystals.
12. Use cap crimper to seal the PTFE-lined cap on GC–MS autosampler vial (see Note 17).
13. The sample is now ready for GC–MS analysis (see Note 18) or storage (see Note 19).

For analysis of less than 1 mg FW of tissue.

3.4. Micro Tissue Extraction

The procedures can be performed at room temperature.

1. Place tissue (less than 1 mg FW, see Note 8) in a microcentrifuge tube.
2. Add 120 μL of 0.01 M HCl and 5 μL of internal standard solution to sample. If using bead mill to grind (see next step), add two 3 mm stainless steel balls to sample.
3. Use either (a) disposable pellet pestles or (b) a bead mill for 3–5 min at 25 s^{-1} frequency to grind the tissue.
4. Shake in vortex mixer with TurboMix attachment for 15 min.
5. Centrifuge samples at 14,000 × *g* for 3 min (see Note 9).

3.5. Micro Solid-Phase Extraction

This procedure should be carried out in a fume hood to avoid breathing fumes from the ammonium hydroxide solution.

1. Condition SPE pipette tip by washing with 100 μL of 0.01 M HCl followed by three washes with 100 μL ddH_2O. Dispense each wash step to waste.
2. Condition SPE pipette tip by aspirating (drawing up) 100 μL of methanol and dispensing to waste, being careful to avoid fully drying out tip bed. Repeat this step four more times, for a total of five methanol washes.
3. Rinse tip by aspirating 100 μL of ddH_2O and dispensing to waste. Repeat this step four more times, for a total of five rinses.
4. Set pipette at 100 μL and aspirate-dispense supernatant 15 times using the convex inset of the cap of the microcentrifuge tube as a spot dish.

5. Wash tip by aspirating 100 μL of methanol/water wash solution and dispensing to waste. Repeat five times.
6. Elute by repeatedly (five times) drawing up and expelling the same 25 μL of elution solution with the tip and then finally expelling the 25 μL of eluent into a 50 μL GC–MS vial insert.
7. Sample is now ready for derivatization or storage (see Note 13).

3.6. Micro Derivatization

This procedure should be conducted in a fume hood. Be careful to avoid breathing fumes from derivatization reagents and chloroform. Use positive displacement pipettes for this section (see Note 14).

1. Derivatize amino acids by adding 2.5 μL of pyridine and 2.5 μL of methyl chloroformate (MCF) to the sample (see Note 15).
2. Briefly vortex to mix.
3. Let stand for 1 min.
4. Add 50 μL of chloroform to sample.
5. Add 50 μL of 50 mM sodium bicarbonate to sample.
6. Vortex well on Maxi Mix, being careful to avoid loss during vortexing (see Note 16).
7. Gently tap until two layers form. If the bottom phase is still turbid or there is not a clean interface between the two liquid phases, vortex again.
8. Transfer the chloroform (bottom) phase to a 250 μL GC–MS vial insert containing a few crystals of anhydrous sodium sulfate.
9. Vortex briefly.
10. Transfer sample, avoiding the sodium sulfate crystals, to a 50 μL GC–MS vial insert in a GC–MS autosampler vial.
11. Use cap crimper to seal GC–MS vial using a PTFE-lined cap (see Note 17).
12. Sample is now ready for GC–MS analysis (see Note 18) or storage (see Note 19).

3.7. GC–MS Analysis

1. Two microliter sample is injected in the splitless mode. The oven temperature is initially held at 70 °C for 3 min. Thereafter the temperature is raised at a rate of 25 °C/min until 280 °C followed by a 5 min. hold at 280 °C. Helium is used as carrier gas and delivered at a constant flow rate at 1 mL/min during the run. The temperature of the inlet is set to 240 °C and the interface temperature at 290 °C.
2. The mass spectra of the MCF-derivatized amino acids and internal standards are obtained in the full-scan mode (50–350 m/z). For larger samples or for microassays and other samples requiring higher sensitivity, data is acquired in the SIM acquisition mode (see Fig. 1). The retention times for each amino acid derivative, the m/z for the molecular ion, and the m/z values for major fragment ions are listed in Table 1.

Table 1
Mass fragment ions of derivatized amino acids (*N*-methoxycarbonyl amino acid methyl esters)

Amino acid	Retention time (min)	Molecular ion (*m*/*z*)	Major fragment ions[b] (*m*/*z*)
Glycine	6.01	147	88*
Alanine	6.04	161	102*, 88
Valine	7.00	189	146, 130*,115, 98
Leucine	7.52	203	144*,115, 102, 88
Isoleucine	7.61	203	144*, 115, 101, 88
Threonine (Thr-OMe)[a]	7.68	205	147*, 115, 100, 88
Proline	7.86	187	128*, 84
Asparagine	7.93	262	146, 127*, 95
Aspartic acid	8.28	219	160*, 128, 118, 101
Serine (Ser-OMe)[a]	8.65	191	176, 144, 114, 100*, 88
Glutamine	8.75	276	141*, 109, 82
Glutamic acid	8.95	233	201, 174*, 142, 114
Methionine	9.06	221	221*, 147, 128, 115
Cysteine	9.54	192	192*, 176, 158, 146, 132
Phenylalanine	9.69	237	178, 162*, 146, 131, 103, 91
Lysine	11.05	276	244, 212, 142*, 88
Histidine	11.37	285	254, 226, 210*, 194, 140, 81
Tyrosine	11.90	267	252, 236*, 220, 192, 165, 146, 121
Tryptophan	13.17	276	130*

[a]Threonine and serine were detected with methyl hydroxyl ether side chains (Thr-OMe and Ser-OMe)
[b]Fragments marked with an asterisk (*) are the ions we primarily monitor for detection, based on the signal-to-noise ratio and specificity of the ions in biological matrices. The most appropriate product ion to monitor may change depending on sample matrix and other factors

4. Notes

1. This variation occurs due to fluctuations in carrier gas flow rate and other factors. For this reason, chromatographic peak heights and spectral abundance values should not be directly compared between analytes or between GC–MS runs.
2. An appropriate concentration of internal standard will depend on several factors, including technique, efficiency of extraction

and derivatization, and sensitivity of GC–MS system. We have found that concentrations of 30 and 100 μg/mL, for tissue samples above 20 mg or below 1 mg, respectively, give very clearly resolved peaks when acquiring GC–MS data in scan mode.

3. Enables shaking of up to twelve 1.5 mL microcentrifuge tubes simultaneously.
4. An alternative to the purchase of commercial mini-spin columns, if desired, is to assemble a column from a standard 200 μL pipette tip, silanized glass wool, and approximately 5 mg (7 mm measured from the glass wool support) of Dowex 50WX2-400 strongly acidic cation exchange (SCX) resin (Sigma-Aldrich, St. Louis, MO, USA). After conditioning with 1 M HCl, the resin is rinsed with ddH_2O until the rinse water is neutral. The H+ form resin is transferred into the pipette tip as a water slurry using a Pasteur pipette (see ref. 3). With filled pipette tips extra care is needed so as not to disrupt the packed resin during the aspiration steps.
5. This solution has a 2-week shelf-life at most, so the solution should be freshly prepared as necessary.
6. To avoid excessive water condensation into the MCF and pyridine, allow bottles to come to room temperature prior to opening them. We usually take these bottles out of the freezer and place them in the hood before beginning the tissue extraction procedure. This way, the reagents are at room temperature by the time the tissue and solid-phase extraction procedures are complete. Since this work was undertaken, the supplier has modified their offerings to make appropriate sized prepackaged bottles available.
7. This is approximately equivalent to two discs cut from leaf tissue using a standard paper punch. This amount may need to be adjusted depending on plant type and tissue source. Our practice is to try 20 mg first, then to increase the amount of tissue if unsatisfactory results are achieved.
8. Frozen or dried tissue may also be used, but the amount of tissue used may need to be adjusted (see Note 7). If using woody, fibrous, or otherwise resilient tissues, such as seeds, consider pulverizing the tissue ahead of time to increase efficiency of extraction. If equipment to pulverize tissue is not available, disposable pellet pestles (Kimble Chase Kontes, Vineland, NJ, USA) may be used to grind the tissue in the microcentrifuge tube prior to bead-mill extraction.
9. This time may need to be adjusted depending on tissue. Also, tissue may not completely pellet. In this case, simply avoid non-pelleted tissue when pipetting in future steps.

10. If processing many samples, a repeater pipette (such as Repeater Plus pipette from Eppendorf AG, Hamburg, Germany) can greatly increase the efficiency.
11. Great care must be taken to avoid drying out the column bed or drawing liquid through the bed too quickly. If processing multiple samples, it may be helpful to use individual valves on the manifold ports (Luer stopcock, Kylar, Item # AP13SML2SXFK; Ark-Plas, Flippin, AR, USA), which allow the manifold operator to start/stop flow and to individually control the flow rate of each column.
12. Be especially careful not to draw liquid through the column too quickly in this step, or the amino acids won't be completely eluted. It is usually possible to start eluting with gravity alone (with the vacuum off). Turn on the vacuum when no more liquid is eluting via gravity alone.
13. Samples can be stored in microcentrifuge tubes for 2 weeks at −80 °C before derivatization.
14. We use positive displacement pipettes for two reasons: (1) they are calibrated for liquids with a range of viscosities; (2) the pistons and capillaries used with these pipettes are not contaminated by phthalates, which are commonly used in the manufacture of traditional pipette tips. Phthalate contamination can result in difficulties in sample analysis and potentially can require expensive cleaning of the GC–MS system.
15. Upon addition of the MCF to the pyridine and extract, there should be gas and heat evolution (bubbling and warming of the mixture). This is an indicator that the derivatization reaction is occurring. This is especially useful when many samples are being processed, where it is easier to lose track of whether both derivatization reagents have been added. If it is not clear, repeat the addition of MCF and pyridine to the samples in question.
16. The amount of vortexing required for sufficient mixing of phases varies depending on tissue, but is typically between 30 and 60 s.
17. If cap does not crimp correctly, or if replacing cap with punctured septum with new cap for long-term storage, use decapping pliers (Chrom Tech Catalog Number: 904371; Wheaton Science Products, Millville, NJ, USA) to carefully remove the old cap.
18. After GC–MS analysis is completed, vials should be recapped (see Note 17) and stored at −80 °C.
19. Derivatized samples can be stored for 12 h at room temperature, 2 days at −20 °C, or longer-term at −80 °C.

Acknowledgments

This work was supported by the U.S. National Science Foundation, Plant Genome Program, grants DBI 0606666 and IOS-0923960, by funds from the Gordon and Margaret Bailey Endowment for Environmental Horticulture and by the Minnesota Agricultural Experiment Station.

References

1. Husek P (1991) Amino acid derivatization and analysis in five minutes. FEBS Lett 280:354–356
2. Kaspar H, Dettmer K, Gronwald W, Oefner PJ (2008) Automated GC-MS analysis of free amino acids in biological fluids. J Chromatogr B 870:222–232
3. Chen W-P, Yang X-Y, Hegeman AD, Gray WM, Cohen JD (2010) Microscale analysis of amino acids using gas chromatography-mass spectrometry after methyl chloroformate derivatization. J Chromatogr B 878:2199–2208
4. Yang X-Y, Chen W-P, Rendahl AK, Hegeman AD, Gray WM, Cohen JD (2010) Measuring the turnover rates of Arabidopsis proteins using deuterium oxide: an auxin signaling case study. Plant J 63:680–695
5. Chen W-P, Yang X-Y, Harms GL, Gray WM, Hegeman AD, Cohen JD (2011) An automated growth enclosure for metabolic labeling of *Arabidopsis thaliana* with ^{13}C-carbon dioxide - An *in vivo* labeling system for proteomics and metabolomics research. Proteome Sci 9:9–23
6. Beynon RJ, Pratt JM (2005) Metabolic labeling of proteins for proteomics. Mol Cell Proteomics 4:857–872
7. Hellerstein MK, Neese RA (1992) Mass isotopomer distribution analysis: a technique for measuring biosynthesis and turnover of polymers. Am J Physiol 263:E988–E1001
8. Lee WN, Byerley LO, Bergner EA, Edmond J (1991) Mass isotopomer analysis: theoretical and practical considerations. Biol Mass Spectrom 20:451–458
9. Barkawi LS, Tam YY, Tillman JA, Normanly J, Cohen JD (2010) A high-throughput method for the quantitative analysis of auxins. Nat Protoc 5:1609–1618

Chapter 13

Arabidopsis thaliana Membrane Lipid Molecular Species and Their Mass Spectral Analysis

Thilani Samarakoon, Sunitha Shiva*, Kaleb Lowe, Pamela Tamura, Mary R. Roth, and Ruth Welti

Abstract

Herein, current approaches to electrospray ionization mass spectrometry-based analyses of membrane lipid molecular species found in *Arabidopsis thaliana* are summarized. Additionally, the identities of over 500 reported membrane lipid molecular species are assembled.

Key words: *Arabidopsis thaliana*, Electrospray ionization, Galactolipids, Lipid profiling, Mass spectrometry, Membrane lipids, Phospholipids, Sphingolipids, Sterols and derivatives, Sulfolipid

1. Introduction

Methods for profiling of plant membrane lipids by electrospray ionization (ESI) mass spectrometry (MS) have been developed and expanded in the past 10 years. The advantages of ESI MS methods over "traditional" lipid analytical approaches have been described (1–7). The most widely adopted ESI MS approaches employ tandem MS in which intact ions are fragmented by collision-induced dissociation. Scanning with a triple quadrupole (QqQ) mass spectrometer in precursor (Pre) and neutral loss (NL) modes provides information (mass/charge ratios, *m/z*s) for each group of lipids that contains a common fragment. One type of common fragment is a lipid head group, and those lipids with a common head group are known as a "class." However, information about other groups of lipids, such as those containing a common fatty acyl chain, can also be collected. Pre or NL scans are a useful starting point for determining the lipid molecular species present within each class or

*The first two authors contributed equally.

Jennifer Normanly (ed.), *High-Throughput Phenotyping in Plants: Methods and Protocols*, Methods in Molecular Biology, vol. 918, DOI 10.1007/978-1-61779-995-2_13, © Springer Science+Business Media, LLC 2012

other group in biological samples. For example, a scan for lipids exhibiting an NL of 141 (NL 141) following fragmentation, i.e., the NL of a phosphoethanolamine fragment, was used to identify lipid molecular species in the phosphatidylethanolamine class in *Arabidopsis thaliana* (8). Lipid classes, groups, and molecular species identified by preliminary scanning can be targeted for quantitative analysis.

Once the target compounds in a group of lipids are known, scans are collected based on the *m/z*s of the intact lipid ion and the characteristic fragment ion and peak intensities are measured. These data may be collected via Pre, NL, or multiple reaction monitoring (MRM) modes on an ESI-QqQ mass spectrometer. Quadrupole time-of-flight (QTOF) MS may also be employed; product ion scans are acquired; and accurate *m/z*s of characteristic fragments of target lipids are identified. Depending on the data type, removal of or adjustment for intensity contributions due to isotopic variants may need to be performed. The intensities of the target lipids are typically compared with intensities for one or more internal standard lipids in order to define the abundance of the target lipid species.

Sample introduction is accomplished either via direct infusion of the sample or via an in-line connection with a chromatographic column. The addition of liquid chromatography (LC) to the protocol provides physical separation of isobaric or near isobaric intact ion–fragment pairs allowing increased compound identification specificity. In direct infusion mode, the isobaric or near isobaric intact ion–fragment pairs may not be resolved. This means that a "target lipid" detected in a direct infusion approach may represent more than one compound. Thus, it is a good idea to subject target compounds defined by direct infusion approaches to further analysis in order to determine their homo- or heterogeneity.

While LC separation reduces ambiguity in compound identification, an advantage of direct infusion is that extended scanning (i.e., extended signal acquisition) for a particular intact ion–fragment pair allows detection and quantification of compounds of low abundance. Extended scanning is possible because scan times are not limited by the elution time of a chromatographic peak. A detailed protocol for direct infusion profiling of polar lipids can be found in another volume in this series (9).

This chapter summarizes ESI MS approaches that have been employed for detection and quantification of *A. thaliana* membrane lipids. Lipid groups and classes, mass spectral scan modes, ions detected, etc. are summarized in Table 1. Additionally, Table 2 includes a list of the individual *A. thaliana* membrane lipid molecular species, the targets for quantitative analyses. It also shows chemical formulas and summarizes mass spectral and other evidence for identification. (Analytical targets not detected in *Arabidopsis* are not listed.)

Table 1
MS scan modes for high-throughput analysis of *Arabidopsis thaliana* membrane lipids

Group	Class	Number of compounds detected in class (Table 2)	Ion analyzed	Scan mode[a]	References
Galactolipids		67[b]			
	DGDG		$[M+Na]^+$	Pre 243.1	(8, 50, 73)
			$[M+NH_4]^+$	NL 341.1	(49)
	DGMG		$[M+NH_4]^+$	NL 341.1	(74)
	MGDG		$[M+Na]^+$	Pre 243.1	(8, 50, 73)
			$[M+NH_4]^+$	NL 179.1	(49)
	MGMG		$[M+NH_4]^+$	NL 179.1	(74)
Phospholipids		159[b]			
	PA		$[M-H]^-$	Pre 153.0	(8, 49)
			$[M+NH_4]^+$	NL 115.0	(52)
	PC and LPC		$[M+H]^+$	Pre 184.1	(8, 49)
	PE and LPE		$[M+H]^+$	NL 141.0	(8, 49)
	PG and LPG		$[M-H]^-$	Pre 153.0	(8)
			$[M-H]^-$	Pre 227.0	(50)
			$[M+NH_4]^+$	NL 189.0	(49)
	3-*trans*-Hexadecenoic acid-containing PG		$[M-H]^-$	NL 236.2	(75)
	PI		$[M-H]^-$	Pre 241.0	(8)
			$[M+NH_4]^+$	NL 277.1	(49)
	PS		$[M-H]^-$	NL 87.0	(8)
			$[M+H]^+$	NL 185.0	(49)
Sulfolipid		17			
	SQDG		$[M-H]^-$	Pre 225.0	(50)

(continued)

Table 1 (continued)

Group	Class	Number of compounds detected in class (Table 2)	Ion analyzed	Scan mode[a]	References
Sterols and derivatives		64			
	Free sterols		$[M+betainyl]^+$	QTOF product ions corresponding to Pre 118.1	(62)
	Steryl glucosides		$[M+NH_4]^+$	QTOF product ions corresponding to NL 197.1	(62)
			$[M+NH_4]^+$	NL 197.1	(63)
	Acyl steryl glucosides				
	Steryl glucoside (16:3)		$[M+NH_4]^+$	QTOF product ions corresponding to NL 429.3	(62)
	Steryl glucoside (16:2)		$[M+NH_4]^+$	QTOF product ions corresponding to NL 431.3	(62)
	Steryl glucoside (16:1)		$[M+NH_4]^+$	QTOF product ions corresponding to NL 433.3	(62)
			$[M+NH_4]^+$	NL 433.3	(63)
	Steryl glucoside (16:0)		$[M+NH_4]^+$	QTOF product ions corresponding to NL 435.3	(62)
			$[M+NH_4]^+$	NL 435.3	(63)
	Steryl glucoside (18:3)		$[M+NH_4]^+$	QTOF product ions corresponding to NL 457.3	(62)
			$[M+NH_4]^+$	NL 457.3	(63)
	Steryl glucoside (18:2)		$[M+NH_4]^+$	QTOF product ions corresponding to NL 459.3	(62)
			$[M+NH_4]^+$	NL 459.3	(63)
	Steryl glucoside (18:1)		$[M+NH_4]^+$	QTOF product ions corresponding to NL 461.3	(62)
			$[M+NH_4]^+$	NL 461.3	(63)
	Steryl glucoside (18:0)		$[M+NH_4]^+$	QTOF product ions corresponding to NL 463.4	(62)
			$[M+NH_4]^+$	NL 463.4	(63)

Oxidized and acylated oxidized lipids		133[b]			
	16:4-O-containing DGDG, MGDG, and acMGDG[c]		$[M-H]^-$, $[M+C_2H_3O_2]^-$	Pre 263.2	(64, 71)
	18:4-O-containing DGDG, MGDG, and PG, and acMGDG		$[M-H]^-$, $[M+C_2H_3O_2]^-$	Pre 291.2	(64, 71)
	18:3-2O-containing DGDG, MGDG, PC, PE, and acMGDG		$[M-H]^-$, $[M+C_2H_3O_2]^-$	Pre 309.2, Pre 291.2	(64, 71)
	18:3-O-containing DGDG, MGDG, PC, PE, and PG		$[M-H]^-$, $[M+C_2H_3O_2]^-$	Pre 293.2	(71)
	18:2-2O-containing PC, PE, and PG		$[M-H]^-$, $[M+C_2H_3O_2]^-$	Pre 293.2	(71)
	18:2-O-containing PC, PE, and PG		$[M-H]^-$, $[M+C_2H_3O_2]^-$	Pre 295.3	(71)
Sphingolipids		62			
	d18:1-containing ceramides, hydroxyceramides, and glucosylceramides		$[M+H]^+$	MRMs corresponding to Pre 264.3	(53)
	d18:0-containing ceramides, hydroxyceramides, and glucosylceramides		$[M+H]^+$	MRMs corresponding to Pre 266.3	(53)
	t18:1-containing ceramides, hydroxyceramides, and glucosylceramides		$[M+H]^+$	MRMs corresponding to Pre 298.3	(53)
	t18:0-containing ceramides, hydroxyceramides, and glucosylceramides		$[M+H]^+$	MRMs corresponding to Pre 300.3	(53)
	Glucosylceramides		$[M+H]^+$	NL 162.1	(7)
	Glycosylinositolphospho ceramides		$[M+H]^+$	MRMs corresponding to NL 598.1	(53)
	Glycosylinositolphospho ceramides		$[M+NH_4]^+$	NL of 615.1 or 179.1	(7)

(continued)

Table 1 (continued)

Group	Class	Number of compounds detected in class (Table 2)	Ion analyzed	Scan mode[a]	References
Lipid A pathway intermediate		4			
	UDP-2,3-diacyl-GlcN		$[M-H]^-$	MRM 1016.5→385.0	(72)
	Lipid X		$[M-H]^-$	MRM 710.4→240.0	(72)
	Disaccharide-1-P		$[M-H]^-$	MRM 1323.8→1097.7	(72)
	Lipid IVA		$[M-2H]^{2-}$	MRM 701.4→853.5	(72)

acMGDG acylated monogalactosyldiacylglycerol, *d18:0* dihydroxy 18:0 sphingoid base, *d18:1* dihydroxy 18:1 sphingoid base, *DGDG* digalactosyldiacylglycerol, *DGMG* digalactosylmonoacylglycerol, *LPC* lysophosphatidylcholine, *LPE* lysophosphatidylethanolamine, *LPG* lysophosphatidylglycerol, *MGDG* monogalactosyldiacylglycerol, *MGMG* monogalactosylmonoacylglycerol, *PA* phosphatidic acid, *PC* phosphatidylcholine, *PE* phosphatidylethanolamine, *PG* phosphatidylglycerol, *PI* phosphatidylinositol, *PS* phosphatidylserine, *SQDG* sulfoquinovosyldiacylglycerol, *t18:0* trihydroxy 18:0 sphingoid base, *t18:1* trihydroxy 18:1 sphingoid base

[a]Unless otherwise indicated, scans were performed on an ESI-QqQ mass spectrometer

[b]Galactolipid and phospholipid totals exclude those with oxidized acyl chains. Among those with oxidized acyl chains, approximately 29 compounds were detected by Q-TOF MS, but not by the indicated scan modes

[c]The "–O" in the oxidized lipid shorthand refers to the number of oxygens beyond the carbonyl, i.e., "extra" oxygens beyond those in normal chain fatty acids

1.1. Galactolipids, Sulfolipid, and Phospholipids

Galactolipids, sulfolipid and phospholipids, or subsets of these from *A. thaliana* have been extensively analyzed by direct infusion ESI MS (10–45) and by LC–MS (46). Analysis is performed after extraction of these lipids by standard protocols such as described by Bligh and Dyer (47). Direct infusion profiling to detect galactolipids and phospholipids, using a combination of positive and negative Pre and NL scans, on an ESI-QqQ mass spectrometer is based on the method of Brügger et al. (48) and has been described by Welti et al. (8) and in the supplementary information of Xiao et al. (49). The scan for sulfolipid is described in Welti et al. (50). Detailed directions for extraction and sample preparation are available (9). A mixture of internal standards that includes two compounds from most lipid classes is employed. In recent applications, the infusion solvent is chloroform/methanol/ammonium acetate (300 mM) in water (300:665:35, v/v/v) (9). For high-throughput lipid profiling by direct infusion ESI-QqQ MS, data processing is available online at http://lipidome.bcf.ku.edu:9000/Lipidomics/ (51); detailed documentation is available at the web site.

In general, the fragments in the intact ion–fragment *m/z* pairs used for Pre and NL scanning of galactolipids, sulfolipid, and phospholipids are head group fragments (8, 49). For diacyl glycerolipid species, head group-specific scans provide information only at the level of total acyl carbons:total acyl double bonds; information on individual acyl chains is not determined directly. To identify individual fatty acyl chains, the membrane glycerolipids are subjected to product ion fragmentation in negative mode, providing detail on the molecular species represented by each intact ion–fragment *m/z* pair (8, 52).

Burgos et al. (46) have employed ultraperformance LC with a C_8 reverse-phase column and an advanced ESI-QTOF mass spectrometer for the analysis of galactolipids, sulfolipid, and phospholipids. Negative ionization mode was used. Using the compounds identified by Devaiah et al. (52) as target compounds, lipid species were identified based on their retention times and accurate *m/z*. Burgos and coworkers (46) also fragmented the compounds to determine the fatty acyl composition and identified several previously undefined acyl combinations in target lipid species. These compounds, species previously determined, and a few new ones identified for the current paper are listed in Table 2 below (also available as an .xls file from the Springer Extras site http://extras.springer.com).

1.2. Sphingolipids

Markham et al. (53) have analyzed a large number of ceramides, hydroxyceramides, glucosylceramides, and glycosylinositolphosphoceramides. Importantly Markham and coworkers (54) have determined that an extraction procedure utilizing the lower phase of an isopropanol/hexane/water (55:20:25, v/v/v) mixture is optimal for extracting sphingolipids. To increase the relative signal for sphingolipids in comparison to other components of the lipid extract, glycerolipids in the lipid extract were hydrolyzed prior to

Table 2
Arabidopsis membrane lipid species identified and analyzed by ESI-MS

Group	Class	Abbreviation	Mass of uncharged compound	Additional steps for compound identification and analyses listed in Table 13.1
Galactolipid	Acylated monogalactosydiacylglycerol	acMGDG (18:3/18:3/16:3)	1,006.7	QTOF MS (product ion) of [M + C2H3O2]-
Galactolipid	Acylated monogalactosydiacylglycerol	acMGDG (18:3/18:3/16:1)	1,010.7	QTOF MS (product ion) of [M + C2H3O2]-
Galactolipid	Acylated monogalactosydiacylglycerol	acMGDG (18:3/18:3/16:0)	1,012.8	QTOF MS (product ion) of [M + C2H3O2]-
Galactolipid	Acylated monogalactosydiacylglycerol	acMGDG (18:3/18:3/18:3)	1,034.7	QTOF MS (product ion) of [M + C2H3O2]-
Galactolipid	Acylated monogalactosydiacylglycerol	acMGDG (18:3/18:3/18:1)	1,038.8	QTOF MS (product ion) of [M + C2H3O2]-
Galactolipid	Acylated monogalactosydiacylglycerol	acMGDG (18:3/18:3/18:0)	1,040.8	QTOF MS (product ion) of [M + C2H3O2]-
Galactolipid	Acylated monogalactosydiacylglycerol (oxidized acyl chain)	acMGDG (18:4-O/32:6)	992.7	QTOF MS (product ion) of [M + C2H3O2]-
Galactolipid	Acylated monogalactosydiacylglycerol (oxidized acyl chain)	acMGDG (16:4-O/18:3/16:3)	992.7	QTOF MS (product ion) of [M + C2H3O2]-
Galactolipid	Acylated monogalactosydiacylglycerol (oxidized acyl chain)	acMGDG (18:4-O/32:3)	998.7	QTOF MS (product ion) of [M + C2H3O2]-

Molecular species identified by additional steps beyond the analyses listed in Table 13.1 (order of acyl chain names does not infer position)	Chemical formula of identified molecular species	References for identification	References for high-throughput analysis	Comments
	C61H98O11	[71]		Detected by product ion analysis but not by the precursor scans (Pre 291.2, Pre 293.2, Pre 295.2) used by Vu et al. (2012)
	C61H102O11	[71]		Detected by product ion analysis but not by the precursor scans (Pre 291.2, Pre 293.2, Pre 295.2) used by Vu et al. (2012)
	C61H104O11	[71]		Detected by product ion analysis but not by the precursor scans (Pre 291.2, Pre 293.2, Pre 295.2) used by Vu et al. (2012)
	C63H102O11	[71]		Detected by product ion analysis but not by the precursor scans (Pre 291.2, Pre 293.2, Pre 295.2) used by Vu et al. (2012)
	C63H106O11	[71]		Detected by product ion analysis but not by the precursor scans (Pre 291.2, Pre 293.2, Pre 295.2) used by Vu et al. (2012)
	C63H108O11	[71]		Detected by product ion analysis but not by the precursor scans (Pre 291.2, Pre 293.2, Pre 295.2) used by Vu et al. (2012)
18:4-O/16:3/16:3	C59H92O12	[71]	[71]	
	C59H92O12	[71]		Detected by product ion analysis but not by the precursor scans (Pre 291.2, Pre 293.2, Pre 295.2) used by Vu et al. (2012)
18:4-O/16:3/16:0	C59H98O12	[71]	[71]	

(continued)

Table 2 (continued)

Group	Class	Abbreviation	Mass of uncharged compound	Additional steps for compound identification and analyses listed in Table 13.1
Galactolipid	Acylated monogalactosydiacylglycerol (oxidized acyl chain)	acMGDG (16:4-O/18:3/16:0)	998.7	QTOF MS (product ion) of [M+C2H3O2]-
Galactolipid	Acylated monogalactosydiacylglycerol (oxidized acyl chain)	acMGDG (16:4-O/16:3/18:0)	998.7	QTOF MS (product ion) of [M+C2H3O2]-
Galactolipid	Acylated monogalactosydiacylglycerol (oxidized acyl chain)	acMGDG (18:4-O/32:7-O)	1,006.6	QTOF MS (product ion) of [M+C2H3O2]-
Galactolipid	Acylated monogalactosydiacylglycerol (oxidized acyl chain)	acMGDG (16:4-O/16:4-O/18:3)	1,006.6	QTOF MS (product ion) of [M+C2H3O2]-
Galactolipid	Acylated monogalactosydiacylglycerol (oxidized acyl chain)	acMGDG (18:4-O/32:5-O)	1,010.7	QTOF MS (product ion) of [M+C2H3O2]-
Galactolipid	Acylated monogalactosydiacylglycerol (oxidized acyl chain)	acMGDG (16:4-O/16:4-O/18:1)	1,010.7	QTOF MS (product ion) of [M+C2H3O2]-
Galactolipid	Acylated monogalactosydiacylglycerol (oxidized acyl chain)	acMGDG (18:4-O/32:4-O)	1,012.7	QTOF MS (product ion) of [M+C2H3O2]-
Galactolipid	Acylated monogalactosydiacylglycerol (oxidized acyl chain)	acMGDG (16:4-O/16:4-O/18:0)	1,012.7	QTOF MS (product ion) of [M+C2H3O2]-
Galactolipid	Acylated monogalactosydiacylglycerol (oxidized acyl chain)	acMGDG (18:3-2O/32:3)	1,016.7	NMR, QTOF MS (product ion) of [M+C2H3O2]-
Galactolipid	Acylated monogalactosydiacylglycerol (oxidized acyl chain)	acMGDG (16:3-2O/16:0/18:3)	1,016.7	QTOF MS (product ion) of [M+C2H3O2]-

Molecular species identified by additional steps beyond the analyses listed in Table 13.1 (order of acyl chain names does not infer position)	Chemical formula of identified molecular species	References for identification	References for high-throughput analysis	Comments
	C59H98O12	[71]		Detected by product ion analysis but not by the precursor scans (Pre 291.2, Pre 293.2, Pre 295.2) used by Vu et al. (2012)
	C59H98O12	[71]		Detected by product ion analysis but not by the precursor scans (Pre 291.2, Pre 293.2, Pre 295.2) used by Vu et al. (2012)
18:4-O/16:4-O/16:3	C59H90O13	[71]	[71]	
	C59H90O13	[71]		Detected by product ion analysis but not by the precursor scans (Pre 291.2, Pre 293.2, Pre 295.2) used by Vu et al. (2012)
18:4-O/16:4-O/16:1	C59H94O13	[71]	[71]	
	C59H94O13	[71]		Detected by product ion analysis but not by the precursor scans (Pre 291.2, Pre 293.2, Pre 295.2) used by Vu et al. (2012)
18:4-O/16:4-O/16:0	C59H96O13	[71]	[71]	
	C59H96O13	[71]		Detected by product ion analysis but not by the precursor scans (Pre 291.2, Pre 293.2, Pre 295.2) used by Vu et al. (2012)
18:3-2O/16:0/16:3	C59H100O13	[71]	[71]	Scanning for Pre 291, i.e., Pre 18:4-O, can be used to detect species containing 18:3-2O also, because 18:3-2O undergoes a water loss to 18:4-O
	C59H100O13	[71]		Detected by product ion analysis but not by the precursor scans (Pre 291.2, Pre 293.2, Pre 295.2) used by Vu et al. (2012)

(continued)

Table 2 (continued)

Group	Class	Abbreviation	Mass of uncharged compound	Additional steps for compound identification and analyses listed in Table 13.1
Galactolipid	Acylated monogalactosydiacylglycerol (oxidized acyl chain)	acMGDG (18:4-O/34:6)	1,020.7	QTOF MS (product ion) of [M+C2H3O2]-
Galactolipid	Acylated monogalactosydiacylglycerol (oxidized acyl chain)	acMGDG (18:3/18:3/16:4-O)	1,020.7	QTOF MS (product ion) of [M+C2H3O2]-
Galactolipid	Acylated monogalactosydiacylglycerol (oxidized acyl chain)	acMGDG (18:4-O/32:8-2O)	1,020.7	QTOF MS (product ion) of [M+C2H3O2]-
Galactolipid	Acylated monogalactosydiacylglycerol (oxidized acyl chain)	acMGDG (18:3-2O/32:7-O)	1,024.6	QTOF MS (product ion) of [M+C2H3O2]-
Galactolipid	Acylated monogalactosydiacylglycerol (oxidized acyl chain)	acMGDG(18:4-O/32:6-2O)	1,024.6	QTOF MS (product ion) of [M+C2H3O2]-
Galactolipid	Acylated monogalactosydiacylglycerol (oxidized acyl chain)	acMGDG (16:3-2O/16:4-O/18:3)	1,024.6	QTOF MS (product ion) of [M+C2H3O2]-
Galactolipid	Acylated monogalactosydiacylglycerol (oxidized acyl chain)	acMGDG (18:4-O/34:3)	1,026.7	QTOF MS (product ion) of [M+C2H3O2]-
Galactolipid	Acylated monogalactosydiacylglycerol (oxidized acyl chain)	acMGDG (18:3-2O/32:4-O)	1,030.7	QTOF MS (product ion) of [M+C2H3O2]-
Galactolipid	Acylated monogalactosydiacylglycerol (oxidized acyl chain)	acMGDG (18:4-O/32:3-2O)	1,030.7	QTOF MS (product ion) of [M+C2H3O2]-

Molecular species identified by additional steps beyond the analyses listed in Table 13.1 (order of acyl chain names does not infer position)	Chemical formula of identified molecular species	References for identification	References for high-throughput analysis	Comments
18:4-O/18:3/16:3	C61H96O12	[71]	[71]	
	C61H96O12	[71]		Detected by product ion analysis but not by the precursor scans (Pre 291.2, Pre 293.2, Pre 295.2) used by Vu et al. (2012)
18:4-O/16:4-O/16:4-O	C59H88O14	[71]	[71]	
18:3-2O/16:4-O/16:3	C59H92O14	[71]	[71]	Scanning for Pre 291, i.e., Pre 18:4-O, can be used to detect species containing 18:3-2O also, because 18:3-2O undergoes a water loss to 18:4-O
18:4-O/16:3-2O/16:3	C59H92O14	[71]	[71]	
	C59H92O14	[71]		Scanning for Pre 263, i.e., Pre 16:4-O, can be used to detect species containing 16:3-2O also, because 16:3-2O undergoes a water loss to 16:4-O; Detected by product ion analysis but not by the precursor scans (Pre 291.2, Pre 293.2, Pre 295.2) used by Vu et al. (2012)
18:4-O/18:3/16:0	C61H102O12	[71]	[71]	
18:3-2O/16:4-O/16:0	C59H98O14	[71]	[71]	Scanning for Pre 291, i.e., Pre 18:4-O, can be used to detect species containing 18:3-2O also, because 18:3-2O undergoes a water loss to 18:4-O
18:4-O/16:3-2O/16:0	C59H98O14	[71]	[71]	

(continued)

Table 2 (continued)

Group	Class	Abbreviation	Mass of uncharged compound	Additional steps for compound identification and analyses listed in Table 13.1
Galactolipid	Acylated monogalactosydiacylglycerol (oxidized acyl chain)	acMGDG (16:3-2O/16:4-O/18:0)	1,030.7	QTOF MS (product ion) of [M+C2H3O2]-
Galactolipid	Acylated monogalactosydiacylglycerol (oxidized acyl chain)	acMGDG (18:4-O/34:7-O)	1,034.7	QTOF MS (product ion) of [M+C2H3O2]-
Galactolipid	Acylated monogalactosydiacylglycerol (oxidized acyl chain)	acMGDG (18:4-O/34:7-O)	1,034.7	QTOF MS (product ion) of [M+C2H3O2]-
Galactolipid	Acylated monogalactosydiacylglycerol (oxidized acyl chain)	acMGDG (18:4-O/34:5-O)	1,038.7	QTOF MS (product ion) of [M+C2H3O2]-
Galactolipid	Acylated monogalactosydiacylglycerol (oxidized acyl chain)	acMGDG (18:4-O/34:5-O)	1,038.7	QTOF MS (product ion) of [M+C2H3O2]-
Galactolipid	Acylated monogalactosydiacylglycerol (oxidized acyl chain)	acMGDG(18:4-O/34:4-O)	1,040.7	QTOF MS (product ion) of [M+C2H3O2]-
Galactolipid	Acylated monogalactosydiacylglycerol (oxidized acyl chain)	acMGDG(18:4-O/34:4-O)	1,040.7	QTOF MS (product ion) of [M+C2H3O2]-
Galactolipid	Acylated monogalactosydiacylglycerol (oxidized acyl chain)	acMGDG (18:3-2O/34:3)	1,044.7	QTOF MS (product ion) of [M+C2H3O2]-
Galactolipid	Acylated monogalactosydiacylglycerol (oxidized acyl chain)	acMGDG (18:3-2O/34:3)	1,044.7	QTOF MS (product ion) of [M+C2H3O2]-
Galactolipid	Acylated monogalactosydiacylglycerol (oxidized acyl chain)	acMGDG(16:3--2O/18:3/18:0)	1,044.7	QTOF MS (product ion) of [M+C2H3O2]-

Molecular species identified by additional steps beyond the analyses listed in Table 13.1 (order of acyl chain names does not infer position)	Chemical formula of identified molecular species	References for identification	References for high-throughput analysis	Comments
	C59H98O14	[71]		Detected by product ion analysis but not by the precursor scans (Pre 291.2, Pre 293.2, Pre 295.2) used by Vu et al. (2012)
18:4-O/16:4-O/18:3	C61H94O13	[71]	[71]	
18:4-O/18:4-O/16:3	C61H94O13	[71]	[71]	
18:4-O/16:4-O/18:1	C61H98O13	[71]	[71]	
18:4-O/18:4-O/16:1	C61H98O13	[71]	[71]	
18:4-O/16:4-O/18:0	C61H100O13	[71]	[71]	
18:4-O/18:4-O/16:0	C61H100O13	[71]	[71]	
18:3-2O/18:3/16:0	C61H104O13	[71]	[71]	Scanning for Pre 291, i.e., Pre 18:4-O, can be used to detect species containing 18:3-2O also, because 18:3-2O undergoes a water loss to 18:4-O
18:3-2O/18:0/16:3	C61H104O13	[71]	[71]	Scanning for Pre 291, i.e., Pre 18:4-O, can be used to detect species containing 18:3-2O also, because 18:3-2O undergoes a water loss to 18:4-O
	C61H104O13	[71]		Scanning for Pre 263, i.e., Pre 16:4-O, can be used to detect species containing 16:3-2O also, because 16:3-2O undergoes a water loss to 16:4-O; Detected by product ion analysis but not by the precursor scans (Pre 291.2, Pre 293.2, Pre 295.2) used by Vu et al. (2012)

(continued)

Table 2 (continued)

Group	Class	Abbreviation	Mass of uncharged compound	Additional steps for compound identification and analyses listed in Table 13.1
Galactolipid	Acylated monogalactosydiacylglycerol (oxidized acyl chain)	acMGDG(18:4-O/34:8-2O)	1,048.6	NMR, QTOF MS (product ion) of [M+C2H3O2]-
Galactolipid	Acylated monogalactosydiacylglycerol (oxidized acyl chain)	acMGDG (18:4-O/36:6)	1,048.7	QTOF MS (product ion) of [M+C2H3O2]-
Galactolipid	Acylated monogalactosydiacylglycerol (oxidized acyl chain)	acMGDG(18:3-2O/34:7-O)	1,052.7	QTOF MS (product ion) of [M+C2H3O2]-
Galactolipid	Acylated monogalactosydiacylglycerol (oxidized acyl chain)	acMGDG(18:4-O/34:6-2O)	1,052.7	QTOF MS (product ion) of [M+C2H3O2]-
Galactolipid	Acylated monogalactosydiacylglycerol (oxidized acyl chain)	acMGDG(18:3-2O/34:7-O)	1,052.7	QTOF MS (product ion) of [M+C2H3O2]-
Galactolipid	Acylated monogalactosydiacylglycerol (oxidized acyl chain)	acMGDG (18:4-O/36:3)	1,054.8	QTOF MS (product ion) of [M+C2H3O2]-
Galactolipid	Acylated monogalactosydiacylglycerol (oxidized acyl chain)	acMGDG (18:4-O/36:3)	1,054.8	QTOF MS (product ion) of [M+C2H3O2]-
Galactolipid	Acylated monogalactosydiacylglycerol (oxidized acyl chain)	acMGDG (18:3-2O/34:4-O)	1,058.7	QTOF MS (product ion) of [M+C2H3O2]-
Galactolipid	Acylated monogalactosydiacylglycerol (oxidized acyl chain)	acMGDG(18:3-2O/34:4-O)	1,058.7	QTOF MS (product ion) of [M+C2H3O2]-
Galactolipid	Acylated monogalactosydiacylglycerol (oxidized acyl chain)	acMGDG(18:4-O/34:9-3O)	1,062.6	QTOF MS (product ion) of [M+C2H3O2]-
Galactolipid	Acylated monogalactosydiacylglycerol (oxidized acyl chain)	acMGDG(18:4-O/36:7-O)	1,062.7	QTOF MS (product ion) of [M+C2H3O2]-

Molecular species identified by additional steps beyond the analyses listed in Table 13.1 (order of acyl chain names does not infer position)	Chemical formula of identified molecular species	References for identification	References for high-throughput analysis	Comments
18:4-O/18:4-O/16:4-O	$C_{61}H_{92}O_{14}$	[71, 65]	[71]	
18:4-O/18:3/18:3	$C_{63}H_{100}O_{12}$	[71]	[71]	
18:3-2O/16:4-O/18:3	$C_{61}H_{96}O_{14}$	[71]	[71]	Scanning for Pre 291, i.e., Pre 18:4-O, can be used to detect species containing 18:3-2O also, because 18:3-2O undergoes a water loss to 18:4-O
18:4-O/16:3-2O/18:3	$C_{61}H_{96}O_{14}$	[71]	[71]	
18:3-2O/18:4-O/16:3	$C_{61}H_{96}O_{14}$	[71]	[71]	Scanning for Pre 291, i.e., Pre 18:4-O, can be used to detect species containing 18:3-2O also, because 18:3-2O undergoes a water loss to 18:4-O
18:4-O/18:3/18:0	$C_{63}H_{106}O_{12}$	[71]	[71]	
18:4-O/18:2/18:1	$C_{63}H_{106}O_{12}$	[71]	[71]	
18:3-2O/18:4-O/16:0	$C_{61}H_{102}O_{14}$	[71]	[71]	Scanning for Pre 291, i.e., Pre 18:4-O, can be used to detect species containing 18:3-2O also, because 18:3-2O undergoes a water loss to 18:4-O
18:3-2O/16:4-O/18:0	$C_{61}H_{102}O_{14}$	[71]	[71]	Scanning for Pre 291, i.e., Pre 18:4-O, can be used to detect species containing 18:3-2O also, because 18:3-2O undergoes a water loss to 18:4-O
18:4-O/16:4-O/18:5-2O	$C_{61}H_{90}O_{15}$	[71]	[71]	
18:4-O/18:4-O/18:3	$C_{63}H_{98}O_{13}$	[71]	[71]	

(continued)

Table 2 (continued)

Group	Class	Abbreviation	Mass of uncharged compound	Additional steps for compound identification and analyses listed in Table 13.1
Galactolipid	Acylated monogalactosydiacylglycerol (oxidized acyl chain)	acMGDG(18:3-2O/34:8-2O)	1,066.7	QTOF MS (product ion) of [M + C2H3O2]-
Galactolipid	Acylated monogalactosydiacylglycerol (oxidized acyl chain)	acMGDG (18:3-2O/36:6)	1,066.7	QTOF MS (product ion) of [M + C2H3O2]-
Galactolipid	Acylated monogalactosydiacylglycerol (oxidized acyl chain)	acMGDG(18:4-O/36:4-O)	1,068.7	QTOF MS (product ion) of [M + C2H3O2]-
Galactolipid	Acylated monogalactosydiacylglycerol (oxidized acyl chain)	acMGDG(18:3-2O/34:6-2O)	1,070.7	QTOF MS (product ion) of [M + C2H3O2]-
Galactolipid	Acylated monogalactosydiacylglycerol (oxidized acyl chain)	acMGDG(18:3-2O/34:6-2O)	1,070.7	QTOF MS (product ion) of [M + C2H3O2]-
Galactolipid	Acylated monogalactosydiacylglycerol (oxidized acyl chain)	acMGDG (18:3-2O/36:4)	1,070.8	QTOF MS (product ion) of [M + C2H3O2]-
Galactolipid	Acylated monogalactosydiacylglycerol (oxidized acyl chain)	acMGDG(18:3-2O/34:5-2O)	1,072.7	QTOF MS (product ion) of [M + C2H3O2]-

Molecular species identified by additional steps beyond the analyses listed in Table 13.1 (order of acyl chain names does not infer position)	Chemical formula of identified molecular species	References for identification	References for high-throughput analysis	Comments
18:3-2O/18:4-O/16:4-O	C61H94O15	[71]	[71]	Scanning for Pre 291, i.e., Pre 18:4-O, can be used to detect species containing 18:3-2O also, because 18:3-2O undergoes a water loss to 18:4-O
18:3-2O/18:3/18:3	C63H102O13	[71]	[71]	Scanning for Pre 291, i.e., Pre 18:4-O, can be used to detect species containing 18:3-2O also, because 18:3-2O undergoes a water loss to 18:4-O
18:4-O/18:4-O/18:0	C63H104O13	[71]	[71]	
18:3-2O/16:3-2O/18:3	C61H98O15	[71]	[71]	Scanning for Pre 291, i.e., Pre 18:4-O, can be used to detect species containing 18:3-2O also, because 18:3-2O undergoes a water loss to 18:4-O
18:3-2O/18:3-2O/16:3	C61H98O15	[71]	[71]	Scanning for Pre 291, i.e., Pre 18:4-O, can be used to detect species containing 18:3-2O also, because 18:3-2O undergoes a water loss to 18:4-O
18:3-2O/18:2/18:2	C63H106O13	[71]	[71]	Scanning for Pre 291, i.e., Pre 18:4-O, can be used to detect species containing 18:3-2O also, because 18:3-2O undergoes a water loss to 18:4-O
18:3-2O/18:2/16:3-2O	C61H100O15	[71]	[71]	Scanning for Pre 291, i.e., Pre 18:4-O, can be used to detect species containing 18:3-2O also, because 18:3-2O undergoes a water loss to 18:4-O

(continued)

Table 2 (continued)

Group	Class	Abbreviation	Mass of uncharged compound	Additional steps for compound identification and analyses listed in Table 13.1
Galactolipid	Acylated monogalactosydiacylglycerol (oxidized acyl chain)	acMGDG (18:3-2O/36:3)	1,072.8	QTOF MS (product ion) of [M+C2H3O2]-
Galactolipid	Acylated monogalactosydiacylglycerol (oxidized acyl chain)	acMGDG (18:3-2O/36:3)	1,072.8	QTOF MS (product ion) of [M+C2H3O2]-
Galactolipid	Acylated monogalactosydiacylglycerol (oxidized acyl chain)	acMGDG(18:4-O/36:8-2O)	1,076.7	NMR, QTOF MS (product ion) of [M+C2H3O2]-
Galactolipid	Acylated monogalactosydiacylglycerol (oxidized acyl chain)	acMGDG(18:3-2O/36:7-O)	1,080.7	QTOF MS (product ion) of [M+C2H3O2]-
Galactolipid	Acylated monogalactosydiacylglycerol (oxidized acyl chain)	acMGDG(18:3-2O/34:7-3O)	1,084.7	QTOF MS (product ion) of [M+C2H3O2]-
Galactolipid	Acylated monogalactosydiacylglycerol (oxidized acyl chain)	acMGDG(18:3-2O/34:7-3O)	1,084.7	QTOF MS (product ion) of [M+C2H3O2]-
Galactolipid	Acylated monogalactosydiacylglycerol (oxidized acyl chain)	acMGDG(18:3-2O/36:4-O)	1,086.8	QTOF MS (product ion) of [M+C2H3O2]-
Galactolipid	Digalactosyldiacylglycerol	DGDG(34:6)	908.5	Product ion of [M+C2H3O2]-

Molecular species identified by additional steps beyond the analyses listed in Table 13.1 (order of acyl chain names does not infer position)	Chemical formula of identified molecular species	References for identification	References for high-throughput analysis	Comments
18:3-2O/18:3/18:0	C63H108O13	[71]	[71]	Scanning for Pre 291, i.e., Pre 18:4-O, can be used to detect species containing 18:3-2O also, because 18:3-2O undergoes a water loss to 18:4-O
18:3-2O/18:2/18:1	C63H108O13	[71]	[71]	Scanning for Pre 291, i.e., Pre 18:4-O, can be used to detect species containing 18:3-2O also, because 18:3-2O undergoes a water loss to 18:4-O
18:4-O/18:4-O/18:4-O	C63H96O14	[69, 71]	[71]	
18:3-2O/18:4-O/18:3	C63H100O14	[71]	[71]	Scanning for Pre 291, i.e., Pre 18:4-O, can be used to detect species containing 18:3-2O also, because 18:3-2O undergoes a water loss to 18:4-O
18:3-2O/16:3-2O/18:4-O	C61H96O16	[71]	[71]	Scanning for Pre 291, i.e., Pre 18:4-O, can be used to detect species containing 18:3-2O also, because 18:3-2O undergoes a water loss to 18:4-O
18:3-2O/18:3-2O/16:4-O	C61H96O16	[71]	[71]	Scanning for Pre 291, i.e., Pre 18:4-O, can be used to detect species containing 18:3-2O also, because 18:3-2O undergoes a water loss to 18:4-O
18:3-2O/18:4-O/18:0	C63H106O14	[71]	[71]	Scanning for Pre 291, i.e., Pre 18:4-O, can be used to detect species containing 18:3-2O also, because 18:3-2O undergoes a water loss to 18:4-O
18:3-16:3	C49H80O15	[52, 46]	[8, 49, 52, 46]	

(continued)

Table 2 (continued)

Group	Class	Abbreviation	Mass of uncharged compound	Additional steps for compound identification and analyses listed in Table 13.1
Galactolipid	Digalactosyldiacylglycerol	DGDG(34:5)a	910.6	Product ion of [M+C2H3O2]−
Galactolipid	Digalactosyldiacylglycerol	DGDG(34:5)b	910.6	Product ion of [M+C2H3O2]−
Galactolipid	Digalactosyldiacylglycerol	DGDG(34:4)	912.6	
Galactolipid	Digalactosyldiacylglycerol	DGDG(34:3)a	914.6	Product ion of [M+C2H3O2]−
Galactolipid	Digalactosyldiacylglycerol	DGDG(34:3)b	914.6	Product ion of [M+C2H3O2]−
Galactolipid	Digalactosyldiacylglycerol	DGDG(34:2)a	916.6	Product ion of [M+C2H3O2]−
Galactolipid	Digalactosyldiacylglycerol	DGDG(34:2)b	916.6	Product ion of [M+C2H3O2]−
Galactolipid	Digalactosyldiacylglycerol	DGDG(34:1)	918.6	Product ion of [M+C2H3O2]−
Galactolipid	Digalactosyldiacylglycerol	DGDG(34:0)	920.6	Product ion of [M+C2H3O2]−
Galactolipid	Digalactosyldiacylglycerol	DGDG(36:6)	936.6	Product ion of [M+C2H3O2]−
Galactolipid	Digalactosyldiacylglycerol	DGDG(36:5)	938.6	Product ion of [M+C2H3O2]−
Galactolipid	Digalactosyldiacylglycerol	DGDG(36:4)a	940.6	Product ion of [M+C2H3O2]−
Galactolipid	Digalactosyldiacylglycerol	DGDG(36:4)b	940.6	Product ion of [M+C2H3O2]−
Galactolipid	Digalactosyldiacylglycerol	DGDG(36:3)a	942.6	Product ion of [M+C2H3O2]−
Galactolipid	Digalactosyldiacylglycerol	DGDG(36:3)b	942.6	Product ion of [M+C2H3O2]−
Galactolipid	Digalactosyldiacylglycerol	DGDG(36:2)a	944.6	Product ion of [M+C2H3O2]−
Galactolipid	Digalactosyldiacylglycerol	DGDG(36:2)b	944.6	Product ion of [M+C2H3O2]−
Galactolipid	Digalactosyldiacylglycerol	DGDG(36:1)	946.6	

Molecular species identified by additional steps beyond the analyses listed in Table 13.1 (order of acyl chain names does not infer position)	Chemical formula of identified molecular species	References for identification	References for high-throughput analysis	Comments
18:3-16:2	C49H82O15	[52, 46]	[8, 49, 52, 46]	
18:2-16:3	C49H82O15	[52]	[8, 49, 52]	
	C49H84O15		[8, 49, 52]	
16:0-18:3	C49H86O15	[52, 46]	[8, 49, 52, 46]	
18:2-16:1	C49H86O15	[52]	[8, 49, 52]	
18:2-16:0	C49H88O15	[52, 46]	[8, 49, 52, 46]	
18:1-16:1	C49H88O15	[52]	[8, 49, 52]	
18:1-16:0	C49H90O15	[52, 46]	[8, 49, 52, 46]	
18:0–16:0	C49H92O15	[46]	[46]	
18:3-18:3	C51H84O15	[52, 46]	[8, 49, 52, 46]	
18:3-18:2	C51H86O15	[52]	[8, 49, 52]	
18:2-18:2	C51H88O15	[52, 46]	[8, 49, 52, 46]	
18:1-18:3	C51H88O15	[52]	[8, 49, 52]	
18:3-18:0	C51H90O15	[52, 46]	[8, 49, 52, 46]	
18:2-18:1	C51H90O15	[52]	[8, 49, 52]	
18:2-18:0	C51H92O15	[52]	[8, 49, 52]	
18:1-18:1	C51H92O15	[52]	[8, 49, 52]	
	C51H94O15		[8, 49, 52]	

(continued)

Table 2 (continued)

Group	Class	Abbreviation	Mass of uncharged compound	Additional steps for compound identification and analyses listed in Table 13.1
Galactolipid	Digalactosyldiacylglycerol	DGDG(38:6)	964.6	
Galactolipid	Digalactosyldiacylglycerol	DGDG(38:5)	966.6	
Galactolipid	Digalactosyldiacylglycerol	DGDG(38:4)	968.6	
Galactolipid	Digalactosyldiacylglycerol	DGDG(38:3)	970.6	
Galactolipid	Digalactosyldiacylglycerol (oxidized acyl chain)	DGDG (18:4-O/16:3)	922.5	QTOF MS (product ion) of [M−H]−
Galactolipid	Digalactosyldiacylglycerol (oxidized acyl chain)	DGDG (16:4-O/18:3)	922.5	QTOF MS (product ion) of [M−H]−
Galactolipid	Digalactosyldiacylglycerol (oxidized acyl chain)	DGDG (18:3-O/16:3)	924.5	QTOF MS (product ion) of [M+C2H3O2]−
Galactolipid	Digalactosyldiacylglycerol (oxidized acyl chain)	DGDG (16:4-O/18:2)	924.5	QTOF MS (product ion) of [M+C2H3O2]−
Galactolipid	Digalactosyldiacylglycerol (oxidized acyl chain)	DGDG (18:4-O/16:0)	928.6	QTOF MS (product ion) of [M+C2H3O2]−
Galactolipid	Digalactosyldiacylglycerol (oxidized acyl chain)	DGDG (18:3-O/16:0)	930.6	QTOF MS (product ion) of [M+C2H3O2]−
Galactolipid	Digalactosyldiacylglycerol (oxidized acyl chain)	DGDG (18:4-O/16:4-O)	936.5	NMR, QTOF MS (product ion) of [M+C2H3O2]−

Molecular species identified by additional steps beyond the analyses listed in Table 13.1 (order of acyl chain names does not infer position)	Chemical formula of identified molecular species	References for identification	References for high-throughput analysis	Comments
	C53H88O15		[8, 49, 52]	May represent a DGDG with one or two oxidized acyl chains instead of the normal chains indicated
	C53H90O15		[8, 49, 52]	May represent a DGDG with one or two oxidized acyl chains instead of the normal chains indicated
	C53H92O15		[8, 49, 52]	May represent a DGDG with one or two oxidized acyl chains instead of the normal chains indicated
	C53H94O15		[8, 49, 52]	May represent a DGDG with one or two oxidized acyl chains instead of the normal chains indicated
	C49H78O16	[71]	[71]	
	C49H78O16	[71]	[71]	Detected by product ion analysis but not by the precursor scans (Pre 291.2, Pre 293.2, Pre 295.2) used by Vu et al. (2012)
	C49H80O16	[71]	[71]	
	C49H80O16	[71]	[71]	Detected by product ion analysis but not by the precursor scans (Pre 291.2, Pre 293.2, Pre 295.2) used by Vu et al. (2012)
	C49H84O16	[71]	[71]	
	C49H86O16	[71]	[71]	
	C49H76O17	[68]	[71]	

(continued)

Table 2 (continued)

Group	Class	Abbreviation	Mass of uncharged compound	Additional steps for compound identification and analyses listed in Table 13.1
Galactolipid	Digalactosyldiacylglycerol (oxidized acyl chain)	DGDG (18:3-O/16:4-O)	938.5	QTOF MS (product ion) of [M+C2H3O2]-
Galactolipid	Digalactosyldiacylglycerol (oxidized acyl chain)	DGDG (18:3-2O/16:3)	940.5	QTOF MS (product ion) of [M+C2H3O2]-
Galactolipid	Digalactosyldiacylglycerol (oxidized acyl chain)	DGDG (16:3-2O/18:3)	940.5	QTOF MS (product ion) of [M+C2H3O2]-
Galactolipid	Digalactosyldiacylglycerol (oxidized acyl chain)	DGDG (18:3-2O/16:0)	946.6	QTOF MS (product ion) of [M−H]-
Galactolipid	Digalactosyldiacylglycerol (oxidized acyl chain)	DGDG (18:4-O/18:3)	950.6	QTOF MS (product ion) of [M+C2H3O2]-
Galactolipid	Digalactosyldiacylglycerol (oxidized acyl chain)	DGDG (18:3-O/18:3)	952.6	QTOF MS (product ion) of [M+C2H3O2]-
Galactolipid	Digalactosyldiacylglycerol (oxidized acyl chain)	DGDG (18:4-O/18:2)	952.6	QTOF MS (product ion) of [M+C2H3O2]-
Galactolipid	Digalactosyldiacylglycerol (oxidized acyl chain)	DGDG (18:3-2O/16:4-O)	954.5	QTOF MS (product ion) of [M+C2H3O2]-
Galactolipid	Digalactosyldiacylglycerol (oxidized acyl chain)	DGDG (18:4-O/16:3-2O)	954.5	QTOF MS (product ion) of [M+C2H3O2]-
Galactolipid	Digalactosyldiacylglycerol (oxidized acyl chain)	DGDG (18:4-O/18:4-O)	964.5	NMR, QTOF MS (product ion) of [M+C2H3O2]-

Molecular species identified by additional steps beyond the analyses listed in Table 13.1 (order of acyl chain names does not infer position)	Chemical formula of identified molecular species	References for identification	References for high-throughput analysis	Comments
	C49H78O17	[71]	[71]	
	C49H80O17	[71]	[71]	Scanning for Pre 291, i.e., Pre 18:4-O, can be used to detect species containing 18:3-2O also, because 18:3-2O undergoes a water loss to 18:4-O
	C49H80O17	[71]	[71]	Detected by product ion analysis but not by the precursor scans (Pre 291.2, Pre 293.2, Pre 295.2) used by Vu et al. (2012)
	C49H86O17	[71]	[71]	Scanning for Pre 291, i.e., Pre 18:4-O, can be used to detect species containing 18:3-2O also, because 18:3-2O undergoes a water loss to 18:4-O
	C51H82O16	[64]	[71]	
	C51H84O16	[71]	[71]	
	C51H84O16	[71]	[71]	Detected by product ion analysis but not by the precursor scans (Pre 291.2, Pre 293.2, Pre 295.2) used by Vu et al. (2012)
	C49H78O18	[71]	[71]	Scanning for Pre 291, i.e., Pre 18:4-O, can be used to detect species containing 18:3-2O also, because 18:3-2O undergoes a water loss to 18:4-O
	C49H78O18	[71]	[71]	
	C51H80O17	[68]	[71]	

(continued)

Table 2 (continued)

Group	Class	Abbreviation	Mass of uncharged compound	Additional steps for compound identification and analyses listed in Table 13.1
Galactolipid	Digalactosyldiacylglycerol (oxidized acyl chain)	DGDG (18:3-O/18:4-O)	966.6	QTOF MS (product ion) of [M+C2H3O2]-
Galactolipid	Digalactosyldiacylglycerol (oxidized acyl chain)	DGDG (18:3-2O/18:3)	968.6	QTOF MS (product ion) of [M+C2H3O2]-
Galactolipid	Digalactosyldiacylglycerol (oxidized acyl chain)	DGDG (18:4-O/18:3-2O)	982.5	QTOF MS (product ion) of [M-H]-
Galactolipid	Digalactosyldiacylglycerol (oxidized acyl chain)	DGDG(18:3-2O/18:4-O)	982.5	QTOF MS (product ion) of [M+C2H3O2]-
Galactolipid	Digalactosylmonoacylglycerol	DGMG(16:3)	648.3	
Galactolipid	Digalactosylmonoacylglycerol	DGMG (16:4-O)	662.3	TOF product ion of [M-H]-and[M+HCOO]-
Galactolipid	Digalactosylmonoacylglycerol	DGMG(18:3)	676.4	
Galactolipid	Digalactosylmonoacylglycerol	DGMG(18:2)	678.4	
Galactolipid	Digalactosylmonoacylglycerol	DGMG(18:1)	680.4	
Galactolipid	Digalactosylmonoacylglycerol	DGMG(18:4-O)	690.3	TOF product ion of [M-H]-and[M+HCOO]-
Galactolipid	Monogalactosyldiacylglycerol	MGDG(34:6)	746.5	Product ion of [M+C2H3O2]-
Galactolipid	Monogalactosyldiacylglycerol	MGDG(34:5)a	748.5	Product ion of [M+C2H3O2]-
Galactolipid	Monogalactosyldiacylglycerol	MGDG(34:5)b	748.5	Product ion of [M+C2H3O2]-
Galactolipid	Monogalactosyldiacylglycerol	MGDG(34:4)a	750.5	Product ion of [M+C2H3O2]-
Galactolipid	Monogalactosyldiacylglycerol	MGDG(34:4)b	750.5	Product ion of [M+C2H3O2]-
Galactolipid	Monogalactosyldiacylglycerol	MGDG(34:4)c	750.5	Product ion of [M+C2H3O2]-

Molecular species identified by additional steps beyond the analyses listed in Table 13.1 (order of acyl chain names does not infer position)	Chemical formula of identified molecular species	References for identification	References for high-throughput analysis	Comments
	C51H82O17	[71]	[71]	
	C51H84O17	[71]	[71]	Scanning for Pre 291, i.e., Pre 18:4-O, can be used to detect species containing 18:3-2O also, because 18:3-2O undergoes a water loss to 18:4-O
	C51H82O18	[64]		
	C51H82O18	[64]	[71]	Scanning for Pre 291, i.e., Pre 18:4-O, can be used to detect species containing 18:3-2O also, because 18:3-2O undergoes a water loss to 18:4-O
	C31H52O14		[74]	
16:4-O	C31H50O15		[74, 66]	
	C33H56O14		[74]	
	C33H58O14		[74]	
	C33H60O14		[74]	
18:4-O	C33H54O15		[74, 66]	
18:3-16:3	C43H70O10	[52, 46]	[8, 49, 52, 46]	
18:3-16:2	C43H72O10	[52, 46]	[8, 49, 52, 46]	
18:2-16:3	C43H72O10	[52]	[8, 49, 52]	
18:3-16:1	C43H74O10	[52, 46]	[8, 49, 52, 46]	
18:2-16:2	C43H74O10	[52]	[8, 49, 52]	
18:1-16:3	C43H74O10	[52]	[8, 49, 52]	

(continued)

Table 2 (continued)

Group	Class	Abbreviation	Mass of uncharged compound	Additional steps for compound identification and analyses listed in Table 13.1
Galactolipid	Monogalactosyldiacylglycerol	MGDG(34:3)a	752.5	Product ion of [M+C2H3O2]-
Galactolipid	Monogalactosyldiacylglycerol	MGDG(34:3)b	752.5	Product ion of [M+C2H3O2]-
Galactolipid	Monogalactosyldiacylglycerol	MGDG(34:3)c	752.5	Product ion of [M+C2H3O2]-
Galactolipid	Monogalactosyldiacylglycerol	MGDG(34:3)d	752.5	Product ion of [M+C2H3O2]-
Galactolipid	Monogalactosyldiacylglycerol	MGDG(34:2)a	754.6	Product ion of [M+C2H3O2]-
Galactolipid	Monogalactosyldiacylglycerol	MGDG(34:2)b	754.6	Product ion of [M+C2H3O2]-
Galactolipid	Monogalactosyldiacylglycerol	MGDG(34:1)a	756.6	Product ion of [M+C2H3O2]-
Galactolipid	Monogalactosyldiacylglycerol	MGDG(34:1)b	756.6	Product ion of [M+C2H3O2]-
Galactolipid	Monogalactosyldiacylglycerol	MGDG(36:6)	774.5	Product ion of [M+C2H3O2]-
Galactolipid	Monogalactosyldiacylglycerol	MGDG(36:5)	776.5	Product ion of [M+C2H3O2]-
Galactolipid	Monogalactosyldiacylglycerol	MGDG(36:4)	778.6	Product ion of [M+C2H3O2]-
Galactolipid	Monogalactosyldiacylglycerol	MGDG(36:3)	780.6	Product ion of [M+C2H3O2]-
Galactolipid	Monogalactosyldiacylglycerol	MGDG(36:2)	782.6	
Galactolipid	Monogalactosyldiacylglycerol	MGDG(36:1)	784.6	
Galactolipid	Monogalactosyldiacylglycerol	MGDG(38:6)	802.6	

Molecular species identified by additional steps beyond the analyses listed in Table 13.1 (order of acyl chain names does not infer position)	Chemical formula of identified molecular species	References for identification	References for high-throughput analysis	Comments
16:0-18:3	C43H76O10	[52, 46]	[8, 49, 52, 46]	
18:2-16:1	C43H76O10	[52]	[8, 49, 52]	
18:0-16:3	C43H76O10	[52]	[8, 49, 52]	
18:1-16:2	C43H76O10	[52]	[8, 49, 52]	
18:1-16:1	C43H78O10	[52, 46]	[8, 49, 52, 46]	
16:0-18:2	C43H78O10	[52]	[8, 49, 52]	
18:0-16:1	C43H80O10	[52]	[8, 49, 52]	
16:0-18:1	C43H80O10	[52, 46]	[8, 49, 52, 46]	
18:3-18:3	C45H74O10	[52, 46]	[8, 49, 52, 46]	In scanning for head group with nominal m/z resolution, may include an MGDG with one or two oxidized acyl chains instead of the normal chains indicated, e.g., MGDG(34:8-2O)
18:3-18:2	C45H76O10	[52, 46]	[8, 49, 52, 46]	
18:2-18:2	C45H78O10	[46]	[8, 49, 52, 46]	
18:3-18:0	C45H80O10	[46]	[8, 49, 52, 46]	
	C45H82O10		[8, 49, 52]	
	C45H84O10		[8, 49, 52]	
20:3-18:3	C47H78O10	This paper	[8, 49, 52]	In scanning for head group with nominal m/z resolution, may include/represent an MGDG with one or two oxidized acyl chains instead of the normal chains indicated, e.g., 36:8-2O

(continued)

Table 2
(continued)

Group	Class	Abbreviation	Mass of uncharged compound	Additional steps for compound identification and analyses listed in Table 13.1
Galactolipid	Monogalactosyldiacylglycerol	MGDG(38:5)	804.6	
Galactolipid	Monogalactosyldiacylglycerol	MGDG(38:4)	806.6	
Galactolipid	Monogalactosyldiacylglycerol	MGDG(38:3)	808.6	
Galactolipid	Monogalactosyldiacylglycerol (oxidized acyl chain)	MGDG (18:4-O/16:3)	760.5	NMR & LC MS/MS, QTOF MS (product ion) of [M−H]-
Galactolipid	Monogalactosyldiacylglycerol (oxidized acyl chain)	MGDG (16:4-O/18:3)	760.5	NMR, QTOF MS (product ion) of [M+C2H3O2]-
Galactolipid	Monogalactosyldiacylglycerol (oxidized acyl chain)	MGDG (18:3-O/16:3)	762.5	QTOF MS (product ion) of [M+C2H3O2]-
Galactolipid	Monogalactosyldiacylglycerol (oxidized acyl chain)	MGDG (16:3-O/18:3)	762.5	QTOF MS (product ion) of [M+C2H3O2]-
Galactolipid	Monogalactosyldiacylglycerol (oxidized acyl chain)	MGDG (18:4-O/16:4-O)	774.5	NMR, QTOF MS (product ion) of [M−H]-
Galactolipid	Monogalactosyldiacylglycerol (oxidized acyl chain)	MGDG (18:3-O/16:4-O)	776.5	QTOF MS (product ion) of [M+C2H3O2]-
Galactolipid	Monogalactosyldiacylglycerol (oxidized acyl chain)	MGDG(18:3-O/16:4-O)	776.5	QTOF MS (product ion) of [M−H]-
Galactolipid	Monogalactosyldiacylglycerol (oxidized acyl chain)	MGDG (18:3-2O/16:3)	778.5	QTOF MS (product ion) of [M−H]-

Molecular species identified by additional steps beyond the analyses listed in Table 13.1 (order of acyl chain names does not infer position)	Chemical formula of identified molecular species	References for identification	References for high-throughput analysis	Comments
	C47H80O10		[8, 49, 52]	May represent an MGDG with one or two oxidized acyl chains instead of the normal chains indicated
	C47H82O10		[8, 49, 52]	May represent an MGDG with one or two oxidized acyl chains instead of the normal chains indicated
	C47H84O10		[8, 49, 52]	May represent an MGDG with one or two oxidized acyl chains instead of the normal chains indicated
	C43H68O11	[64, 70]	[71]	
	C43H68O11	[76]		
	C43H70O11	[71]	[71]	
	C43H70O12	[71]	[71]	Detected by product ion analysis but not by the precursor scans (Pre 291.2, Pre 293.2, Pre 295.2) used by Vu et al. (2012)
	C43H66O12	[64, 67]	[71]	
	C43H6812	[71]	[71]	
	C43H68O12	[71]	[71]	
	C43H70O12	[64]	[71]	Scanning for Pre 291, i.e., Pre 18:4-O, can be used to detect species containing 18:3-2O also, because 18:3-2O undergoes a water loss to 18:4-O

(continued)

Table 2 (continued)

Group	Class	Abbreviation	Mass of uncharged compound	Additional steps for compound identification and analyses listed in Table 13.1
Galactolipid	Monogalactosyldiacylglycerol (oxidized acyl chain)	MGDG (18:3/16:3-2O)	778.5	QTOF MS (product ion) of [M−H]−
Galactolipid	Monogalactosyldiacylglycerol (oxidized acyl chain)	MGDG (18:4-O/18:3)	788.5	QTOF MS (product ion) of [M−H]−
Galactolipid	Monogalactosyldiacylglycerol (oxidized acyl chain)	MGDG (18:3-O/18:3)	790.5	QTOF MS (product ion) of [M−H]−
Galactolipid	Monogalactosyldiacylglycerol (oxidized acyl chain)	MGDG (18:4-O/18:2)	790.5	QTOF MS (product ion) of [M−H]−
Galactolipid	Monogalactosyldiacylglycerol (oxidized acyl chain)	MGDG (18:3-2O/16:4-O)	792.5	QTOF MS (product ion) of [M−H]−
Galactolipid	Monogalactosyldiacylglycerol (oxidized acyl chain)	MGDG (18:4-O/18:4-O)	802.5	NMR, QTOF MS (product ion) of [M−H]−
Galactolipid	Monogalactosyldiacylglycerol (oxidized acyl chain)	MGDG (18:3-O/18:4-O)	804.5	QTOF MS (product ion) of [M+C2H3O2]−
Galactolipid	Monogalactosyldiacylglycerol (oxidized acyl chain)	MGDG (18:3-2O/18:3)	806.5	QTOF MS (product ion) of [M−H]−
Galactolipid	Monogalactosyldiacylglycerol (oxidized acyl chain)	MGDG (18:3-2O/18:4-O)	820.5	QTOF MS (product ion) of [M−H]−
Galactolipid	Monogalactosylmonoacylglycerol	MGMG(16:3)	486.3	
Galactolipid	Monogalactosylmonoacylglycerol	MGMG(16:2)	488.3	
Galactolipid	Monogalactosylmonoacylglycerol	MGMG(16:1)	490.3	
Galactolipid	Monogalactosylmonoacylglycerol	MGMG(16:4-O)	500.3	TOF product ion of [M−H]−and [M+HCOO]−

Molecular species identified by additional steps beyond the analyses listed in Table 13.1 (order of acyl chain names does not infer position)	Chemical formula of identified molecular species	References for identification	References for high-throughput analysis	Comments
	C43H70O12	[64]		
	C45H72O11	[64]	[71]	
	C45H74O11	[71]	[71]	
	C45H74O11	[71]	[71]	Detected by product ion analysis but not by the precursor scans (Pre 291.2, Pre 293.2, Pre 295.2) used by Vu et al. (2012)
	C43H68O13	[64]	[71]	Scanning for Pre 291, i.e., Pre 18:4-O, can be used to detect species containing 18:3-2O also, because 18:3-2O undergoes a water loss to 18:4-O
	C45H70O12	[64, 67]	[71]	
	C45H72O12	[71]	[71]	
	C45H74O12	[64]	[71]	Scanning for Pre 291, i.e., Pre 18:4-O, can be used to detect species containing 18:3-2O also, because 18:3-2O undergoes a water loss to 18:4-O
	C45H72O13	[64]	[71]	Scanning for Pre 291, i.e., Pre 18:4-O, can be used to detect species containing 18:3-2O also, because 18:3-2O undergoes a water loss to 18:4-O
	C25H42O9		[74]	
	C25H44O9		[74]	
	C25H46O9		[74]	
16:4-O	C25H40O10		[74, 66]	

(continued)

Table 2 (continued)

Group	Class	Abbreviation	Mass of uncharged compound	Additional steps for compound identification and analyses listed in Table 13.1
Galactolipid	Monogalactosylmonoacylglycerol	MGMG(18:3)	514.3	
Galactolipid	Monogalactosylmonoacylglycerol	MGMG(18:2)	516.3	
Galactolipid	Monogalactosylmonoacylglycerol	MGMG(18:1)	518.3	
Galactolipid	Monogalactosylmonoacylglycerol	MGMG(18:4-O)	528.3	TOF product ion of [M−H]− and [M+HCOO]−
Lipid A pathway intermediate	Lipid A pathway intermediate	Lipid X	711.4	QTOF MS (product ion) of [M−H]−
Lipid A pathway intermediate	Lipid A pathway intermediate	UDP-2,3-diacyl-GlcN	1,017.5	QTOF MS (product ion) of [M−H]−
Lipid A pathway intermediate	Lipid A pathway intermediate	Disaccharide-1-P	1,324.9	QTOF MS (product ion) of [M−H]−
Lipid A pathway intermediate	Lipid A pathway intermediate	Lipid IVA	1,404.9	QTOF MS (product ion) of [M−2H]2−
Phospholipid	Lysophosphatidylcholine	LPC(16:1)	493.3	Product ion of [M+C2H3O2]−
Phospholipid	Lysophosphatidylcholine	LPC(16:0)	495.3	Product ion of [M+C2H3O2]−
Phospholipid	Lysophosphatidylcholine	LPC(18:3)	517.3	Product ion of [M+C2H3O2]−
Phospholipid	Lysophosphatidylcholine	LPC(18:2)	519.3	Product ion of [M+C2H3O2]−
Phospholipid	Lysophosphatidylcholine	LPC(18:1)	521.3	Product ion of [M+C2H3O2]−
Phospholipid	Lysophosphatidylcholine	LPC(18:0)	523.4	
Phospholipid	Lysophosphatidylethanolamine	LPE(16:1)	451.3	
Phospholipid	Lysophosphatidylethanolamine	LPE(16:0)	453.3	Product ion of [M−H]−
Phospholipid	Lysophosphatidylethanolamine	LPE(18:3)	475.3	Product ion of [M−H]−

Molecular species identified by additional steps beyond the analyses listed in Table 13.1 (order of acyl chain names does not infer position)	Chemical formula of identified molecular species	References for identification	References for high-throughput analysis	Comments
	$C_{27}H_{46}O_9$		[74]	
	$C_{27}H_{48}O_9$		[74]	
	$C_{27}H_{50}O_9$		[74]	
18:4-O	$C_{27}H_{44}O_{10}$		[74]	
	$C_{34}H_{66}O_{12}NP$		[72]	
	$C_{43}H_{77}O_{20}N_3P_2$		[72]	
	$C_{68}H_{129}O_{20}N_2P$		[72]	
	$C_{68}H_{130}O_{23}N_2P_2$		[72]	
16:1	$C_{24}H_{48}O_7PN$	[52]	[8, 49, 52]	
16:0	$C_{24}H_{50}O_7PN$	[52]	[8, 49, 52]	
18:3	$C_{26}H_{48}O_7PN$	[52]	[8, 49, 52]	
18:2	$C_{26}H_{50}O_7PN$	[52]	[8, 49, 52]	
18:1	$C_{26}H_{52}O_7PN$	[52]	[8, 49, 52]	
18:0	$C_{26}H_{54}O_7PN$		[8, 49, 52]	
16:1	$C_{21}H_{42}O_7PN$	This paper	[8, 49, 52]	
16:0	$C_{21}H_{44}O_7PN$	[52]	[8, 49, 52]	
18:3	$C_{23}H_{42}O_7PN$	[52]	[8, 49, 52]	

(continued)

Table 2 (continued)

Group	Class	Abbreviation	Mass of uncharged compound	Additional steps for compound identification and analyses listed in Table 13.1
Phospholipid	Lysophosphatidylethanolamine	LPE(18:2)	477.3	Product ion of [M-H]-
Phospholipid	Lysophosphatidylethanolamine	LPE(18:1)	479.3	Product ion of [M-H]-
Phospholipid	Lysophosphatidylglycerol	LPG(16:1)	482.3	
Phospholipid	Lysophosphatidylglycerol	LPG(16:0)	484.3	Product ion of [M-H]-
Phospholipid	Lysophosphatidylglycerol	LPG(18:3)	506.3	Product ion of [M-H]-
Phospholipid	Lysophosphatidylglycerol	LPG(18:2)	508.3	Product ion of [M-H]-
Phospholipid	Lysophosphatidylglycerol	LPG(18:1)	510.3	Product ion of [M-H]-
Phospholipid	Phosphatidic acid	PA(32:0)	648.5	Product ion of [M-H]-
Phospholipid	Phosphatidic acid	PA(34:6)	664.4	Product ion of [M-H]-
Phospholipid	Phosphatidic acid	PA(34:5)	666.4	Product ion of [M-H]-
Phospholipid	Phosphatidic acid	PA(34:4)	668.4	Product ion of [M-H]-
Phospholipid	Phosphatidic acid	PA(34:3)a	670.5	Product ion of [M-H]-
Phospholipid	Phosphatidic acid	PA(34:3)b	670.5	Product ion of [M-H]-
Phospholipid	Phosphatidic acid	PA(34:2)	672.5	Product ion of [M-H]-
Phospholipid	Phosphatidic acid	PA(34:1)	674.5	Product ion of [M-H]-
Phospholipid	Phosphatidic acid	PA(36:6)	692.4	Product ion of [M-H]-

Molecular species identified by additional steps beyond the analyses listed in Table 13.1 (order of acyl chain names does not infer position)	Chemical formula of identified molecular species	References for identification	References for high-throughput analysis	Comments
18:2	C23H44O7PN	[52]	[8, 49, 52]	
18:1	C23H46O7PN	[52]	[8, 49, 52]	
18:0	C22H43O9P		[8, 49, 52]	
16:0	C22H45O9P	[52]	[8, 49, 52]	
18:3	C24H43O9P	[52]	[8, 49, 52]	
18:2	C24H45O9P	[52]	[8, 49, 52]	
18:1	C24H47O9P	[52]	[8, 49, 52]	
16:0-16:0	C35H69O8P	This paper		
18:3-16:3	C37H61O8P	[52]	[8, 49, 52]	
18:2-16:3	C37H63O8P	This paper		
18:3-16:1	C37H65O8P	[52]	[8, 49, 52]	
18:3-16:0	C37H67O8P	[52]	[8, 49, 52]	
18:2-16:1	C37H67O8P	[52]	[8, 49, 52]	Signals from the component compounds of a species labeled with letters (e.g., a or b) in the Abbreviations column are combined when a head group-specific scan is employed in direct-infusion mode
18:2-16:0	C37H69O8P		[8, 49, 52]	
18:1-16:0	C37H71O8P			
18:3-18:3	C39H65O8P	[52]	[8, 49, 52]	

(continued)

Table 2 (continued)

Group	Class	Abbreviation	Mass of uncharged compound	Additional steps for compound identification and analyses listed in Table 13.1
Phospholipid	Phosphatidic acid	PA(36:5)	694.5	Product ion of [M−H]-
Phospholipid	Phosphatidic acid	PA(36:4)a	696.5	Product ion of [M−H]-
Phospholipid	Phosphatidic acid	PA(36:4)b	696.5	Product ion of [M−H]-
Phospholipid	Phosphatidic acid	PA(36:3)a	698.5	Product ion of [M−H]-
Phospholipid	Phosphatidic acid	PA(36:3)b	698.5	Product ion of [M−H]-
Phospholipid	Phosphatidic acid	PA(36:2)a	700.5	Product ion of [M−H]-
Phospholipid	Phosphatidic acid	PA(36:2)b	700.5	Product ion of [M−H]-
Phospholipid	Phosphatidylcholine	PC(32:0)	733.6	Product ion of [M+C2H3O2]-
Phospholipid	Phosphatidylcholine	PC(34:6)	749.5	Product ion of [M+C2H3O2]-
Phospholipid	Phosphatidylcholine	PC(34:4)	753.5	Product ion of [M+C2H3O2]-
Phospholipid	Phosphatidylcholine	PC(34:3)	755.5	Product ion of [M+C2H3O2]-

Molecular species identified by additional steps beyond the analyses listed in Table 13.1 (order of acyl chain names does not infer position)	Chemical formula of identified molecular species	References for identification	References for high-throughput analysis	Comments
18:2-18:3	C39H67O8P	[52]	[8, 49, 52]	
18:2-18:2	C39H69O8P	[52]	[8, 49, 52]	
18:3-18:1	C39H69O8P	[52]	[8, 49, 52]	Signals from the component compounds of a species labeled with letters (e.g., a or b) in the Abbreviations column are combined when a head group-specific scan is employed in direct-infusion mode
18:2-18:1	C39H71O8P	[52]	[8, 49, 52]	
18:0-18:3	C39H71O8P	[52]	[8, 49, 52]	Signals from the component compounds of a species labeled with letters (e.g., a or b) in the Abbreviations column are combined when a head group-specific scan is employed in direct-infusion mode
18:0-18:2	C39H73O8P	[52]	[8, 49, 52]	
18:1-18:1	C39H73O8P	[52]	[8, 49, 52]	Signals from the component compounds of a species labeled with letters (e.g., a or b) in the Abbreviations column are combined when a head group-specific scan is employed in direct-infusion mode
	C40H80O8PN		[8, 49, 52]	
	C42H72O8PN		[46]	
16:1-18:3	C42H76O8PN	[52]	[8, 49, 52]	
16:0-18:3	C42H78O8PN	[52, 46]	[8, 49, 52, 46]	

(continued)

Table 2 (continued)

Group	Class	Abbreviation	Mass of uncharged compound	Additional steps for compound identification and analyses listed in Table 13.1
Phospholipid	Phosphatidylcholine	PC(34:2)	757.6	Product ion of [M+C2H3O2]-
Phospholipid	Phosphatidylcholine	PC(34:1)	759.6	Product ion of [M+C2H3O2]-
Phospholipid	Phosphatidylcholine	PC(36:6)	777.5	Product ion of [M+C2H3O2]-
Phospholipid	Phosphatidylcholine	PC(36:5)	779.5	Product ion of [M+C2H3O2]-
Phospholipid	Phosphatidylcholine	PC(36:4)a	781.6	Product ion of [M+C2H3O2]-
Phospholipid	Phosphatidylcholine	PC(36:4)b	781.6	Product ion of [M+C2H3O2]-
Phospholipid	Phosphatidylcholine	PC(36:3)a	783.6	Product ion of [M+C2H3O2]-
Phospholipid	Phosphatidylcholine	PC(36:3)b	783.6	Product ion of [M+C2H3O2]-
Phospholipid	Phosphatidylcholine	PC(36:2)a	785.6	Product ion of [M+C2H3O2]-
Phospholipid	Phosphatidylcholine	PC(36:2)b	785.6	Product ion of [M+C2H3O2]-

Molecular species identified by additional steps beyond the analyses listed in Table 13.1 (order of acyl chain names does not infer position)	Chemical formula of identified molecular species	References for identification	References for high-throughput analysis	Comments
16:0-18:2	C42H80O8PN	[52, 46]	[8, 49, 52, 46]	
16:0-18:1	C42H82O8PN	[46]	[8, 49, 52, 46]	
18:3-18:3	C44H76O8PN	[52, 46]	[8, 49, 52, 46]	
18:3-18:2	C44H78O8PN	[52]	[8, 49, 52, 46]	
18:2-18:2	C44H80O8PN	[52, 46]	[8, 49, 52, 46]	
18:1-18:3	C44H80O8PN	[52]	[8, 49, 52]	Signals from the component compounds of a species labeled with letters (e.g., a or b) in the Abbreviations column are combined when a head group-specific scan is employed in direct-infusion mode
18:1-18:2	C44H82O8PN	[52, 46]	[8, 49, 52, 46]	
18:0-18:3	C44H82O8PN	[52]	[8, 49, 52]	Signals from the component compounds of a species labeled with letters (e.g., a or b) in the Abbreviations column are combined when a head group-specific scan is employed in direct-infusion mode
18:0-18:2	C44H84O8PN	[52]	[8, 49, 52]	
18:1-18:1	C44H84O8PN	[52, 46]	[8, 49, 52, 46]	Signals from the component compounds of a species labeled with letters (e.g., a or b) in the Abbreviations column are combined when a head group-specific scan is employed in direct-infusion mode

(continued)

Table 2 (continued)

Group	Class	Abbreviation	Mass of uncharged compound	Additional steps for compound identification and analyses listed in Table 13.1
Phospholipid	Phosphatidylcholine	PC(36:1)	787.6	Product ion of [M+C2H3O2]-
Phospholipid	Phosphatidylcholine	PC(38:6)	805.6	Product ion of [M+C2H3O2]-
Phospholipid	Phosphatidylcholine	PC(38:5)	807.6	Product ion of [M+C2H3O2]-
Phospholipid	Phosphatidylcholine	PC(38:4)a	809.6	Product ion of [M+C2H3O2]-
Phospholipid	Phosphatidylcholine	PC(38:4)b	809.6	Product ion of [M+C2H3O2]-
Phospholipid	Phosphatidylcholine	PC(38:3)a	811.6	Product ion of [M+C2H3O2]-
Phospholipid	Phosphatidylcholine	PC(38:3)b	811.6	Product ion of [M+C2H3O2]-
Phospholipid	Phosphatidylcholine	PC(38:2)a	813.6	
Phospholipid	Phosphatidylcholine	PC(38:2)b	813.6	
Phospholipid	Phosphatidylcholine	PC(40:5)	835.6	
Phospholipid	Phosphatidylcholine	PC(40:4)	837.6	
Phospholipid	Phosphatidylcholine	PC(40:3)	839.6	Product ion of [M+C2H3O2]-

Molecular species identified by additional steps beyond the analyses listed in Table 13.1 (order of acyl chain names does not infer position)	Chemical formula of identified molecular species	References for identification	References for high-throughput analysis	Comments
18:0-18:1	C44H86O8PN	[46]	[8, 49, 52, 46]	
	C46H80O8PN		[8, 49, 52, 46]	
20:2-18:3	C46H82O8PN	[52]	[8, 49, 52, 46]	
20:1-18:3	C46H84O8PN	[52]	[8, 49, 52, 46]	
20:2-18:2	C46H84O8PN	[52]	[8, 49, 52, 46]	Signals from the component compounds of a species labeled with letters (e.g., a or b) in the Abbreviations column are combined when a head group-specific scan is employed in direct-infusion mode
20:1-18:2	C46H86O8PN	[52]	[8, 49, 52, 46]	
20:0-18:3	C46H86O8PN	[52]	[8, 49, 52, 46]	Signals from the component compounds of a species labeled with letters (e.g., a or b) in the Abbreviations column are combined when a head group-specific scan is employed in direct-infusion mode
	C46H88O8PN		[8, 49, 52]	
	C46H88O8PN		[8, 49, 52]	Signals from the component compounds of a species labeled with letters (e.g., a or b) in the Abbreviations column are combined when a head group-specific scan is employed in direct-infusion mode
	C48H86O8PN		[8, 49, 52]	
	C48H88O8PN		[8, 49, 52]	
22:0-18:3	C48H90O8PN	[52]	[8, 49, 52]	

(continued)

Table 2 (continued)

Group	Class	Abbreviation	Mass of uncharged compound	Additional steps for compound identification and analyses listed in Table 13.1
Phospholipid	Phosphatidylcholine	PC(40:2)a	841.7	Product ion of [M+C2H3O2]-
Phospholipid	Phosphatidylcholine	PC(40:2)b	841.7	Product ion of [M+C2H3O2]-
Phospholipid	Phosphatidylcholine (oxidized acyl chain)	PC(18:3-O/16:0)	771.5	QTOF MS (product ion) of [M+C2H3O2]-
Phospholipid	Phosphatidylcholine (oxidized acyl chain)	PC(18:2-O/16:0)	773.6	QTOF MS (product ion) of [M+C2H3O2]-
Phospholipid	Phosphatidylcholine (oxidized acyl chain)	PC(18:3-2O/16:0)	787.5	QTOF MS (product ion) of [M+C2H3O2]-
Phospholipid	Phosphatidylcholine (oxidized acyl chain)	PC(18:2-2O/16:0)	789.5	QTOF MS (product ion) of [M+C2H3O2]-
Phospholipid	Phosphatidylcholine (oxidized acyl chain)	PC(18:4-O/18:2)	793.5	QTOF MS (product ion) of [M+C2H3O2]-
Phospholipid	Phosphatidylcholine (oxidized acyl chain)	PC(18:3-O/18:3)	793.5	QTOF MS (product ion) of [M+C2H3O2]-
Phospholipid	Phosphatidylcholine (oxidized acyl chain)	PC(18:3-O/18:2)	795.5	QTOF MS (product ion) of [M+C2H3O2]-
Phospholipid	Phosphatidylcholine (oxidized acyl chain)	PC(18:2-O/18:3)	795.5	QTOF MS (product ion) of [M+C2H3O2]-
Phospholipid	Phosphatidylcholine (oxidized acyl chain)	PC(18:2-O/18:2)	797.6	QTOF MS (product ion) of [M+C2H3O2]-
Phospholipid	Phosphatidylcholine (oxidized acyl chain)	PC(18:3-2O/18:3)	809.5	QTOF MS (product ion) of [M+C2H3O2]-

Molecular species identified by additional steps beyond the analyses listed in Table 13.1 (order of acyl chain names does not infer position)	Chemical formula of identified molecular species	References for identification	References for high-throughput analysis	Comments
22:1-18:1	C48H92O8PN	[52]	[8, 49, 52]	
22:0-18:2	C48H92O8PN	[52]	[8, 49, 52]	Signals from the component compounds of a species labeled with letters (e.g., a or b) in the Abbreviations column are combined when a head group-specific scan is employed in direct-infusion mode
	C42H78O9PN	[71]	[71]	
	C42H80O9PN	[71]	[71]	
	C42H78O10PN	[71]	[71]	
	C42H80O10PN	[71]	[71]	Scanning for Pre 293, i.e., Pre 18:3-O, can be used to detect species containing 18:2-2O also, because 18:2-2O undergoes a water loss to 18:3-O
	C44H76O9PN	[71]		Detected by product ion analysis but not by the precursor scans (Pre 291.2, Pre 293.2, Pre 295.2) used by Vu et al. (2012)
	C44H76O9PN	[71]	[71]	
	C44H78O9PN	[71]	[71]	
	C44H78O9PN	[71]	[71]	
	C44H80O9PN	[71]	[71]	
	C44H76O10PN	[71]	[71]	

(continued)

Table 2 (continued)

Group	Class	Abbreviation	Mass of uncharged compound	Additional steps for compound identification and analyses listed in Table 13.1
Phospholipid	Phosphatidylcholine (oxidized acyl chain)	PC(18:2-2O/18:3)	811.5	QTOF MS (product ion) of [M+C2H3O2]-
Phospholipid	Phosphatidylcholine (oxidized acyl chain)	PC(18:3-2O/18:2)	811.5	QTOF MS (product ion) of [M+C2H3O2]-
Phospholipid	Phosphatidylcholine (oxidized acyl chain)	PC(18:2-2O/18:3)	811.5	QTOF MS (product ion) of [M+C2H3O2]-
Phospholipid	Phosphatidylcholine (oxidized acyl chain)	PC(18:2-2O/18:2)	813.5	QTOF MS (product ion) of [M+C2H3O2]-
Phospholipid	Phosphatidylcholine (oxidized acyl chain)	PC(18:3-2O/18:1)	813.5	QTOF MS (product ion) of [M+C2H3O2]-
Phospholipid	Phosphatidylethanolamine	PE(32:3)	685.5	
Phospholipid	Phosphatidylethanolamine	PE(32:2)	687.5	
Phospholipid	Phosphatidylethanolamine	PE(32:1)	689.5	
Phospholipid	Phosphatidylethanolamine	PE(32:0)	691.5	
Phospholipid	Phosphatidylethanolamine	PE(34:6)	709.5	Product ion of [M-H]-
Phospholipid	Phosphatidylethanolamine	PE(34:4)	711.5	Product ion of [M-H]-

Molecular species identified by additional steps beyond the analyses listed in Table 13.1 (order of acyl chain names does not infer position)	Chemical formula of identified molecular species	References for identification	References for high-throughput analysis	Comments
	C44H78O10PN	[71]	[71]	Scanning for Pre 293, i.e., Pre 18:3-O, can be used to detect species containing 18:2-2O also, because 18:2-2O undergoes a water loss to 18:3-O
	C44H78O10PN	[71]	[71]	Scanning for Pre 291, i.e., Pre 18:4-O, can be used to detect species containing 18:3-2O also, because 18:3-2O undergoes a water loss to 18:4-O
	C44H78O10PN	[71]	[71]	Detected by product ion analysis but not by the precursor scans (Pre 291.2, Pre 293.2, Pre 295.2) used by Vu et al. (2012)
	C44H80O10PN	[71]	[71]	Scanning for Pre 293, i.e., Pre 18:3-O, can be used to detect species containing 18:2-2O also, because 18:2-2O undergoes a water loss to 18:3-O
	C44H80O10PN	[71]	[71]	Detected by product ion analysis but not by the precursor scans (Pre 291.2, Pre 293.2, Pre 295.2) used by Vu et al. (2012)
16:0-16:3	C37H68O8PN	This paper		
16:1-16:1	C37H70O8PN	This paper		
16:0-16:1	C37H72O8PN	This paper		
16:0-16:0	C37H74O8PN	This paper		
16:3-18:3	C39H68O8PN	[46]	[46]	
16:1-18:3	C39H70O8PN	[52]	[8, 49, 52, 46]	

(continued)

Table 2 (continued)

Group	Class	Abbreviation	Mass of uncharged compound	Additional steps for compound identification and analyses listed in Table 13.1
Phospholipid	Phosphatidylethanolamine	PE(34:3)a	713.5	Product ion of [M–H]-
Phospholipid	Phosphatidylethanolamine	PE(34:3)b	713.5	Product ion of [M–H]-
Phospholipid	Phosphatidylethanolamine	PE(34:2)	715.5	Product ion of [M–H]-
Phospholipid	Phosphatidylethanolamine	PE(34:1)	717.5	Product ion of [M–H]-
Phospholipid	Phosphatidylethanolamine	PE(36:6)	735.5	Product ion of [M–H]-
Phospholipid	Phosphatidylethanolamine	PE(36:5)	737.5	Product ion of [M–H]-
Phospholipid	Phosphatidylethanolamine	PE(36:4)a	739.5	Product ion of [M–H]-
Phospholipid	Phosphatidylethanolamine	PE(36:4)b	739.5	Product ion of [M–H]-
Phospholipid	Phosphatidylethanolamine	PE(36:3)a	741.5	Product ion of [M–H]-

Molecular species identified by additional steps beyond the analyses listed in Table 13.1 (order of acyl chain names does not infer position)	Chemical formula of identified molecular species	References for identification	References for high-throughput analysis	Comments
16:0-18:3	C39H72O8PN	[52, 46]	[8, 49, 52, 46]	Signals from the component compounds of a species labeled with letters (e.g., a or b) in the Abbreviations column are combined when a head group-specific scan is employed in direct-infusion mode
16:1-18:2	C39H72O8PN	[52]	[8, 49, 52]	
16:0-18:2	C39H74O8PN	[52, 46]	[8, 49, 52, 46]	
16:0-18:1	C39H76O8PN	[52, 46]	[8, 49, 52, 46]	
18:3-18:3	C41H70O8PN	[52, 46]	[8, 49, 52, 46]	
18:3-18:2	C41H72O8PN	[52, 46]	[8, 49, 52, 46]	
18:2-18:2	C41H74O8PN	[52, 46]	[8, 49, 52, 46]	Signals from the component compounds of a species labeled with letters (e.g., a or b) in the Abbreviations column are combined when a head group-specific scan is employed in direct-infusion mode
18:1-18:3	C41H74O8PN	[52]	[8, 49, 52]	Signals from the component compounds of a species labeled with letters (e.g., a or b) in the Abbreviations column are combined when a head group-specific scan is employed in direct-infusion mode
18:0-18:3	C41H76O8PN	[52]	[8, 49, 52]	

(continued)

Table 2 (continued)

Group	Class	Abbreviation	Mass of uncharged compound	Additional steps for compound identification and analyses listed in Table 13.1
Phospholipid	Phosphatidylethanolamine	PE(36:3)b	741.5	Product ion of [M-H]-
Phospholipid	Phosphatidylethanolamine	PE(36:2)a	743.5	Product ion of [M-H]-
Phospholipid	Phosphatidylethanolamine	PE(36:2)b	743.5	Product ion of [M-H]-
Phospholipid	Phosphatidylethanolamine	PE(38:6)	763.5	Product ion of [M-H]-
Phospholipid	Phosphatidylethanolamine	PE(38:5)a	765.5	Product ion of [M-H]-
Phospholipid	Phosphatidylethanolamine	PE(38:5)b	765.5	Product ion of [M-H]-
Phospholipid	Phosphatidylethanolamine	PE(38:4)a	767.5	Product ion of [M-H]-
Phospholipid	Phosphatidylethanolamine	PE(38:4)b	767.5	Product ion of [M-H]-

Molecular species identified by additional steps beyond the analyses listed in Table 13.1 (order of acyl chain names does not infer position)	Chemical formula of identified molecular species	References for identification	References for high-throughput analysis	Comments
18:1-18:2	$C_{41}H_{76}O_8PN$	[52, 46]	[8, 49, 52, 46]	Signals from the component compounds of a species labeled with letters (e.g., a or b) in the Abbreviations column are combined when a head group-specific scan is employed in direct-infusion mode
18:0-18:2	$C_{41}H_{78}O_8PN$	[52, 46]	[8, 49, 52, 46]	
18:1-18:1	$C_{41}H_{78}O_8PN$	[52]	[8, 49, 52]	Signals from the component compounds of a species labeled with letters (e.g., a or b) in the Abbreviations column are combined when a head group-specific scan is employed in direct-infusion mode
20:3-18:3	$C_{43}H_{74}O_8PN$	[46]	[58]	
20:2-18:3	$C_{43}H_{76}O_8PN$	[52]	[8, 49, 52]	Signals from the component compounds of a species labeled with letters (e.g., a or b) in the Abbreviations column are combined when a head group-specific scan is employed in direct-infusion mode
20:3-18:2	$C_{43}H_{76}O_8PN$	[46]	[8, 49, 52, 46]	
20:1-18:3	$C_{43}H_{78}O_8PN$	[52]	[8, 49, 52]	
20:2-18:2	$C_{43}H_{78}O_8PN$	[52, 46]	[8, 49, 52, 46]	Signals from the component compounds of a species labeled with letters (e.g., a or b) in the Abbreviations column are combined when a head group-specific scan is employed in direct-infusion mode

(continued)

Table 2 (continued)

Group	Class	Abbreviation	Mass of uncharged compound	Additional steps for compound identification and analyses listed in Table 13.1
Phospholipid	Phosphatidylethanolamine	PE(38:3)a	769.6	Product ion of [M−H]−
Phospholipid	Phosphatidylethanolamine	PE(38:3)b	769.6	Product ion of [M−H]−
Phospholipid	Phosphatidylethanolamine	PE(38:2)	771.6	Product ion of [M−H]−
Phospholipid	Phosphatidylethanolamine	PE(40:3)	797.6	Product ion of [M−H]−
Phospholipid	Phosphatidylethanolamine	PE(40:2)	799.6	Product ion of [M−H]−
Phospholipid	Phosphatidylethanolamine	PE(42:4)	823.6	Product ion of [M−H]−
Phospholipid	Phosphatidylethanolamine	PE(42:3)a	825.6	Product ion of [M−H]−
Phospholipid	Phosphatidylethanolamine	PE(42:3)b	825.6	Product ion of [M−H]−
Phospholipid	Phosphatidylethanolamine	PE(42:2)	827.6	Product ion of [M−H]−
Phospholipid	Phosphatidylethanolamine (oxidized acyl chain)	PE(18:3-O/16:0)	729.5	QTOF MS (product ion) of [M−H]−
Phospholipid	Phosphatidylethanolamine (oxidized acyl chain)	PE(18:2-O/16:0)	731.5	QTOF MS (product ion) of [M−H]−
Phospholipid	Phosphatidylethanolamine (oxidized acyl chain)	PE(18:3-2O/16:0)	745.5	QTOF MS (product ion) of [M−H]−

Molecular species identified by additional steps beyond the analyses listed in Table 13.1 (order of acyl chain names does not infer position)	Chemical formula of identified molecular species	References for identification	References for high-throughput analysis	Comments
20:1-18:2	C43H80O8PN	[52]	[8, 49, 52]	
20:0-18:3	C43H80O8PN	[52]	[8, 49, 52]	Signals from the component compounds of a species labeled with letters (e.g., a or b) in the Abbreviations column are combined when a head group-specific scan is employed in direct-infusion mode
20:0-18:2	C43H82O8PN	[52]	[8, 49, 52, 46]	
22:0-18:3	C45H84O8PN	[52, 46]	[8, 49, 52, 46]	
22:0-18:2	C45H86O8PN	[52, 46]	[8, 49, 52, 46]	
24:1-18:3	C47H86O8PN	[52, 46]	[8, 49, 52, 46]	
24:1-18:2	C47H88O8PN	[52, 46]	[8, 49, 52, 46]	
24:0-18:3	C47H88O8PN	[52]	[8, 49, 52]	Signals from the component compounds of a species labeled with letters (e.g., a or b) in the Abbreviations column are combined when a head group-specific scan is employed in direct-infusion mode
24:0-18:2	C47H90O8PN	[52, 46]	[8, 49, 52, 46]	
	C39H72O9PN	[71]	[71]	
	C39H74O9PN	[71]	[71]	
	C39H72O10PN	[71]	[71]	Scanning for Pre 291, i.e., Pre 18:4-O, can be used to detect species containing 18:3-2O also, because 18:3-2O undergoes a water loss to 18:4-O

(continued)

Table 2 (continued)

Group	Class	Abbreviation	Mass of uncharged compound	Additional steps for compound identification and analyses listed in Table 13.1
Phospholipid	Phosphatidylethanolamine (oxidized acyl chain)	PE(18:2-2O/16:0)	747.5	QTOF MS (product ion) of [M - H]-
Phospholipid	Phosphatidylethanolamine (oxidized acyl chain)	PE(18:4-O/18:2)	751.0	QTOF MS (product ion) of [M - H]-
Phospholipid	Phosphatidylethanolamine (oxidized acyl chain)	PE(18:3-O/18:3)	751.5	QTOF MS (product ion) of [M - H]-
Phospholipid	Phosphatidylethanolamine (oxidized acyl chain)	PE(18:3-O/18:2)	753.5	QTOF MS (product ion) of [M - H]-
Phospholipid	Phosphatidylethanolamine (oxidized acyl chain)	PE(18:2-O/18:3)	753.5	QTOF MS (product ion) of [M - H]-
Phospholipid	Phosphatidylethanolamine (oxidized acyl chain)	PE(18:3-O/18:2)	753.5	QTOF MS (product ion) of [M - H]-
Phospholipid	Phosphatidylethanolamine (oxidized acyl chain)	PE(18:2-O/18:2)	755.5	QTOF MS (product ion) of [M - H]-
Phospholipid	Phosphatidylethanolamine (oxidized acyl chain)	PE(18:3-2O/18:3)	767.5	QTOF MS (product ion) of [M - H]-
Phospholipid	Phosphatidylethanolamine (oxidized acyl chain)	PE(18:3-2O/18:2)	769.5	QTOF MS (product ion) of [M - H]-

Molecular species identified by additional steps beyond the analyses listed in Table 13.1 (order of acyl chain names does not infer position)	Chemical formula of identified molecular species	References for identification	References for high-throughput analysis	Comments
	C39H74O10PN	[71]	[71]	Scanning for Pre 293, i.e., Pre 18:3-O, can be used to detect species containing 18:2-2O also, because 18:2-2O undergoes a water loss to 18:3-O
	C41H70O9PN	[71]	[71]	Detected by product ion analysis but not by the precursor scans (Pre 291.2, Pre 293.2, Pre 295.2) used by Vu et al. (2012)
	C41H70O9PN	[71]	[71]	
	C41H72O9PN	[71]	[71]	
	C41H72O9PN	[71]	[71]	
	C41H72O9PN	[71]	[71]	Detected by product ion analysis but not by the precursor scans (Pre 291.2, Pre 293.2, Pre 295.2) used by Vu et al. (2012)
	C41H74O9PN	[71]	[71]	
	C41H70O10PN	[71]	[71]	Scanning for Pre 291, i.e., Pre 18:4-O, can be used to detect species containing 18:3-2O also, because 18:3-2O undergoes a water loss to 18:4-O
	C41H72O10PN	[71]	[71]	Scanning for Pre 291, i.e., Pre 18:4-O, can be used to detect species containing 18:3-2O also, because 18:3-2O undergoes a water loss to 18:4-O

(continued)

Table 2 (continued)

Group	Class	Abbreviation	Mass of uncharged compound	Additional steps for compound identification and analyses listed in Table 13.1
Phospholipid	Phosphatidylethanolamine (oxidized acyl chain)	PE(18:2-2O/18:3)	769.5	QTOF MS (product ion) of [M−H]−
Phospholipid	Phosphatidylethanolamine (oxidized acyl chain)	PE(18:2-2O/18:2)	771.5	QTOF MS (product ion) of [M−H]−
Phospholipid	Phosphatidylglycerol	PG(32:1)	720.5	Product ion of [M−H]−
Phospholipid	Phosphatidylglycerol	PG(32:0)	722.5	Product ion of [M−H]−
Phospholipid	Phosphatidylglycerol	PG(34:5)	740.5	Product ion of [M−H]−
Phospholipid	Phosphatidylglycerol	PG(34:4)	742.5	Product ion of [M−H]−
Phospholipid	Phosphatidylglycerol	PG(34:3)a	744.5	Product ion of [M−H]−
Phospholipid	Phosphatidylglycerol	PG(34:3)b	744.5	Product ion of [M−H]−
Phospholipid	Phosphatidylglycerol	PG(34:2)a	746.5	Product ion of [M−H]−
Phospholipid	Phosphatidylglycerol	PG(34:2)b	746.5	Product ion of [M−H]−

Molecular species identified by additional steps beyond the analyses listed in Table 13.1 (order of acyl chain names does not infer position)	Chemical formula of identified molecular species	References for identification	References for high-throughput analysis	Comments
	C41H72O10PN	[71]	[71]	Scanning for Pre 293, i.e., Pre 18:3-O, can be used to detect species containing 18:2-2O also, because 18:2-2O undergoes a water loss to 18:3-O
	C41H74O10PN	[71]	[71]	Scanning for Pre 293, i.e., Pre 18:3-O, can be used to detect species containing 18:2-2O also, because 18:2-2O undergoes a water loss to 18:3-O
16:1-16:0	C38H73O10P	[52, 75, 46]	[8, 49, 52, 75, 46]	
16:0-16:0	C38H75O10P	[52, 46]	[8, 49, 52, 46]	
	C40H69O10P		[46]	
16:1-18:3	C40H71O10P	[52, 75, 46]	[8, 49, 52, 75, 46]	
18:3-16:0	C40H73O10P	[52, 46]	[8, 49, 52, 46]	
16:1-18:2	C40H73O10P	[52]	[8, 49, 52]	Signals from the component compounds of a species labeled with letters (e.g., a or b) in the Abbreviations column are combined when a head group-specific scan is employed in direct-infusion mode
18:2-16:0	C40H75O10P	[52, 46]	[8, 49, 52, 46]	
16:1-18:1	C40H75O10P	[52]	[8, 49, 52]	Signals from the component compounds of a species labeled with letters (e.g., a or b) in the Abbreviations column are combined when a head group-specific scan is employed in direct-infusion mode

(continued)

Table 2 (continued)

Group	Class	Abbreviation	Mass of uncharged compound	Additional steps for compound identification and analyses listed in Table 13.1
Phospholipid	Phosphatidylglycerol	PG(34:1)a	748.5	Product ion of [M-H]-
Phospholipid	Phosphatidylglycerol	PG(34:1)b	748.5	Product ion of [M-H]-
Phospholipid	Phosphatidylglycerol	PG(34:0)	750.5	Product ion of [M-H]-
Phospholipid	Phosphatidylglycerol	PG(36:7)	764.5	Product ion of [M-H]-
Phospholipid	Phosphatidylglycerol	PG(36:6)	766.5	Product ion of [M-H]-
Phospholipid	Phosphatidylglycerol	PG(36:5)	768.5	Product ion of [M-H]-
Phospholipid	Phosphatidylglycerol	PG(36:4)a	770.5	Product ion of [M-H]-
Phospholipid	Phosphatidylglycerol	PG(36:4)b	770.5	Product ion of [M-H]-
Phospholipid	Phosphatidylglycerol	PG(36:3)	772.5	
Phospholipid	Phosphatidylglycerol	PG(36:2)	774.5	Product ion of [M-H]-
Phospholipid	Phosphatidylglycerol	PG(36:1)	776.5	
Phospholipid	Phosphatidylglycerol (oxidized acyl chain)	PG(18:4-O/16:1)	756.5	QTOF MS (product ion) of [M-H]-
Phospholipid	Phosphatidylglycerol (oxidized acyl chain)	PG(18:4-O/16:0)	758.5	QTOF MS (product ion) of [M-H]-

Molecular species identified by additional steps beyond the analyses listed in Table 13.1 (order of acyl chain names does not infer position)	Chemical formula of identified molecular species	References for identification	References for high-throughput analysis	Comments
18:1-16:0	C40H77O10P	[52, 58]	[8, 49, 52, 46]	
16:1-18:0	C40H77O10P	[52]	[8, 49, 52]	Signals from the component compounds of a species labeled with letters (e.g., a or b) in the Abbreviations column are combined when a head group-specific scan is employed in direct-infusion mode
18:0-16:0	C40H79O10P	[52, 58]	[8, 49, 52, 46]	
	C42H69O10P	[58]	[58]	
18:3-18:3	C42H71O10P	[58]	[8, 49, 52, 46]	
18:3-18:2	C42H73O10P	This paper	[8, 49, 52, 46]	
18:3-18:1	C42H75O10P	This paper	[8, 49, 52, 46]	
18:2-18:2	C42H75O10P	This paper	[8, 49, 52, 46]	Signals from the component compounds of a species labeled with letters (e.g., a or b) in the Abbreviations column are combined when a head group-specific scan is employed in direct-infusion mode
	C42H77O10P		[8, 49, 52, 46]	
18:2-18:0	C42H79O10P	This paper	[8, 49, 52, 46]	
	C42H81O10P		[8, 49, 52]	
	C40H69O11P	[64]	[71]	
	C40H71O11P	[64]	[71]	

(continued)

Table 2 (continued)

Group	Class	Abbreviation	Mass of uncharged compound	Additional steps for compound identification and analyses listed in Table 13.1
Phospholipid	Phosphatidylglycerol (oxidized acyl chain)	PG(18:3-O/16:1)	758.5	QTOF MS (product ion) of [M–H]–
Phospholipid	Phosphatidylglycerol (oxidized acyl chain)	PG(18:3-O/16:0)	760.5	QTOF MS (product ion) of [M–H]–
Phospholipid	Phosphatidylglycerol (oxidized acyl chain)	PG(18:2-O/16:1)	760.5	QTOF MS (product ion) of [M–H]–
Phospholipid	Phosphatidylglycerol (oxidized acyl chain)	PG(18:2-O/16:0)	762.5	QTOF MS (product ion) of [M–H]–
Phospholipid	Phosphatidylglycerol (oxidized acyl chain)	PG(18:3-2O/16:0)	776.0	QTOF MS (product ion) of [M–H]–
Phospholipid	Phosphatidylglycerol (oxidized acyl chain)	PG(18:2-2O/16:1)	776.5	QTOF MS (product ion) of [M–H]–
Phospholipid	Phosphatidylglycerol (oxidized acyl chain)	PG(18:2-2O/16:0)	778.5	QTOF MS (product ion) of [M–H]–
Phospholipid	Phosphatidylinositol	PI(32:3)	804.5	
Phospholipid	Phosphatidylinositol	PI(32:1)	808.5	
Phospholipid	Phosphatidylinositol	PI(32:0)	810.5	
Phospholipid	Phosphatidylinositol	PI(34:5)	828.5	Product ion of [M–H]–
Phospholipid	Phosphatidylinositol	PI(34:4)	830.5	Product ion of [M–H]–
Phospholipid	Phosphatidylinositol	PI(34:3)	832.5	Product ion of [M–H]–
Phospholipid	Phosphatidylinositol	PI(34:2)	834.5	Product ion of [M–H]–
Phospholipid	Phosphatidylinositol	PI(34:1)	836.5	Product ion of [M–H]–

Molecular species identified by additional steps beyond the analyses listed in Table 13.1 (order of acyl chain names does not infer position)	Chemical formula of identified molecular species	References for identification	References for high-throughput analysis	Comments
	C40H71O11P	[71]	[75, 71]	
	C40H73O11P	[71]	[71]	
	C40H73O11P	[71]	[71]	
	C40H75O11P	[71]	[71]	
	C40H73O12P	[71]	[71]	Detected by product ion analysis but not by the precursor scans (Pre 291.2, Pre 293.2, Pre 295.2) used by Vu et al. (2012)
	C40H73O12P	[71]	[71]	Scanning for Pre 293, i.e., Pre 18:3-O, can be used to detect species containing 18:2-2O also, because 18:2-2O undergoes a water loss to 18:3-O
	C40H75O12P	[71]	[71]	Scanning for Pre 293, i.e., Pre 18:3-O, can be used to detect species containing 18:2-2O also, because 18:2-2O undergoes a water loss to 18:3-O
	C41H73O13P		[46]	
	C41H77O13P		[49]	
	C41H79O13P		[49]	
	C43H73O13P	[46]	[46]	
16:1-18:3	C43H75O13P	[52, 46]	[8, 49, 52, 46]	
18:3-16:0	C43H77O13P	[52, 46]	[8, 49, 52, 46]	
16:0-18:2	C43H79O13P	[52]	[8, 49, 52]	
16:0-18:1	C43H81O13P	[52]	[8, 49, 52]	

(continued)

Table 2 (continued)

Group	Class	Abbreviation	Mass of uncharged compound	Additional steps for compound identification and analyses listed in Table 13.1
Phospholipid	Phosphatidylinositol	PI(36:6)	854.5	Product ion of [M-H]-
Phospholipid	Phosphatidylinositol	PI(36:5)	856.5	Product ion of [M-H]-
Phospholipid	Phosphatidylinositol	PI(36:4)a	858.5	Product ion of [M-H]-
Phospholipid	Phosphatidylinositol	PI(36:4)b	858.5	Product ion of [M-H]-
Phospholipid	Phosphatidylinositol	PI(36:3)a	860.5	Product ion of [M-H]-
Phospholipid	Phosphatidylinositol	PI(36:3)b	860.5	Product ion of [M-H]-
Phospholipid	Phosphatidylinositol	PI(36:2)	862.5	Product ion of [M-H]-
Phospholipid	Phosphatidylinositol	PI(36:1)	864.6	Product ion of [M-H]-
Phospholipid	Phosphatidylserine	PS(34:4)	755.5	
Phospholipid	Phosphatidylserine	PS(34:3)	757.5	
Phospholipid	Phosphatidylserine	PS(34:2)	759.5	
Phospholipid	Phosphatidylserine	PS(34:1)	761.5	
Phospholipid	Phosphatidylserine	PS(36:6)	779.5	
Phospholipid	Phosphatidylserine	PS(36:5)	781.5	Product ion of [M-H]-
Phospholipid	Phosphatidylserine	PS(36:4)a	783.5	Product ion of [M-H]-

Molecular species identified by additional steps beyond the analyses listed in Table 13.1 (order of acyl chain names does not infer position)	Chemical formula of identified molecular species	References for identification	References for high-throughput analysis	Comments
18:3-18:3	C45H75O13P	[52]	[8, 49, 52]	
18:2-18:3	C45H77O13P	[52]	[8, 49, 52, 46]	
18:2-18:2	C45H79O13P	[52]	[8, 49, 52]	
18:1-18:3	C45H79O13P	[52]	[8, 49, 52]	Signals from the component compounds of a species labeled with letters (e.g., a or b) in the Abbreviations column are combined when a head group-specific scan is employed in direct-infusion mode
18:1-18:2	C45H81O13P	[52]	[8, 49, 52]	
18:0-18:3	C45H81O13P	[52]	[8, 49, 52]	Signals from the component compounds of a species labeled with letters (e.g., a or b) in the Abbreviations column are combined when a head group-specific scan is employed in direct-infusion mode
18:2-18:0	C45H83O13P	[52]	[8, 49, 52]	
18:0-18:1	C45H85O13P	[52]	[8, 49, 52]	
	C40H70O10PN		[49, 52]	
	C40H72O10PN		[49, 52]	
	C40H74O10PN		[49, 52]	
	C40H76O10PN		[49, 52]	
	C42H70O10PN		[49, 52]	
18:3-18:2	C42H72O10PN	This paper	[49, 52]	
18:3-18:1	C42H74O10PN	This paper	[49, 52]	

(continued)

Table 2 (continued)

Group	Class	Abbreviation	Mass of uncharged compound	Additional steps for compound identification and analyses listed in Table 13.1
Phospholipid	Phosphatidylserine	PS(36:4)b	783.5	Product ion of [M-H]-
Phospholipid	Phosphatidylserine	PS(36:3)a	785.5	Product ion of [M-H]-
Phospholipid	Phosphatidylserine	PS(36:3)b	785.5	Product ion of [M-H]-
Phospholipid	Phosphatidylserine	PS(36:2)	787.5	Product ion of [M-H]-
Phospholipid	Phosphatidylserine	PS(36:1)	789.6	
Phospholipid	Phosphatidylserine	PS(38:6)	807.5	
Phospholipid	Phosphatidylserine	PS(38:5)a	809.5	
Phospholipid	Phosphatidylserine	PS(38:5)b	809.5	
Phospholipid	Phosphatidylserine	PS(38:4)a	811.5	
Phospholipid	Phosphatidylserine	PS(38:4)b	811.5	
Phospholipid	Phosphatidylserine	PS(38:3)a	813.6	
Phospholipid	Phosphatidylserine	PS(38:3)b	813.6	
Phospholipid	Phosphatidylserine	PS(38:2)	815.6	
Phospholipid	Phosphatidylserine	PS(38:1)	817.6	
Phospholipid	Phosphatidylserine	PS(40:4)	839.6	
Phospholipid	Phosphatidylserine	PS(40:3)	841.6	Product ion of [M-H]-
Phospholipid	Phosphatidylserine	PS(40:2)	843.6	Product ion of [M-H]-

Molecular species identified by additional steps beyond the analyses listed in Table 13.1 (order of acyl chain names does not infer position)	Chemical formula of identified molecular species	References for identification	References for high-throughput analysis	Comments
18:2-18:2	$C_{42}H_{74}O_{10}PN$	This paper	[49, 52]	Signals from the component compounds of a species labeled with letters (e.g., a or b) in the Abbreviations column are combined when a head group-specific scan is employed in direct-infusion mode
18:3-18:0	$C_{42}H_{76}O_{10}PN$	This paper	[49, 52]	
18:2-18:1	$C_{42}H_{76}O_{10}PN$	This paper	[49, 52]	Signals from the component compounds of a species labeled with letters (e.g., a or b) in the Abbreviations column are combined when a head group-specific scan is employed in direct-infusion mode
18:2-18:0	$C_{42}H_{78}O_{10}PN$	This paper	[49, 52]	
	$C_{42}H_{80}O_{10}PN$		[49, 52]	
	$C_{44}H_{74}O_{10}PN$		[49, 52]	
	$C_{44}H_{76}O_{10}PN$		[49, 52]	
	$C_{44}H_{76}O_{10}PN$		[49, 52]	
	$C_{44}H_{78}O_{10}PN$		[49, 52]	
	$C_{44}H_{78}O_{10}PN$		[49, 52]	
	$C_{44}H_{80}O_{10}PN$		[49, 52, 46]	
	$C_{44}H_{80}O_{10}PN$		[49, 52, 46]	
	$C_{44}H_{82}O_{10}PN$		[49, 52, 46]	
	$C_{44}H_{84}O_{10}PN$		[49, 52]	
	$C_{46}H_{82}O_{10}PN$		[49, 52]	
22:0-18:3	$C_{46}H_{84}O_{10}PN$	This paper	[49, 52]	
22:0-18:2	$C_{46}H_{86}O_{10}PN$	This paper	[49, 52, 58]	

(continued)

Table 2 (continued)

Group	Class	Abbreviation	Mass of uncharged compound	Additional steps for compound identification and analyses listed in Table 13.1
Phospholipid	Phosphatidylserine	PS(40:1)	845.6	Product ion of [M–H]-
Phospholipid	Phosphatidylserine	PS(42:4)	867.6	Product ion of [M–H]-
Phospholipid	Phosphatidylserine	PS(42:3)a	869.6	Product ion of [M–H]-
Phospholipid	Phosphatidylserine	PS(42:3)b	869.6	Product ion of [M–H]-
Phospholipid	Phosphatidylserine	PS(42:2)	871.6	Product ion of [M–H]-
Phospholipid	Phosphatidylserine	PS(42:1)	873.7	
Phospholipid	Phosphatidylserine	PS(44:3)	897.6	Product ion of [M–H]-
Phospholipid	Phosphatidylserine	PS(44:2)	899.7	Product ion of [M–H]-
Sphingolipid	Ceramide	Cer(16:0-d18:1)	537.5	
Sphingolipid	Ceramide	Cer(16:0-d18:0)	539.5	
Sphingolipid	Ceramide	Cer(16:0-t18:1)	553.5	
Sphingolipid	Ceramide	Cer(16:0-t18:0)	555.5	
Sphingolipid	Ceramide	Cer(18:0-d18:1)	565.6	
Sphingolipid	Ceramide	Cer(18:0-t18:1)	581.6	
Sphingolipid	Ceramide	Cer(20:0-d18:1)	593.6	
Sphingolipid	Ceramide	Cer(20:0-t18:1)	609.6	
Sphingolipid	Ceramide	Cer(20:0-t18:0)	611.6	
Sphingolipid	Ceramide	Cer(22:0-d18:1)	621.6	
Sphingolipid	Ceramide	Cer(22:0-t18:1)	637.6	
Sphingolipid	Ceramide	Cer(22:0-t18:0)	639.6	
Sphingolipid	Ceramide	Cer(24:1-d18:1)	647.7	

Molecular species identified by additional steps beyond the analyses listed in Table 13.1 (order of acyl chain names does not infer position)	Chemical formula of identified molecular species	References for identification	References for high-throughput analysis	Comments
22:0-18:1	C46H88O10PN	This paper	[49, 52]	
24:1-18:3	C48H86O10PN	[52]	[49, 52, 46]	
24:0-18:3	C48H88O10PN	[52]	[49, 52]	
24:1-18:2	C48H88O10PN	[52, 46]	[49, 52, 46]	Signals from the component compounds of a species labeled with letters (e.g., a or b) in the Abbreviations column are combined when a head group-specific scan is employed in direct-infusion mode
24:0-18:2	C48H90O10PN	[52, 46]	[49, 52, 46]	
	C48H92O10PN		[49, 52]	
26:0-18:3	C50H92O10PN	[52]	[49, 52]	
26:0-18:2	C50H94O10PN	[52]	[49, 52]	
	C34H67O3N		[53]	
	C34H69O3N		[53]	
	C34H67O4N		[53]	
	C34H69O4N		[53]	
	C36H71O3N		[53]	
	C36H71O4N		[53]	
	C38H75O3N		[53]	
	C38H75O4N		[53]	
	C38H77O4N		[53]	
	C40H79O3N		[53]	
	C40H79O4N		[53]	
	C40H81O4N		[53]	
	C42H81O3N		[53]	

(continued)

Table 2 (continued)

Group	Class	Abbreviation	Mass of uncharged compound	Additional steps for compound identification and analyses listed in Table 13.1
Sphingolipid	Ceramide	Cer(24:1-d18:0)	649.6	
Sphingolipid	Ceramide	Cer(24:0-d18:1)	649.6	
Sphingolipid	Ceramide	Cer(24:0-d18:0)	651.7	
Sphingolipid	Ceramide	Cer(24:1-t18:1)	663.6	
Sphingolipid	Ceramide	Cer(24:1-t18:0)	665.6	
Sphingolipid	Ceramide	Cer(24:0-t18:1)	665.6	
Sphingolipid	Ceramide	Cer(24:0-t18:0)	667.7	
Sphingolipid	Ceramide	Cer(26:0-d18:1)	677.7	
Sphingolipid	Ceramide	Cer(26:1-t18:1)	691.6	
Sphingolipid	Ceramide	Cer(26:1-t18:0)	693.7	
Sphingolipid	Ceramide	Cer(26:0-t18:1)	693.7	
Sphingolipid	Ceramide	Cer(26:0-t18:0)	695.7	
Sphingolipid	Glucosylceramide	GlcCer(h16:0-d18:1)	715.6	
Sphingolipid	Glucosylceramide	GlcCer(h16:0-d18:0)	717.6	
Sphingolipid	Glucosylceramide	GlcCer(h16:0-t18:1)	731.6	
Sphingolipid	Glucosylceramide	GlcCer(h18:0-t18:0)	761.6	
Sphingolipid	Glucosylceramide	GlcCer(h20:0-t18:1)	787.6	
Sphingolipid	Glucosylceramide	GlcCer(h22:0-d18:1)	799.7	
Sphingolipid	Glucosylceramide	GlcCer(h22:1-t18:1)	813.6	
Sphingolipid	Glucosylceramide	GlcCer(h22:0-t18:1)	815.6	
Sphingolipid	Glucosylceramide	GlcCer(h24:1-d18:1)	825.7	
Sphingolipid	Glucosylceramide	GlcCer(h24:0-d18:1)	827.7	
Sphingolipid	Glucosylceramide	GlcCer(h24:1-t18:1)	841.7	
Sphingolipid	Glucosylceramide	GlcCer(h24:0-t18:1)	843.7	
Sphingolipid	Glucosylceramide	GlcCer(h24:1-t18:0)	843.7	
Sphingolipid	Glucosylceramide	GlcCer(h26:1-d18:1)	853.7	
Sphingolipid	Glucosylceramide	GlcCer(h26:0-d18:1)	855.7	
Sphingolipid	Glucosylceramide	GlcCer(h26:1-t18:1)	869.7	
Sphingolipid	Glucosylceramide	GlcCer(h26:0-t18:1)	871.7	

Molecular species identified by additional steps beyond the analyses listed in Table 13.1 (order of acyl chain names does not infer position)	Chemical formula of identified molecular species	References for identification	References for high-throughput analysis	Comments
	C42H83O3N		[53]	
	C42H83O3N		[53]	
	C42H85O3N		[53]	
	C42H81O4N		[53]	
	C42H83O4N		[53]	
	C42H83O4N		[53]	
	C42H85O4N		[53]	
	C44H87O3N		[53]	
	C44H85O4N		[53]	
	C44H87O4N		[53]	
	C44H87O4N		[53]	
	C44H89O4N		[53]	
	C40H77O9N		[53]	
	C40H79O9N		[53]	
	C40H77O10N		[53]	
	C42H83O10N		[53]	
	C44H85O10N		[53]	
	C46H89O9N		[53]	
	C46H87O10N		[53]	
	C46H89O10N		[53]	
	C48H91O9N		[53]	
	C48H93O9N		[53]	
	C48H91O10N		[53]	
	C48H93O10N		[53]	
	C48H93O10N		[53]	
	C50H95O9N		[53]	
	C50H97O9N		[53]	
	C50H95O10N		[53]	
	C50H97O10N		[53]	

(continued)

Table 2 (continued)

Group	Class	Abbreviation	Mass of uncharged compound	Additional steps for compound identification and analyses listed in Table 13.1
Sphingolipid	Glycosylated inositolphosphoceramide	GIPC(h16:0-d18:0)	1,135.6	
Sphingolipid	Glycosylated inositolphosphoceramide	GIPC(h16:0-t18:1)	1,149.6	
Sphingolipid	Glycosylated inositolphosphoceramide	GIPC(h16:0-t18:0)	1,151.6	
Sphingolipid	Glycosylated inositolphosphoceramide	GIPC(h18:0-t18:1)	1,177.6	
Sphingolipid	Glycosylated inositolphosphoceramide	GIPC(h20:0-t18:1)	1,205.7	
Sphingolipid	Glycosylated inositolphosphoceramide	GIPC(h22:0-d18:0)	1,219.7	
Sphingolipid	Glycosylated inositolphosphoceramide	GIPC(h22:1-t18:1)	1,231.7	
Sphingolipid	Glycosylated inositolphosphoceramide	GIPC(h22:0-t18:1)	1,233.7	
Sphingolipid	Glycosylated inositolphosphoceramide	GIPC(h22:0-t18:0)	1,235.7	
Sphingolipid	Glycosylated inositolphosphoceramide	GIPC(h24:1-d18:0)	1,245.7	
Sphingolipid	Glycosylated inositolphosphoceramide	GIPC(h24:0-d18:0)	1,247.8	
Sphingolipid	Glycosylated inositolphosphoceramide	GIPC(h24:1-t18:1)	1,259.7	
Sphingolipid	Glycosylated inositolphosphoceramide	GIPC(h24:0-t18:1)	1,261.7	
Sphingolipid	Glycosylated inositolphosphoceramide	GIPC(h24:1-t18:0)	1,261.7	
Sphingolipid	Glycosylated inositolphosphoceramide	GIPC(h24:0-t18:0)	1,263.7	
Sphingolipid	Glycosylated inositolphosphoceramide	GIPC(h26:0-d18:0)	1,275.8	
Sphingolipid	Glycosylated inositolphosphoceramide	GIPC(h26:1-t18:1)	1,287.7	
Sphingolipid	Glycosylated inositolphosphoceramide	GIPC(h26:0-t18:1)	1,289.8	

Molecular species identified by additional steps beyond the analyses listed in Table 13.1 (order of acyl chain names does not infer position)	Chemical formula of identified molecular species	References for identification	References for high-throughput analysis	Comments
	C52H98O23NP		[53]	
	C52H96O24NP		[53]	
	C52H98O24NP		[53]	
	C54H100O24NP		[53]	
	C56H104O24NP		[53]	
	C58H110O23NP		[53]	
	C58H106O24NP		[53]	
	C58H108O24NP		[53]	
	C58H110O24NP		[53]	
	C60H112O23NP		[53]	
	C60H114O23NP		[53]	
	C60H110O24NP		[53]	
	C60H112O24NP		[53]	
	C60H112O24NP		[53]	
	C60H114O24NP		[53]	
	C62H118O23NP		[53]	
	C62H114O24NP		[53]	
	C62H116O24NP		[53]	

(continued)

Table 2 (continued)

Group	Class	Abbreviation	Mass of uncharged compound	Additional steps for compound identification and analyses listed in Table 13.1
Sphingolipid	Glycosylated inositolphosphoceramide	GIPC(h26:1-t18:0)	1,289.8	
Sphingolipid	Glycosylated inositolphosphoceramide	GIPC(h26:0-t18:0)	1,291.8	
Sterols and derivatives	Acylated steryl glucoside	Cholesteryl glucoside(16:3)	780.6	QTOF MS (product ion) of [M+NH4]+
Sterols and derivatives	Acylated steryl glucoside	Cholesteryl glucoside(16:2)	782.6	QTOF MS (product ion) of [M+NH4]+
Sterols and derivatives	Acylated steryl glucoside	Cholesteryl glucoside(16:1)	784.7	QTOF MS (product ion) of [M+NH4]+
Sterols and derivatives	Acylated steryl glucoside	Cholesteryl glucoside(16:0)	786.7	QTOF MS (product ion) of [M+NH4]+
Sterols and derivatives	Acylated steryl glucoside	Campesteryl glucoside(16:3)	794.6	QTOF MS (product ion) of [M+NH4]+
Sterols and derivatives	Acylated steryl glucoside	Campesteryl glucoside(16:2)	796.7	QTOF MS (product ion) of [M+NH4]+
Sterols and derivatives	Acylated steryl glucoside	Campesteryl glucoside(16:1)	798.7	QTOF MS (product ion) of [M+NH4]+
Sterols and derivatives	Acylated steryl glucoside	Campesteryl glucoside(16:0)	800.7	QTOF MS (product ion) of [M+NH4]+
Sterols and derivatives	Acylated steryl glucoside	Stigmasteryl glucoside(16:3)	806.6	QTOF MS (product ion) of [M+NH4]+
Sterols and derivatives	Acylated steryl glucoside	Stigmasteryl glucoside(16:2)	808.7	QTOF MS (product ion) of [M+NH4]+
Sterols and derivatives	Acylated steryl glucoside	Sitosteryl glucoside(16:3)	808.7	QTOF MS (product ion) of [M+NH4]+
Sterols and derivatives	Acylated steryl glucoside	Cholesteryl glucoside(18:3)	808.7	QTOF MS (product ion) of [M+NH4]+
Sterols and derivatives	Acylated steryl glucoside	Stigmasteryl glucoside(16:1)	810.7	QTOF MS (product ion) of [M+NH4]+
Sterols and derivatives	Acylated steryl glucoside	Sitosteryl glucoside(16:2)	810.7	QTOF MS (product ion) of [M+NH4]+
Sterols and derivatives	Acylated steryl glucoside	Cholesteryl glucoside(18:2)	810.7	QTOF MS (product ion) of [M+NH4]+
Sterols and derivatives	Acylated steryl glucoside	Stigmasteryl glucoside(16:0)	812.7	QTOF MS (product ion) of [M+NH4]+

Molecular species identified by additional steps beyond the analyses listed in Table 13.1 (order of acyl chain names does not infer position)	Chemical formula of identified molecular species	References for identification	References for high-throughput analysis	Comments
	C62H116O24NP		[53]	
	C62H118O24NP		[53]	
	C49H80O7	[62]	[62]	
	C49H82O7	[62]	[62]	
	C49H84O7	[62]	[62]	
	C49H86O7	[62]	[62]	
	C50H82O7	[62]	[62]	
	C50H84O7	[62]	[62]	
	C50H86O7	[62]	[62, 63]	
	C50H88O7	[62]	[62, 63]	
	C51H82O7	[62]	[62]	
	C51H84O7	[62]	[62]	
	C51H84O7	[62]	[62]	
	C51H84O7	[62]	[62]	
	C51H86O7	[62]	[62, 63]	
	C51H86O7	[62]	[62]	
	C51H86O7	[62]	[62]	
	C51H88O7	[62]	[62, 63]	

(continued)

Table 2 (continued)

Group	Class	Abbreviation	Mass of uncharged compound	Additional steps for compound identification and analyses listed in Table 13.1
Sterols and derivatives	Acylated steryl glucoside	Sitosteryl glucoside(16:1)	812.7	QTOF MS (product ion) of [M+NH4]+
Sterols and derivatives	Acylated steryl glucoside	Cholesteryl glucoside(18:1)	812.7	QTOF MS (product ion) of [M+NH4]+
Sterols and derivatives	Acylated steryl glucoside	Sitosteryl glucoside(16:0)	814.7	QTOF MS (product ion) of [M+NH4]+
Sterols and derivatives	Acylated steryl glucoside	Cholesteryl glucoside(18:0)	814.7	QTOF MS (product ion) of [M+NH4]+
Sterols and derivatives	Acylated steryl glucoside	Campesteryl glucoside(18:3)	822.7	QTOF MS (product ion) of [M+NH4]+
Sterols and derivatives	Acylated steryl glucoside	Campesteryl glucoside(18:2)	824.7	QTOF MS (product ion) of [M+NH4]+
Sterols and derivatives	Acylated steryl glucoside	Campesteryl glucoside(18:1)	826.7	QTOF MS (product ion) of [M+NH4]+
Sterols and derivatives	Acylated steryl glucoside	Campesteryl glucoside(18:0)	828.7	QTOF MS (product ion) of [M+NH4]+
Sterols and derivatives	Acylated steryl glucoside	Stigmasteryl glucoside(18:3)	834.7	QTOF MS (product ion) of [M+NH4]+
Sterols and derivatives	Acylated steryl glucoside	Sitosteryl glucoside(18:3)	836.7	QTOF MS (product ion) of [M+NH4]+
Sterols and derivatives	Acylated steryl glucoside	Stigmasteryl glucoside(18:2)	836.7	QTOF MS (product ion) of [M+NH4]+
Sterols and derivatives	Acylated steryl glucoside	Cholesteryl glucoside(20:3)	836.7	QTOF MS (product ion) of [M+NH4]+
Sterols and derivatives	Acylated steryl glucoside	Sitosteryl glucoside(18:2)	838.7	QTOF MS (product ion) of [M+NH4]+
Sterols and derivatives	Acylated steryl glucoside	Stigmasteryl glucoside(18:1)	838.7	QTOF MS (product ion) of [M+NH4]+
Sterols and derivatives	Acylated steryl glucoside	Stigmasteryl glucoside(18:0)	840.7	QTOF MS (product ion) of [M+NH4]+
Sterols and derivatives	Acylated steryl glucoside	Sitosteryl glucoside(18:1)	840.7	QTOF MS (product ion) of [M+NH4]+
Sterols and derivatives	Acylated steryl glucoside	Cholesteryl glucoside(20:1)	840.7	QTOF MS (product ion) of [M+NH4]+
Sterols and derivatives	Acylated steryl glucoside	Sitosteryl glucoside(18:0)	842.7	QTOF MS (product ion) of [M+NH4]+

Molecular species identified by additional steps beyond the analyses listed in Table 13.1 (order of acyl chain names does not infer position)	Chemical formula of identified molecular species	References for identification	References for high-throughput analysis	Comments
	C51H88O7	[62]	[62, 63]	
	C51H88O7	[62]	[62]	
	C51H90O7	[62]	[62, 63]	
	C51H90O7	[62]	[62]	
	C52H86O7	[62]	[62, 63]	
	C52H88O7	[62]	[62, 63]	
	C52H90O7	[62]	[62, 63]	
	C52H92O7	[62]	[62, 63]	
	C53H86O7	[62]	[62, 63]	
	C53H88O7	[62]	[62, 63]	
	C53H88O7	[62]	[62, 63]	
	C53H88O7	[62]	[62]	
	C53H90O7	[62]	[62, 63]	
	C53H90O7	[62]	[62, 63]	
	C53H92O7	[62]	[62, 63]	
	C53H92O7	[62]	[62, 63]	
	C53H92O7	[62]	[62]	
	C53H94O7	[62]	[62, 63]	

(continued)

Table 2 (continued)

Group	Class	Abbreviation	Mass of uncharged compound	Additional steps for compound identification and analyses listed in Table 13.1
Sterols and derivatives	Acylated steryl glucoside	Campesteryl glucoside(20:1)	854.7	QTOF MS (product ion) of [M+NH4]+
Sterols and derivatives	Acylated steryl glucoside	Campesteryl glucoside(20:0)	856.7	QTOF MS (product ion) of [M+NH4]+
Sterols and derivatives	Acylated steryl glucoside	Stigmasteryl glucoside(20:3)	862.7	QTOF MS (product ion) of [M+NH4]+
Sterols and derivatives	Acylated steryl glucoside	Stigmasteryl glucoside(20:2)	864.7	QTOF MS (product ion) of [M+NH4]+
Sterols and derivatives	Acylated steryl glucoside	Sitosteryl glucoside(20:3)	864.7	QTOF MS (product ion) of [M+NH4]+
Sterols and derivatives	Acylated steryl glucoside	Stigmasteryl glucoside(20:1)	866.7	QTOF MS (product ion) of [M+NH4]+
Sterols and derivatives	Acylated steryl glucoside	Sitosteryl glucoside(20:2)	866.7	QTOF MS (product ion) of [M+NH4]+
Sterols and derivatives	Acylated steryl glucoside	Stigmasteryl glucoside(20:0)	868.7	QTOF MS (product ion) of [M+NH4]+
Sterols and derivatives	Acylated steryl glucoside	Sitosteryl glucoside(20:1)	868.7	QTOF MS (product ion) of [M+NH4]+
Sterols and derivatives	Acylated steryl glucoside	Cholesteryl glucoside(22:1)	868.7	QTOF MS (product ion) of [M+NH4]+
Sterols and derivatives	Acylated steryl glucoside	Sitosteryl glucoside(20:0)	870.8	QTOF MS (product ion) of [M+NH4]+
Sterols and derivatives	Acylated steryl glucoside	Campesteryl glucoside(22:3)	878.6	QTOF MS (product ion) of [M+NH4]+
Sterols and derivatives	Acylated steryl glucoside	Campesteryl glucoside(22:2)	880.7	QTOF MS (product ion) of [M+NH4]+

Molecular species identified by additional steps beyond the analyses listed in Table 13.1 (order of acyl chain names does not infer position)	Chemical formula of identified molecular species	References for identification	References for high-throughput analysis	Comments
	$C54H94O7$	[62]	[62]	
	$C54H96O7$	[62]	[62]	
	$C55H90O7$	[62]	[62]	
	$C55H92O7$	[62]	[62]	
	$C55H92O7$	[62]	[62]	
	$C55H94O7$	[62]	[62]	
	$C55H94O7$	[62]	[62]	
	$C55H96O7$	[62]	[62]	
	$C55H96O7$	[62]	[62]	
	$C55H96O7$	[62]	[62]	
	$C55H98O7$	[62]	[62]	
	$C56H94O7$	[62]	[62]	
	$C56H96O7$	[62]	[62]	

(continued)

Table 2 (continued)

Group	Class	Abbreviation	Mass of uncharged compound	Additional steps for compound identification and analyses listed in Table 13.1
Sterols and derivatives	Acylated steryl glucoside	Campesteryl glucoside(22:0)	884.8	QTOF MS (product ion) of [M+NH4]+
Sterols and derivatives	Acylated steryl glucoside	Stigmasteryl glucoside(22:3)	890.6	QTOF MS (product ion) of [M+NH4]+
Sterols and derivatives	Acylated steryl glucoside	Sitosteryl glucoside(22:3)	892.6	QTOF MS (product ion) of [M+NH4]+
Sterols and derivatives	Acylated steryl glucoside	Stigmasteryl glucoside(22:2)	892.7	QTOF MS (product ion) of [M+NH4]+
Sterols and derivatives	Acylated steryl glucoside	Stigmasteryl glucoside(22:1)	894.8	QTOF MS (product ion) of [M+NH4]+
Sterols and derivatives	Acylated steryl glucoside	Sitosteryl glucoside(22:2)	894.8	QTOF MS (product ion) of [M+NH4]+
Sterols and derivatives	Acylated steryl glucoside	Stigmasteryl glucoside(22:0)	896.8	QTOF MS (product ion) of [M+NH4]+
Sterols and derivatives	Acylated steryl glucoside	Sitosteryl glucoside(22:1)	896.8	QTOF MS (product ion) of [M+NH4]+
Sterols and derivatives	Acylated steryl glucoside	Sitosteryl glucoside(22:0)	898.8	QTOF MS (product ion) of [M+NH4]+
Sterols and derivatives	Sterol	Cholesterol	386.4	QTOF MS (product ion) of [M+betainyl]+
Sterols and derivatives	Sterol	Campesterol	400.4	QTOF MS (product ion) of [M+betainyl]+

Molecular species identified by additional steps beyond the analyses listed in Table 13.1 (order of acyl chain names does not infer position)	Chemical formula of identified molecular species	References for identification	References for high-throughput analysis	Comments
	$C_{56}H_{100}O_7$	[62]	[62]	
	$C_{57}H_{94}O_7$	[62]	[62]	
	$C_{57}H_{96}O_7$	[62]	[62]	
	$C_{57}H_{96}O_7$	[62]	[62]	
	$C_{57}H_{98}O_7$	[62]	[62]	
	$C_{57}H_{98}O_7$	[62]	[62]	
	$C_{57}H_{100}O_7$	[62]	[62]	
	$C_{57}H_{100}O_7$	[62]	[62]	
	$C_{57}H_{102}O_7$	[62]	[62]	
	$C_{27}H_{46}O$	[62]	[62]	The free sterols were betainylated with N-chlorobetainyl chloride prior to analysis
	$C_{28}H_{48}O$	[62]	[62]	The free sterols were betainylated with N-chlorobetainyl chloride prior to analysis

(continued)

Table 2 (continued)

Group	Class	Abbreviation	Mass of uncharged compound	Additional steps for compound identification and analyses listed in Table 13.1
Sterols and derivatives	Sterol	Stigmasterol	412.4	QTOF MS (product ion) of [M + betainyl]+
Sterols and derivatives	Sterol	Sitosterol	414.4	QTOF MS (product ion) of [M + betainyl]+
Sterols and derivatives	Steryl glucoside	Cholesteryl glucoside	548.4	QTOF MS (product ion) of [M + NH4]+
Sterols and derivatives	Steryl glucoside	Campesteryl glucoside	562.4	QTOF MS (product ion) of [M + NH4]+
Sterols and derivatives	Steryl glucoside	Stigmasteryl glucoside	574.4	QTOF MS (product ion) of [M + NH4]+
Sterols and derivatives	Steryl glucoside	Sitosteryl glucoside	576.4	QTOF MS (product ion) of [M + NH4]+
Sulfolipid	Sulfoquinovosyldiacylglycerol	SQDG(32:0)	794.5	
Sulfolipid	Sulfoquinovosyldiacylglycerol	SQDG(34:6)	810.5	Product ion of [M − H]−
Sulfolipid	Sulfoquinovosyldiacylglycerol	SQDG(34:5)	812.5	Product ion of [M − H]−
Sulfolipid	Sulfoquinovosyldiacylglycerol	SQDG(34:4)	814.5	Product ion of [M − H]−
Sulfolipid	Sulfoquinovosyldiacylglycerol	SQDG(34:3)a	816.5	Product ion of [M − H]−
Sulfolipid	Sulfoquinovosyldiacylglycerol	SQDG(34:3)b	816.5	Product ion of [M − H]−

Molecular species identified by additional steps beyond the analyses listed in Table 13.1 (order of acyl chain names does not infer position)	Chemical formula of identified molecular species	References for identification	References for high-throughput analysis	Comments
	C29H48O	[62]	[62]	The free sterols were betainylated with N-chlorobetainyl chloride prior to analysis
	C29H50O	[62]	[62]	The free sterols were betainylated with N-chlorobetainyl chloride prior to analysis
	C33H56O6	[62]	[62]	
	C34H58O6	[62]	[62, 63]	
	C35H58O6	[62]	[62, 63]	
	C35H60O6	[62]	[62, 63]	
	C41H78O12S		[50, 77]	
18:3-16:3	C43H70O12S	[52]	[46]	
16:2-18:3	C43H72O12S	[52]	[46]	
18:3-16:1	C43H74O12S	[52]		
18:3-16:0	C43H76O12S	[52]	[50, 46, 77]	Signals from the component compounds of a species labeled with letters (e.g., a or b) in the Abbreviations column are combined when a head group-specific scan is employed in direct-infusion mode
18:2-16:1	C43H76O12S	[46]	[58, 77]	

(continued)

Group	Class	Abbreviation	Mass of uncharged compound	Additional steps for compound identification and analyses listed in Table 13.1
Sulfolipid	Sulfoquinovosyldiacylglycerol	SQDG(34:2)	818.5	Product ion of [M−H]−
Sulfolipid	Sulfoquinovosyldiacylglycerol	SQDG(34:1)a	820.5	Product ion of [M−H]−
Sulfolipid	Sulfoquinovosyldiacylglycerol	SQDG(34:1)b	820.5	Product ion of [M−H]−
Sulfolipid	Sulfoquinovosyldiacylglycerol	SQDG(34:0)	822.5	Product ion of [M−H]−
Sulfolipid	Sulfoquinovosyldiacylglycerol	SQDG(36:6)	838.5	Product ion of [M−H]−
Sulfolipid	Sulfoquinovosyldiacylglycerol	SQDG(36:5)	840.5	Product ion of [M−H]−
Sulfolipid	Sulfoquinovosyldiacylglycerol	SQDG(36:4)a	842.5	Product ion of [M−H]−
Sulfolipid	Sulfoquinovosyldiacylglycerol	SQDG(36:4)b	842.5	Product ion of [M−H]−
Sulfolipid	Sulfoquinovosyldiacylglycerol	SQDG(36:3)a	844.5	Product ion of [M−H]−
Sulfolipid	Sulfoquinovosyldiacylglycerol	SQDG(36:3)b	844.5	Product ion of [M−H]−
Sulfolipid	Sulfoquinovosyldiacylglycerol	SQDG(36:2)	846.6	Product ion of [M−H]−

Molecular species identified by additional steps beyond the analyses listed in Table 13.1 (order of acyl chain names does not infer position)	Chemical formula of identified molecular species	References for identification	References for high-throughput analysis	Comments
16:0-18:2	C43H78O12S	[52]	[58, 77]	
16:0-18:1	C43H80O12S	[52]	[58]	
16:1-18:0	C43H80O12S	[46]	[58]	
18:0-16:0	C43H78O12S	[46]	[58]	
18:3-18:3	C45H74O12S	[52]	[50, 46, 77]	
18:3-18:2	C45H76O12S	[52, 46]	[46]	
18:2-18:2	C45H78O12S	[52]	[46]	
18:1-18:3	C45H78O12S	[52]	[46]	Signals from the component compounds of a species labeled with letters (e.g., a or b) in the Abbreviations column are combined when a head group-specific scan is employed in direct-infusion mode
18:2-18:1	C45H80O12S	[52, 46]	[46]	
18:3-18:0	C45H80O12S	[52]	[46]	Signals from the component compounds of a species labeled with letters (e.g., a or b) in the Abbreviations column are combined when a head group-specific scan is employed in direct-infusion mode
18:0-18:2	C45H82O12S	[52]		

the lipid analysis by incubation with methylamine (53). For analysis, a reversed-phase LC coupled with an ESI-QqQ mass spectrometer operating in MRM mode was used. The LC provides separation of isobaric and near-isobaric compounds and simplifies isotopic deconvolution, providing data at the molecular species level. The detection modes are indicated in Table 1, and the sphingolipid molecular species detected in *Arabidopsis* are indicated in Table 2. Quantification was in comparison to the signal for a mammalian sphingolipid, G_{M1} (53). Markham and coworkers have utilized sphingolipid profiling to identify the function of several genes involved in sphingolipid metabolism (55–61).

1.3. Sterols and Derivatives

Recently Wewer et al. (62) have used a direct infusion nanospray ionization-QTOF MS method to quantify sterols, sterol glucosides, acyl sterol glucosides, and sterol esters (Table 1). Except for the esters, which are unlikely to be membrane lipids, the detected lipids are listed in Table 2. Wewer and coworkers used a pre-analytical solid-phase extraction (SPE) step; the separation of the sterols and derivatives from more polar lipids increased ionization efficiency compared to that achievable by direct infusion of the total lipid extract (62). Derivatizing free sterols with *N*-chlorobetainyl chloride prior to the analysis provided for good ionization compared to underivatized free sterols, which ionize quite poorly (62, 63). Schrick and coworkers also have analyzed steryl glucosides and acyl steryl glucosides using direct infusion with ESI-QqQ MS (Table 1). Both groups used the new methods to examine variations in *Arabidopsis* sterol metabolism (62, 63).

1.4. Lipids with Oxidized Acyl Chains

Membrane lipids containing oxidized acyl chains are drawing interest due to their possible roles in stress responses. Up until recently, only a handful of oxidized lipid species had been profiled (23, 64–70). Very recently, a direct-infusion ESI-QqQ MS approach was utilized to quantify a larger number of membrane lipids containing oxidized fatty acyl chains (71). Precursor scanning (in negative mode) for three oxidized fatty acyl anions was used to quantify membrane lipids with oxidized fatty acyl chains in an unfractionated lipid extract. Quantification was in comparison to a saturated MGDG species (18:0/16:0). Chemical formulas of acyl chains in target lipids were identified by ESI-QTOF MS and Fourier transform ion cyclotron resonance MS. Lipids containing oxidized acyl chains in phosphatidylcholine, phosphatidylethanolamine, phosphatidylglycerol, digalactosyldiacylglycerol, monogalactosyldiacylglycerol, and acylated monogalactosyldiacylglycerol are listed in Table 2.

1.5. Lipid A Pathway Intermediates

Very interestingly, Li and coworkers (72) have recently employed LC ESI-QqQ MS to measure compounds in the pathway to Lipid A, a hexa-acylated disaccharide of glucosamine. This pathway, previously described in bacteria, was unknown in plants, although genes that encode enzymes related to those in the bacterial Lipid A pathway are present in *Arabidopsis*. Normal-phase LC was used for separation in the quantitative analysis and detection of the Lipid A pathway intermediates was in the MRM mode (Table 1). Quantification was in comparison to an abundant endogenous lipid, PE (16:0/18:2). To confirm the structure of the Lipid A-related compounds (Table 2), QTOF MS product ion scanning was used (72).

2. Overview

Reliable analyses are now available for many lipids. However, for many applications, identifying and quantifying more lipids from a single sample is a remaining goal. Following the Bligh and Dyer (47) extraction with the isopropanol/hexane/water extraction described by Markham et al. (53) extracts more compounds more fully than either method alone (Welti lab, unpublished; combined extraction protocol described in (9)). Expanding analyses to greater numbers of lipids requires maintenance of specificity and sensitivity. Building on the results of Wewer et al. (62), perhaps optimization of a pre-separation approach might provide a compromise retaining the benefits of chromatography (specificity) and direct infusion (sensitivity via ability to adjust analysis times). It is likely that many more *Arabidopsis* lipid species will be discovered, providing new challenges and opportunities for continued method development.

Supplementary material

This chapter contains a supplementary material which can be found at the publisher's website (http://extras.springer.com).

References

1. Welti R, Wang X (2004) Lipid species profiling: a high-throughput approach to identify lipid compositional changes and determine the function of genes involved in lipid metabolism and signaling. Curr Opin Plant Biol 7:337–344
2. Welti R, Shah J, LeVine S et al (2005) High-throughput lipid profiling to identify and characterize genes involved in lipid metabolism, signaling, and stress response. In: Feng L, Prestwich GD (eds) Functional lipidomics. Marcel Dekker, New York
3. Welti R, Roth MR, Deng Y et al (2007) Lipidomics: ESI MS/MS-based profiling to determine the function of genes involved in metabolism of complex lipids. In: Nikolau B (ed) Plant metabolomics. Springer, Dordrecht

4. Isaac G, Jeannotte R, Esch SW et al (2007) New mass-spectrometry-based strategies for lipids. Gen Eng Rev 28:129–157
5. Welti R, Shah J, Li W et al (2007) Plant lipidomics: discerning biological function by profiling plant complex lipids using mass spectrometry. Front Biosci 12:2494–2506
6. Welti R (2010) Plant lipidomics. In: AOCS lipid library. http://lipidlibrary.aocs.org/plantbio/plantlipidomics/index.htm
7. Li-Beisson Y, Shorrosh B, Beisson F et al (2010) Acyl lipid metabolism. Arabidopsis Book 8:1–65. http://aralip.plantbiology.msu.edu/data/tab_methods.pdf
8. Welti R, Li W, Li M et al (2002) Profiling membrane lipids in plant stress responses. Role of phospholipase D alpha in freezing-induced lipid changes in *Arabidopsis*. J Biol Chem 277:31994–32002
9. Shiva S, Vu HS, Roth MR et al Lipidomic analysis of plant membrane lipids by direct infusion tandem mass spectrometry. In: Munnik T, Heilmann I (eds) Plant lipid signaling protocols, methods in molecular biology. Humana Press, New York (in press)
10. Nandi A, Krothapalli K, Buseman C et al (2003) The *Arabidopsis thaliana* sfd mutants affect plastidic lipid composition and suppress dwarfing, cell death and the enhanced disease resistance phenotypes resulting from the deficiency of a fatty acid desaturase. Plant Cell 15:2383–2398
11. Zhang W, Wang C, Qin C et al (2003) The oleate-stimulated phospholipase D, PLDδ, and phosphatidic acid decrease H_2O_2-induced cell death in *Arabidopsis*. Plant Cell 15: 2285–2295
12. Nandi A, Welti R, Shah J (2004) The *Arabidopsis thaliana* dihydroxyacetone phosphate reductase gene *suppressor of fatty acid desaturase deficiency1* is required for glycerolipid metabolism and for the activation of systemic acquired resistance. Plant Cell 16:465–477
13. Li W, Li M, Zhang W et al (2004) The plasma membrane-bound phospholipase Dδ enhances freezing tolerance in *Arabidopsis thaliana*. Nat Biotechnol 22:427–433
14. Li M, Zhang W, Welti R et al (2006) Double knockouts of phospholipase Dζ1 and ζ2 in *Arabidopsis* affect root elongation under phosphate limitation, but do not affect root hair patterning. Plant Physiol 140:761–770
15. Li M, Welti R, Wang X (2006) Quantitative profiling of *Arabidopsis* polar glycerolipids in response to phosphorus starvation: roles of PLDζ1 and PLDζ2 in phosphatidylcholine hydrolysis and digalactosyldiacylglycerol accumulation in phosphorus-starved plants. Plant Physiol 142:750–761
16. Cruz-Ramirez A, Oropeza-Aburto A, Razo-Hernandez F et al (2006) Phospholipase DZ2 plays an important role in extraplastidic galactolipid biosynthesis and phosphate recycling in *Arabidopsis* roots. Proc Natl Acad Sci USA 103: 6765–6770
17. Chen J, Burke JJ, Xin Z et al (2006) Characterization of the *Arabidopsis* thermosensitive mutant *atts02* reveals an important role for galactolipids in thermotolerance. Plant Cell Environ 29:1437–1448
18. Fritz M, Lokstein H, Hackenberg D et al (2007) Chanelling of eukaryotic diacylglycerol into the biosynthesis of plastidial phosphatidylglycerol. J Biol Chem 282:4613–4625
19. Yang W, Devaiah S, Pan X et al (2007) AtPLAI is an LRR-containing acyl hydrolase involved in basal jasmonic acid product ion and *Arabidopsis* resistance to *Botrytis cinerea*. J Biol Chem 282:18116–18128
20. Devaiah S, Pan X, Hong Y et al (2007) Enhancing seed quality and viability by suppressing phospholipase D in *Arabidopsis*. Plant J 50:950–957
21. Kachroo A, Shanklin J, Whittle E et al (2007) The *Arabidopsis* stearoyl-acyl carrier protein-desaturase family and the contribution of leaf isoforms to oleic acid synthesis. Plant Mol Biol 63:257–271
22. Li W, Wang R, Li M et al (2008) Differential degradation of extraplastidic and plastidic lipids during freezing and post-freezing recovery in *Arabidopsis thaliana*. J Biol Chem 283:461–468
23. Maeda H, Sage TL, Isaac G et al (2008) Tocopherols modulate extra-plastidic polyunsaturated fatty acid metabolism in *Arabidopsis* at low temperature. Plant Cell 20:452–470
24. Hong Y, Pan X, Welti R et al (2008) Alterations of phospholipase Dα3 change *Arabidopsis* response to salinity and water deficits. Plant Cell 20:803–816
25. Hong Y, Pan X, Welti R et al (2008) The effect of phospholipase Dα3 on *Arabidopsis* response to hyperosmotic stress and glucose. Plant Signal Behav 3:1099–1100
26. Xiao S, Li HY, Zhang JP et al (2008) *Arabidopsis* acyl-CoA-binding proteins ACBP4 and ACBP5 are subcellularly localized to the cytosol and ACBP4 depletion affects membrane lipid composition. Plant Mol Biol 68:571–583
27. Chen M, Markham JE, Dietrich CR et al (2008) Sphingolipid long-chain base hydroxylation is important for growth and regulation of sphingolipid content and composition in *Arabidopsis*. Plant Cell 20:1862–1878

28. Chen Q-F, Shi X, Chye M-L (2008) Overexpression of the *Arabidopsis* 10-Kilodalton acyl-coenzyme A-binding protein ACBP6 enhances freezing tolerance. Plant Physiol 148: 304–315
29. Hong Y, Devaiah SP, Bahn S et al (2009) Phospholipase Dε and phosphatidic acid enhance *Arabidopsis* growth. Plant J 58:376–387
30. Zhang Y, Zhu H, Zhang Q et al (2009) Phospholipase Dα1 and phosphatidic acid regulate NADPH oxidase activity and production of reactive oxygen species in ABA-mediated stomatal closure in *Arabidopsis*. Plant Cell 21: 2357–2377
31. Zhang Q, Fry J, Rajashekar C et al (2009) Membrane polar lipid changes in zoysiagrass rhizomes and their potential role in freezing tolerance. J Am Soc Hort Sci 134:322–328
32. Xia Y, Gao Q, Yu K et al (2009) An intact cuticle in distal tissues is essential for the induction of systemic acquired resistance in plants. Cell Host Microbe 5:151–165
33. Reina-Pinto J, Voisin D, Kurdyukov S et al (2009) Misexpression of *fatty acid elongation1* in the *Arabidopsis* epidermis induces cell death and suggests a critical role for phospholipase A2 in this process. Plant Cell 21:1252–1272
34. Keogh M, Courtney PD, Kinney AJ et al (2009) Functional characterization of phospholipid N-methyltransferases from *Arabidopsis* and soybean. J Biol Chem 284:15439–15447
35. Kirik A, Mudgett MB (2009) SOBER1 phospholipase activity suppresses phosphatidic acid accumulation and plant immunity in response to bacterial effector. Proc Natl Acad Sci USA 106:20532–20537
36. Bais P, Moon SM, He K et al (2010) PlantMetabolomics.org: a web portal for plant metabolomics experiments. Plant Physiol 152:1807–1816
37. Peters C, Li M, Narasimhan R et al (2010) Nonspecific phospholipase C NPC4 promotes responses to abscisic acid and tolerance to hyperosmotic stress in *Arabidopsis*. Plant Cell 22:2642–2659
38. Yu L, Nie J, Cao C et al (2010) Phosphatidic acid mediates salt stress response by regulation of MPK6 in *Arabidopsis thaliana*. New Phytol 188:762–773
39. Shen W, Li JQ, Dauk M et al (2010) Metabolic and transcriptional responses of glycerolipid pathways to a perturbation of glycerol 3-phosphate metabolism in *Arabidopsis*. J Biol Chem 285:22957–22965
40. Du Z-Y, Xiao S, Chen QF et al (2010) Depletion of the membrane-associated acyl-coenzyme A-binding protein ACBP1 enhances the ability of cold acclimation in *Arabidopsis*. Plant Physiol 152:1585–1597
41. Chen H, Xiong L (2010) *myo*-Inositol-1-phosphate synthase is required for polar auxin transport and organ development. J Biol Chem 285:24238–24247
42. Chen M, Thelen JJ (2010) The plastid isoform of triose phosphate isomerase is required for the postgerminative transition from heterotrophic to autotrophic growth in *Arabidopsis*. Plant Cell 22:77–90
43. Chen QF, Xian S, Qi W et al (2010) The *Arabidopsis* acbp1acbp2 double mutant lacking acyl-CoA-binding proteins ACBP1 and ACBP2 is embryo lethal. New Phytol 186:843–855
44. Kim H, Vijayan P, Carlsson AS et al (2010) A mutation in the *LPAT1* gene suppresses the sensitivity of *fab1* plants to low temperature. Plant Physiol 153:1135–1143
45. Li M, Bahn SC, Guo L et al (2011) Alterations of patatin-related phospholipase pPLAIIIβ reveal effects of membrane lipid metabolism on cellulose content and anisotropic cell expansion in *Arabidopsis*. Plant Cell 23:1107–1123
46. Burgos A, Szymanski J, Seiwert B et al (2011) Analysis of short-term changes in the *Arabidopsis thaliana* glycerolipidome in response to temperature and light. Plant J 66:656–666
47. Bligh EG, Dyer WJ (1959) A rapid method of total lipid extraction and purification. Can J Biochem Physiol 37:911–917
48. Brügger B, Erben G, Sandhoff R et al (1997) Quantitative analysis of biological membrane lipids at the low picomole level by nano-electrospray ionization tandem mass spectrometry. Proc Natl Acad Sci USA 94:2339–2344
49. Xiao S, Gao W, Chen Q-F et al (2010) Overexpression of *Arabidopsis* acyl-CoA binding protein ACBP3 promotes starvation-induced and age-dependent leaf senescence. Plant Cell 22:1463–1482
50. Welti R, Wang X, Williams TD (2003) Electrospray ionization tandem mass spectrometry scan modes for plant chloroplast lipids. Anal Biochem 314:149–152
51. Zhou Z, Marepally SR, Nune DS et al (2011) LipidomeDB data calculation environment: online processing of direct-infusion mass spectral data for lipid profiles. Lipids 46:879–884
52. Devaiah SP, Roth MR, Baughman E et al (2006) Quantitative profiling of polar glycerolipid species and the role of phospholipase Dα1 in defining the lipid species in *Arabidopsis* tissues. Phytochemistry 67:1907–1924
53. Markham JE, Jaworski JG (2007) Rapid measurement of sphingolipids from *Arabidopsis thaliana* by reversed-phase high-performance

liquid chromatography coupled to electrospray ionization tandem mass spectrometry. Rapid Commun Mass Spectrom 21:1304–1314

54. Markham J, Li J, Cahoon EB et al (2006) Separation and identification of major plant sphingolipid classes from leaves. J Biol Chem 281:22684–22694
55. Chao D-Y, Gable K, Chen M et al (2011) Sphingolipids in the root play an important role in regulating the leaf ionome in *Arabidopsis thaliana*. Plant Cell 23:1061–1081
56. Chen M, Markham JE, Dietrich CR et al (2008) Sphingolipid long-chain base hydroxylation is important for growth and regulation of sphingolipid content and composition in *Arabidopsis*. Plant Cell 20:1862–1878
57. Tsegaye Y, Richardson CG, Bravo JE et al (2007) *Arabidopsis* mutants lacking long chain base phosphate lyase are fumonisin-sensitive and accumulate trihydroxy-18:1 long chain base phosphate. J Biol Chem 282:28195–28206
58. Chen M, Markham JE, Cahoon EB (2011) Sphingolipid Δ8 unsaturation is important for glucosylceramide biosynthesis and low temperature performance in *Arabidopsis*. Plant J. doi:10.1111/j.1365-313X.2011.04829.x
59. Saucedo-García M, Guevara-García A, González-Solís A et al (2011) MPK6, sphinganine and the LCB2a gene from serine palmitoyltransferase are required in the signaling pathway that mediates cell death induced by long chain bases in *Arabidopsis*. New Phytol 191:943–957
60. Markham JE, Molino D, Gissot L et al (2011) Sphingolipids containing very-long-chain fatty acids define a secretory pathway for specific polar plasma membrane protein targeting in *Arabidopsis*. Plant Cell 23:2362–2378
61. Roudier F, Gissot L, Beaudoin F et al (2010) Very-long-chain fatty acids are involved in polar auxin transport and developmental patterning in *Arabidopsis*. Plant Cell 22:364–375
62. Wewer V, Dombrick I, vom Dorp K et al (2011) Quantification of sterol lipids in plants by quadrupole time-of-flight mass spectrometry. J Lipid Res 52:1039–1054
63. Schrick K, Shiva S, Arpin J et al (2011) Steryl glucoside and acyl steryl glucoside analysis of *Arabidopsis* seeds by electrospray ionization tandem mass spectrometry. Lipids. doi:10.1007/s11745-011-3602-9
64. Buseman C, Tamura P, Sparks A et al (2006) Wounding stimulates the accumulation of glycerolipids containing oxophytodienoic acid and dinor-oxophytodienoic acid in *Arabidopsis* leaves. Plant Physiol 142:28–39
65. Andersson MX, Hamberg M, Kourtchenko O et al (2006) Oxylipin profiling of the hypersensitive response in *Arabidopsis thaliana*. Formation of a novel oxo-phytodienoic acid-containing galactolipid Arabidopside E. J Biol Chem 281:31528–31537
66. Glauser G, Grata E, Rudaz S et al (2008) High-resolution profiling of oxylipin containing galactolipids in *Arabidopsis* extracts by ultraperformance liquid chromatography/time-of-flight mass spectrometry. Rapid Commun Mass Spectrom 22:3154–3160
67. Hisamatsu Y, Goto N, Hasegawa K et al (2003) *Arabidopsides* A and B, two new oxylipins from *Arabidopsis thaliana*. Tetrahedron Lett 44: 5553–5556
68. Hisamatsu Y, Goto N, Sekiguchi M et al (2005) Oxylipins arabidopsides C and D from *Arabidopsis thaliana*. J Nat Prod 68:600–603
69. Kourtchenko O, Andersson MX, Hamberg M et al (2007) Oxo-phytodienoic acid containing galactolipids in *Arabidopsis*: Jasmonate signaling dependence. Plant Physiol 145:1658–1669
70. Stelmach BA, Muller A, Hennig P et al (2001) A novel class of oxylipins, *sn*1-*O*-(12-oxophytodienoyl)-*sn*2-*O*-(hexadecatrienoyl)-monogalactosyl diglyceride, from *Arabidopsis thaliana*. J Biol Chem 276:12832–12838
71. Vu HS, Tamura P, Galeva NA et al (2012) Direct infusion mass spectrometry of oxylipin-containing *Arabidopsis thaliana* membrane lipids reveals varied patterns in different stress responses. Plant Physiol 158:324–339
72. Li C, Guan Z, Liu D et al (2011) Pathway for lipid A biosynthesis in *Arabidopsis thaliana* resembling that of *Escherichia coli*. Proc Natl Acad Sci USA 108:11387–11392
73. Kim YH, Choi J-S, Yoo JS et al (1999) Structural identification of glycerolipid molecular species isolated from *Cyanobacterium synechocystis* sp. PCC 6803 using fast atom bombardment tandem mass spectrometry. Anal Biochem 267:260–270
74. Yang W, Zheng Y, Bahn SC et al (2012) The patatin-containing phospholipase A pPLAIIα modulates oxylipin formation and water loss in *Arabidopsis thaliana*. Mol Plant 5:452–460
75. Hsu FF, Turk J, Williams TD et al (2007) Electrospray ionization multiple stage quadrupole ion-trap and tandem quadrupole mass spectrometric studies on phosphatidylglycerol from *Arabidopsis* leaves. J Am Soc Mass Spectrom 18:783–790
76. Nakajyo H, Hisamatsu Y, Sekiguchi M et al (2006) Arabidopside F, a new oxylipin from *Arabidopsis thaliana*. Heterocycles 69:295–301
77. Okazaki Y, Shimojima M, Sawada Y et al (2009) A chloroplastic UDP-glucose pyrophosphorylase from *Arabidopsis* is the committed enzyme for the first step of sulfolipid biosynthesis. Plant Cell 21:892–909

Chapter 14

Inductively Coupled Plasma–Mass Spectrometry as a Tool for High-Throughput Analysis of Plants

Javier Seravalli

Abstract

Ionomics is the study of the elemental composition of biological tissues. It complements knowledge acquired by metabolomics, proteomics, bioinformatics, and genomics in elucidating the physiological status of plants as well as the identification of genes involved in the transport and metabolism of individual elements and their interactions. Inductively coupled plasma–mass spectrometry (ICP–MS) technology provides a very sensitive method for the medium- and high-throughput elemental analysis of plant tissues. This chapter introduces the plant physiologist to the ICP–MS technique, a method for sample preparation and analysis.

Key words: ICP–MS, Elemental analysis, High-throughput, Isotope, Mass spectrometry, Ionomics, Plant tissue analysis

1. Introduction

Knowing the elemental composition and the accumulation of trace elements in plants is important for several reasons: (a) the accumulation of toxic elements by plants plays an important role in animal/human nutrition and the role of the environment in agriculture; (b) it enables the evaluation of the nutritional content and growth conditions, which can lead to trace-element fortification of edible crops; (c) it allows the study of the physiological status of plants based on provenance and genomic composition (ionomics). While the use of elemental analysis in toxicology and agriculture is well documented, the analysis of whole ion content and its mapping to localized regions of the genome is very recent. This chapter will introduce the reader to the inductively coupled plasma–mass spectrometry (ICP–MS) methodology and to its application in the analysis of plant tissues and the potential for identifying genes involve in metal homeostasis. First, the concept of ionomics and its

Jennifer Normanly (ed.), *High-Throughput Phenotyping in Plants: Methods and Protocols*, Methods in Molecular Biology, vol. 918, DOI 10.1007/978-1-61779-995-2_14, © Springer Science+Business Media, LLC 2012

applications in plant research will be introduced, followed by an explanation of the principle of operation of an ICP–MS. Then the methods for preparation of samples for sample tissues, yeast and bacterial samples and calibrations will be described in detail.

1.1. Ionomics of Plants

The ionome is the total inventory of elements within a cell, a tissue, or an organism (1–3). The genome supports the functions of living organisms through the combination of the proteome, transcriptome, metabolome, and ionome (4). The ionome overlaps with the metabolome and proteome in that many small molecules and proteins contain metal cofactors. Also, phosphorus is a common element to all the four "omes." Just like the proteome, metabolome, and transcriptome, the ionome responds to environmental cues and stressors. The ionome affects the metabolome and transcriptome through the induction of genes and enzyme activities, which lead to metal uptake, storage, utilization, and excretion. The opposite also holds, as signaling pathways can lead to changes in a particular element concentration. There are also interdependencies between the levels of different elements that compose the ionome. A similar concept is the metallome, which is the inventory of all the metals, a subset of the ionome. The field of ionomics (and metallomics) deals with metal homeostasis, metal utilization, the mechanism of catalysis by metallo enzymes, the role of metals in disease, the structural characteristics of metal binding to biological molecules, metal-based bio-imaging (5), metal-based affinity tagging and purification (MeCat, (6)), the mechanism of action of metal-containing drugs, and metal exchange among organisms within a given ecosystem. This area of investigation is quite new and still expanding, which carries important implications for the development of crops. Since elemental analysis is by definition a multivariate analytical technique, just like nuclear magnetic resonance, mass spectrometry, and chromatography it is amenable to analysis by chemometric multivariate statistical methods such as partial least squares and principal component analysis (7). These methods aim to reduce the number of variables (in the case of ionomics the concentration of individual elements) in order to better visualize the data, uncover relationships and hidden variables, and identify patterns and correlations between multivariate changes. The variations in a single element or combination of them can be studied as a function of the stage of growth of the plant, the soil composition, the provenance of the plant, the genes being transcribed, etc. Another methodology is the use of recombinant inbred lines (RIL), which allow the identification of quantitative trait loci (QTLs) that control the levels of some elements (8).

1.2. The Instrument

ICP–MS is an elemental analysis technique that uses MS to separate and analyze the elemental content of a sample (9, 10). ICP is a gas of ions generated and maintained by a rapidly oscillating

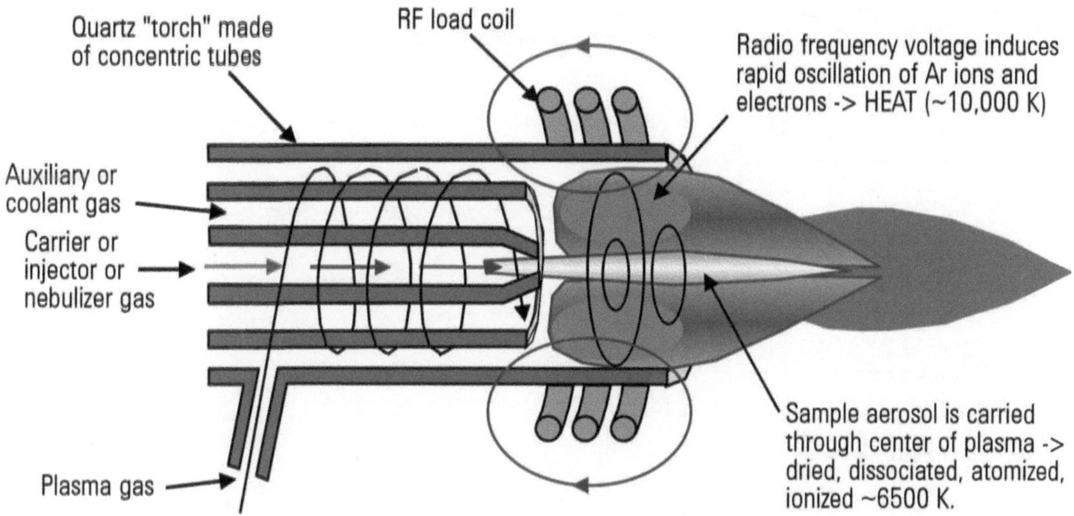

Fig. 1. Diagram of the Fassel-type of torch operating in an ICP instrument. There are three points of Ar entry; the plasma, the sample, and the auxiliary gas. The cross section of the RF coil shows that the oscillating magnetic field causes the dissociation and ionization of the sample atoms within the center of the plasma. Figure courtesy of Agilent Technologies.

magnetic field applied to a cylindrical flow of argon (Ar) gas (11, 12). Typically Ar is fed through a torch made out of quartz (Fig. 1). The torch contains three concentric cylinders, all of which carry Ar gas at separate flow rates. The main plasma flow is usually 16–18 L of Ar per minute and circulates on the outer ring. The sample flow is 0.5–1.5 L/min of Ar and the auxiliary or makeup flow is 0.1–0.5 L/min. The plasma flow is the gas source that provides for the generation of Ar ions, i.e., the plasma itself, while the sample gas pushes the fine aerosol mist into the radio frequency (RF) coil that keeps the plasma running. The auxiliary gas flow has the role of focusing the sample plus Ar gas at the center of the plasma. Ignition is accomplished by a Tesla coil, which provides electrons to the hollow copper coil which maintains the plasma (Fig. 1). After ignition, the rapidly oscillating magnetic field causes the Ar atoms to ionize by means of a loss of one electron ($Ar \rightarrow Ar^{+} + e^{-}$). The frequency of the oscillation is usually 27 MHz (RF range), which prevents the immediate recombination of Ar^{+} ions and electrons, and therefore maintains the plasma as long as the flow of gas continues and the RF is maintained at a constant power. Given the high resulting kinetic energy from the RF radiation at powers between 900 and 1,500 W, typical temperatures of 7,000–10,000 K are achieved in a standard commercial ICP instrument. The high temperatures are advantageous, as chemical species will completely decompose into atoms upon passage through the plasma (Fig. 2). The Ar^{+} ions act then as primary ionizers of the sample ions by collision/electron transfer reactions, thus producing Ar and M^{+} ions. Ions of higher charge can also be present or produced at the plasma,

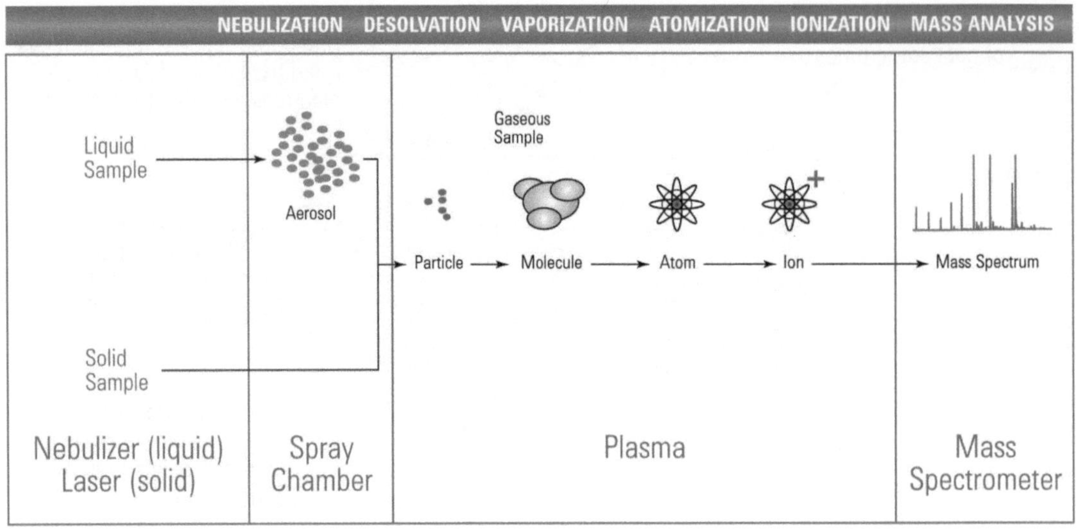

Fig. 2. Physical processes taking place within the plume of the plasma. First the sample is dispersed into fine droplets by the spray chamber. Then the aerosol is successively converted into particles, de-solvated molecules, atoms, and ions before entering the quadrupole accelerator. Figure courtesy of Agilent Technologies.

but are normally discriminated against by a set of electrostatic lenses that focuses the beam of ions and which also eliminates undesired species (anions, photons, doubly charged species) from the mass spectrometer analyzer (Fig. 3). Importantly, since the ionization potential of Ar is much higher than virtually any element (except Ne, F, and a few others), the reactions between Ar^+ and M are heavily favored in the direction of the products Ar and M^+.

The process of separation of the single-charged atoms and detection is very analogous to the standard methods for small molecule and biological mass spectrometry. The ions generated at the plasma are decelerated by a skimmer cone, which is part of the MS interface (Fig. 3) and which collects a small amount of ions in the sample to be analyzed. The skimmer cone is followed by a sample cone, which eliminates anionic species, photons, and uncharged species. A stack of two to four lenses (depending upon the instrument design) focuses the ion beam for optimal detection of the isotopes of interest (Fig. 3). The atoms are then accelerated and selected according to their mass/charge ratios by either a time-of-flight, magnetic sector, or quadrupole magnet and counted by means of an electron multiplier. It is outside the scope of this chapter to explain the differences between the operations and characteristics of the different magnets. By far the most common one is the quadrupole, also the more economical solution. The application of two direct current and oscillating voltages across each pair of magnetic poles (hence the name quadrupole) acts in concert to selected ions of a specific m/z ratio (Fig. 4), while all others ions fall into unstable trajectories that ultimately do not collide with the detector. Since the range of single-charged atomic weights spans

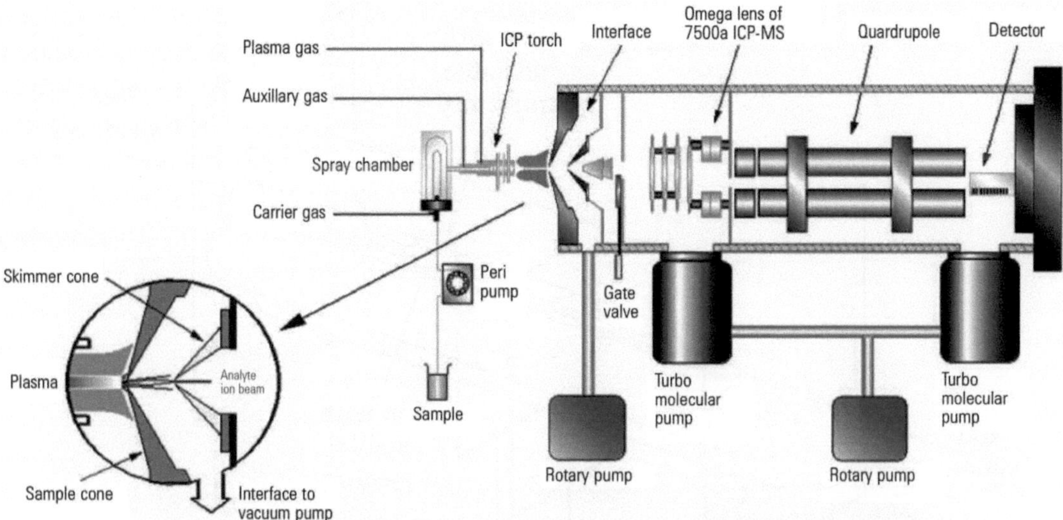

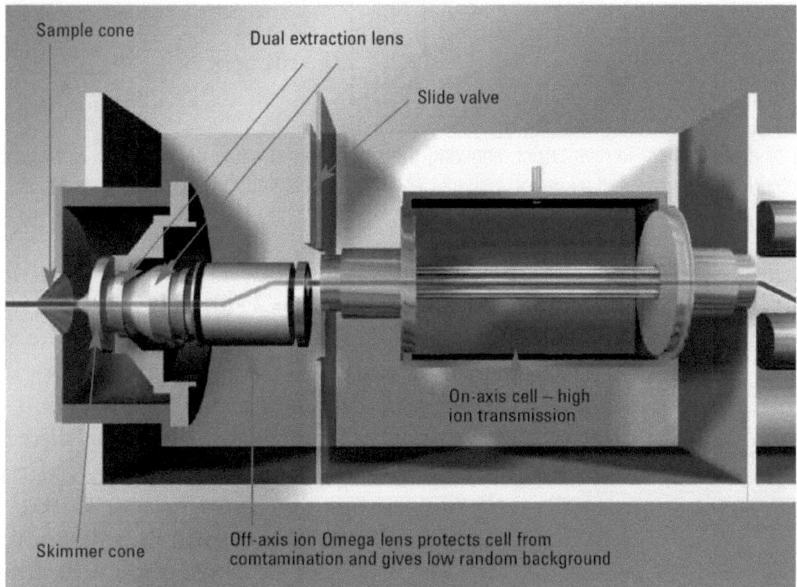

Fig. 3. Schematic diagram of an ICP–MS instrument. The components shown are the nebulizer, torch, plasma, ion lenses, quadrupole separator, and detector. The instrument on *top* does not have a collision cell. The instrument on the *bottom* is a diagram showing the path the ions must travel. The sample and skimmer cones collimate the sample into a set of lenses. The last one is an omega lens, which deflects the ions into an off-axis path which eliminates photons and reduces the background. The octopole collision cell is used to eliminate polyatomic species. Figure courtesy of Agilent technologies.

1–260 a.m.u., a high resolution magnet is not required in routine ICP–MS; quadrupole instruments can achieve resolutions between 0.05 and 0.1 a.u. However, for highly precise determination of isotope ratios a magnetic sector magnet is required, and such instruments accomplish resolutions closer to 0.01 a.m.u. Since only singly charged ions are selected at the lenses at the point of entry into the quadrupole analyzer (Fig. 2), when the frequency is set to a certain mass it essentially detects only one isotope. The

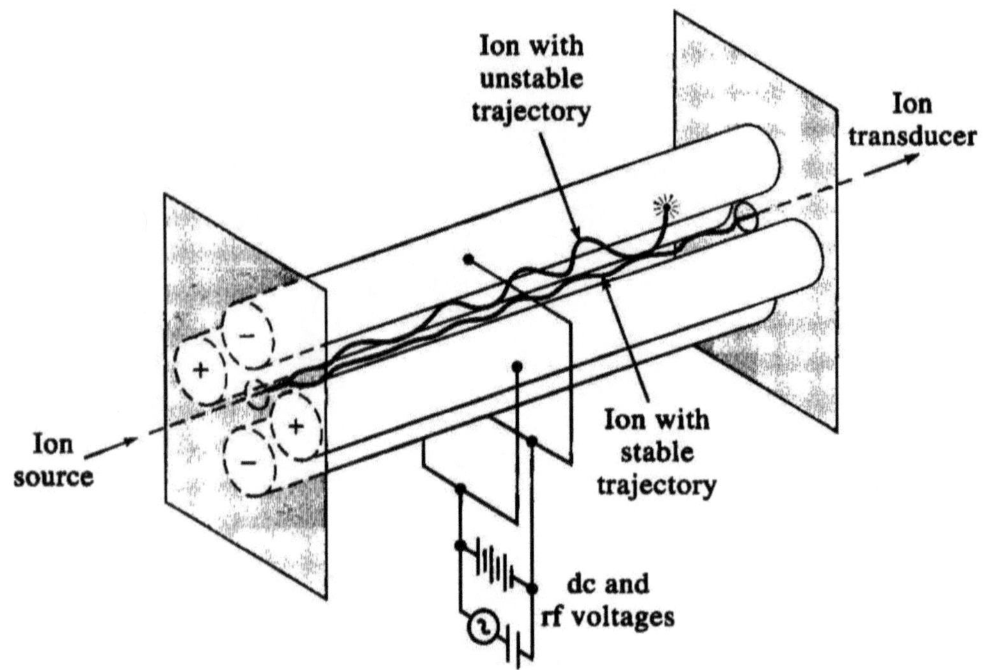

Fig. 4. Drawing of a quadrupole ion separator, showing the ortho-normal pairs of poles that apply voltages of opposite signs. The selected ions are collected by an ion transducer, while unselected ions go on off-axis paths.

instrument scans one m/z value at a time, so for instance if the quadrupole is set to read $m/z = 24.0$, only Mg^{+} ions will be detected (13). A single scan can take as little as 1 ms, and the averaging time for a single peak is usually ~100 ms, while each peak is scanned three times in order to calculate a standard deviation. Thus a single element can be read in about 0.3 s. Since the distribution of isotopes for most samples is pretty close to their natural abundance, normally only one isotope per element is measured, though more than one isotope can be measured. Elements with multiple isotopes have more "peaks" that can be chosen from for the purposes of quantitation, a fact that is useful if one suspects the presence of other elements in a given peak (e.g., $^{90}Zr^{2+}$ in a ^{45}Sc peak and $^{138}Ba^{2+}$, $^{138}La^{2+}$, and $^{138}Ce^{2+}$ in a ^{69}Ga peak).

Another crucial component of the ICP–MS is the sample introduction system. Liquid samples are introduced via the nebulizer, which is a device constructed from perfluoro alkoxy polymer (PFA, tradename Teflon). Teflon is used due to its resistance to attack by strong acids and bases and the fact that it contains very few trace elements that can slow leach into the sample or carrier solution. The nebulizer is essentially a capillary nozzle attached to a quartz device known as a spray chamber, which mixes the Ar carrier with the sample aerosol generated by the nebulizer (Fig. 5a, d). The spray chamber and nebulizer assembly (Figs. 3 and 5d) are in turn attached to the torch leading to the ICP generator (RF coil) generator. As the name indicates, the nebulizer creates a rather fine

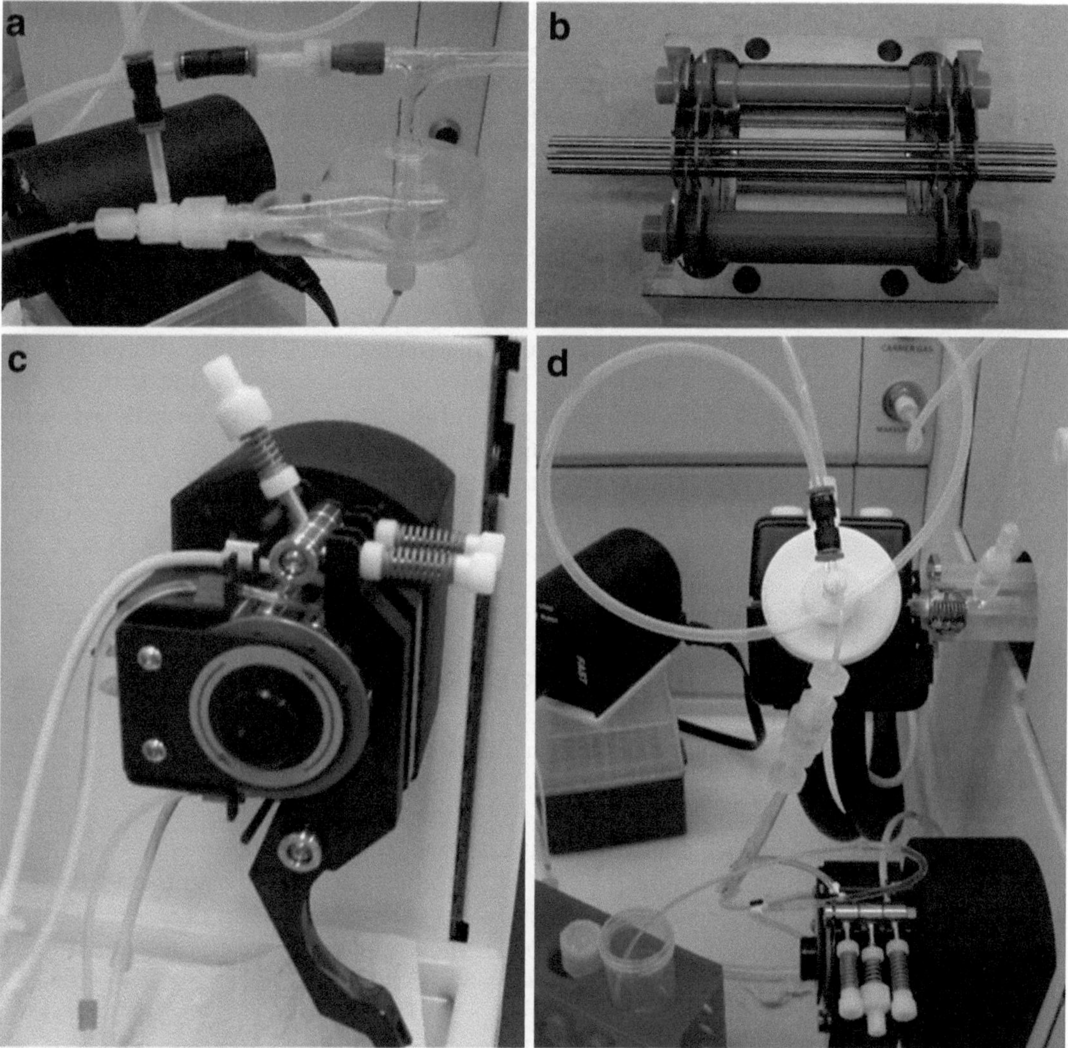

Fig. 5. Photographs of the sample introduction of an ICP–MS. (**a**) Quartz cyclonic spray chamber with a PFA nebulizer, which has two quick-connect inlets for the sample and make up Ar gas lines. (**b**) An actual octopole cell for an Agilent 7500cx instrument. (**c**) Peristaltic pump with two lines for uptake of carrier solution (tygon tubing) and waste line from the spray chamber (Pharmed tubing). (**d**) Frontal view of a double-pass spray chamber mounted next to the entry to the torch.

sample mist while larger drops are eliminated. The spray chamber is necessary to insure that the sample that will end up in the torch is composed only of the finely dispersed small droplets (<10 μm diameter) plus Ar and to minimize the content of elements from the sample matrix (mainly hydrogen, nitrogen, oxygen plus traces of chlorine and sulfur). The efficiency of the ionization process is inversely related to the size of the droplets, so both these factors, the droplet size and sample desolvation, lead to increased accuracy and lower limits of detection. The larger drops are collected at the bottom of the spray chamber and pumped away by a peristaltic pump that also injects the sample. The liquid samples are usually "pushed" into the nebulizer/spray chamber assembly by either the highly accurate peristaltic pump (Fig. 5c) or an injection system driven by motorized-glass syringes. A popular spray chamber that can quickly introduce the samples into the torch (less than 30 s) is the cyclonic spray chamber (Fig. 5a). This spray chamber is also ideal for reducing sample matrix elements and for handling very small amounts of sample (less than 1 mL total volume).

The final component of the ICP–MS is the detector. The ICP–MS uses an electron multiplier, which works similarly to a photomultiplier detector used in photon-based spectroscopies. Ions that are separated at the quadrupole magnet are collected by a dynode and further amplified if necessary. The level of amplification can be separately set for each individual isotope measured; so the dynamic range of detection can span detection levels between five and six orders of magnitude (for most elements). For biological samples this means that routinely trace levels of some elements such as Mo, Cd, As can be measured simultaneously in a sample containing high levels of main elements, such as Na, Mg, K, Ca, P, and S. The efficiency of detection can be as high as 1 in 10^6 ions, so depending upon the element of interest detection limits are from parts per million (mg/L or ppm) down to parts per quadrillion (pg/L or ppq). One more important point is that the detection limit is a function of the ionization potential of a particular element. Elements that easily ionize at the plasma such as metals give higher signals and, correspondingly, lower limits of detection (defined as three times the standard deviation on the background signal). Elements that are harder to ionize such as S, P, Cl, Br, I, and Se (non-metals) will produce lower ion counts and result in higher levels of detection. In addition, some of these elements are volatile, so preparation of the samples is a crucial consideration in their analysis as will be described below.

1.3. Interferences and Elements Analyzed

One problem with the process of desolvation, atomization, and ionization that occurs within the ICP is the formation of polyatomic interferences. Some of the polyatomics that are generated at or after the plasma come from atoms of the carrier gas ($^{36}Ar^+$, $^{38}Ar^+$, $^{40}Ar^+$, and the dimers $^{36}Ar^{40}Ar^+$, $^{38}Ar^{40}Ar^+$, and $^{80}Ar^{40}Ar^+$), the main components of the sample matrix, (NO^+, NOH^+, O_2^+, CO_2^+, N_2^+) or both (ArO^+, $ArOH^+$, ArN^+, $ArNH^+$) (14–16). These polyatomic species are generated in the region of the plasma plume where the cooling is first occurring and ions are entering the sample lenses (Fig. 1). Manufacturers of ICP–MS and analysts have dealt with solutions for this problem for the last 30 years. The most frequently used methods involve collision dissociation with nonreactive gases and reactive neutralization. For collisional dissociation, the principle is that species such as ArO^+ ($m/z=56$) or $ArNH^+$ ($m/z=55$), which interfere with the detection of ^{56}Fe and ^{55}Mn, respectively, have a higher molecular volume than the metal ions of interest. If the ion beam emerging from the lenses is subjected to a small steady flow of an inert gas such as He or Xe, these polyatomic species will dissociate ArO^+ and $ArNH^+$ into Ar and O^+ or NH^+, which will not interfere with the correct masses for iron and manganese, respectively. Collisions of He or Xe with $^{56}Fe^+$ and $^{55}Mn^+$ in this example do still occur, though much less frequently than with the polyatomics. The collision gas is usually fed into a chamber positioned after the ion lenses (but before the quadrupole magnet; Fig. 5c), which maintains the beam focus by means of an octopole magnetic field. Reactive dissociation gases like NH_3 or H_2 are introduced in the collision cell to react with ArO^+ to produce ArOH, a species that is not charged and therefore will not enter the quadrupole analyzer. Hydrogen is particularly useful for the analysis of ^{40}Ca, ^{55}Mn, ^{56}Fe, and ^{78}Se, all of which are important in elemental analysis of plants and which are interfered with by $^{40}Ar^+$, $ArNH^+$, ArO^+, and $^{38}Ar^{40}Ar^+$, respectively. An advantage of the collisional mode is that it is applicable to almost all polyatomic species, so the analyst does not need to know the types of potential polyatomics formed in the sample, but it results in loss of ions of interest and decrease sensitivity. Hydrogen has the advantage that it removes positively charged polyatomics with less collisions, so a lower background and higher sensitivity than He is obtained, but it can also react with metal ions resulting in hydrides, which increase their mass by 1 and can give loss of signal. A list of interferences for most biological elements is shown in Table 1. The choice of an isotope for elemental analysis is dependent upon whatever else is present in the sample. For example, while ^{63}Cu is the major isotope of copper, it is heavily interfered with by $^{40}Ar^{23}Na^+$, so in samples that are rich in sodium, ^{65}Cu might be a better choice. However, ^{65}Cu can be interfered with by $^{14}N^{16}O^{35}Cl$ in samples high in chlorine. In either case the use of an octopole chamber such as the one shown in Fig. 5c with as little as 2.5 mL/min of He gas is enough to eliminate most polyatomic interferences. For the purpose of ionomic analysis of plant tissues, there is no difference in the results between these two interference removal methods. However, one must keep in mind that external calibrations have to be performed under identical conditions of collision gases and Ar flows as the samples.

Figure 6 shows an ICP–MS periodic chart, where the bars correspond to the stable isotopes for each element and the colors correlate with the approximate limits of quantitation that can be achieved. The highest sensitivity (1 ng/L or ppt) is possible for rare earth elements since they are not common contaminants in the environment and are ionized easily by Ar. Next are the halogens, which are rare in nature, easy to ionize but exhibit interferences, followed by the transition metals (yellow), which in most cases can be reliably measured in the low ppt to high ppt range; although they are not present in the environment in high quantities. The first row of transition metals exhibit interferences by polyatomic species. The alkaline metals do not have interferences

Table 1
Summary of the most important interferences for elements commonly measured in plants

Element	Recommended isotope	Interferences
Li	^{7}Li	$^{14}N^{2+}$
B	^{11}B	None
Na	^{23}Na ([a])	$^{46}Ti^{2+}$, $^{46}Ca^{2+}$
Mg	^{24}Mg	$^{7}Li^{16}O$, $^{48}Ti^{2+}$, $^{48}Ca^{2+}$
P	^{31}P ([a])	$^{14}N_2{}^{1}H$, $^{15}N^{16}O$, $^{14}N^{17}O$, $^{13}C^{18}O$, $^{12}C^{18}O^{1}H$, $^{62}Ni^{2+}$
S	^{34}S	$^{16}O^{18}O$, $^{15}N^{18}O^{1}H$, $^{16}O^{17}O^{1}H$, $^{17}O^{17}O$
K	^{39}K ([a])	^{38}ArH, $^{23}Na^{16}O$, $^{78}Se^{2+}$
Ca	^{44}Ca	$^{16}O_2{}^{12}C$, $^{28}Si^{16}O$, $^{88}Sr^{2+}$
Mn	^{55}Mn ([a])	$^{40}Ar^{14}N^{1}H$, $^{39}K^{16}O$, $^{37}Cl^{18}O$, $^{40}Ar^{15}N$, $^{38}Ar^{17}O$, $^{36}Ar^{18}O^{1}H$, $^{38}Ar^{16}O^{1}H$, $^{37}Cl^{17}O^{1}H$, $^{23}Na^{32}S$
Fe	^{57}Fe	$^{38}Ar^{18}O^{1}H$, $^{40}Ar^{17}O$, $^{40}Ar^{16}O^{1}H$, $^{40}Ca^{17}O$
Co	^{59}Co ([a])	$^{42}Ca^{16}O^{1}H$, $^{40}Ar^{18}O^{1}H$, $^{36}Ar^{23}Na$, $^{43}Ca^{16}O$, $^{24}Mg^{35}Cl$
Ni	^{60}Ni	$^{43}Ca^{16}O^{1}H$, $^{44}Ca^{16}O$, $^{23}Na^{37}Cl$
Cu	^{63}Cu	$^{40}Ar^{23}Na$, $^{47}Ti^{16}O$, $^{14}N^{12}C^{37}Cl$, $^{16}O^{12}C^{35}Cl$, $^{23}Na^{40}Ca$
Zn	^{66}Zn	$^{50}Ti^{16}O$, $^{50}Cr^{16}O$, $^{50}V^{16}O$, $^{34}S^{16}O_2$, $^{32}S^{16}O^{18}O$, $^{32}S^{17}O_2$, $^{33}S^{16}O^{17}O$, $^{32}S^{34}S$, $^{33}S_2$
As	^{75}As ([a])	$^{40}Ar^{35}Cl$, $^{59}Co^{16}O$, $^{36}Ar^{38}Ar^{1}H$, $^{38}Ar^{37}Cl$, $^{36}Ar^{39}K$, $^{150}Nd^{2+}$, $^{150}Sm^{2+}$
Se	^{78}Se	$^{40}Ar^{38}Ar$
Mo	^{95}Mo	$^{40}Ar^{39}K^{16}O$, $^{79}Br^{16}O$, $^{190}Os^{2+}$, $^{190}Pt^{2+}$
Cd	^{111}Cd	$^{95}Mo^{16}O$

[a]Monoisotopic elements

(except for potassium) but can be measured in the low ppb range. Most nonmetals can be analyzed by ICP–MS, though with less sensitivity since they have higher ionization potentials and most of them can have interferences. Of particular interest are B, Si, P, S, As, and Se. Each of those elements has its own challenges (Si is insoluble, S and Se are volatile in biological samples, all of them have interferences).

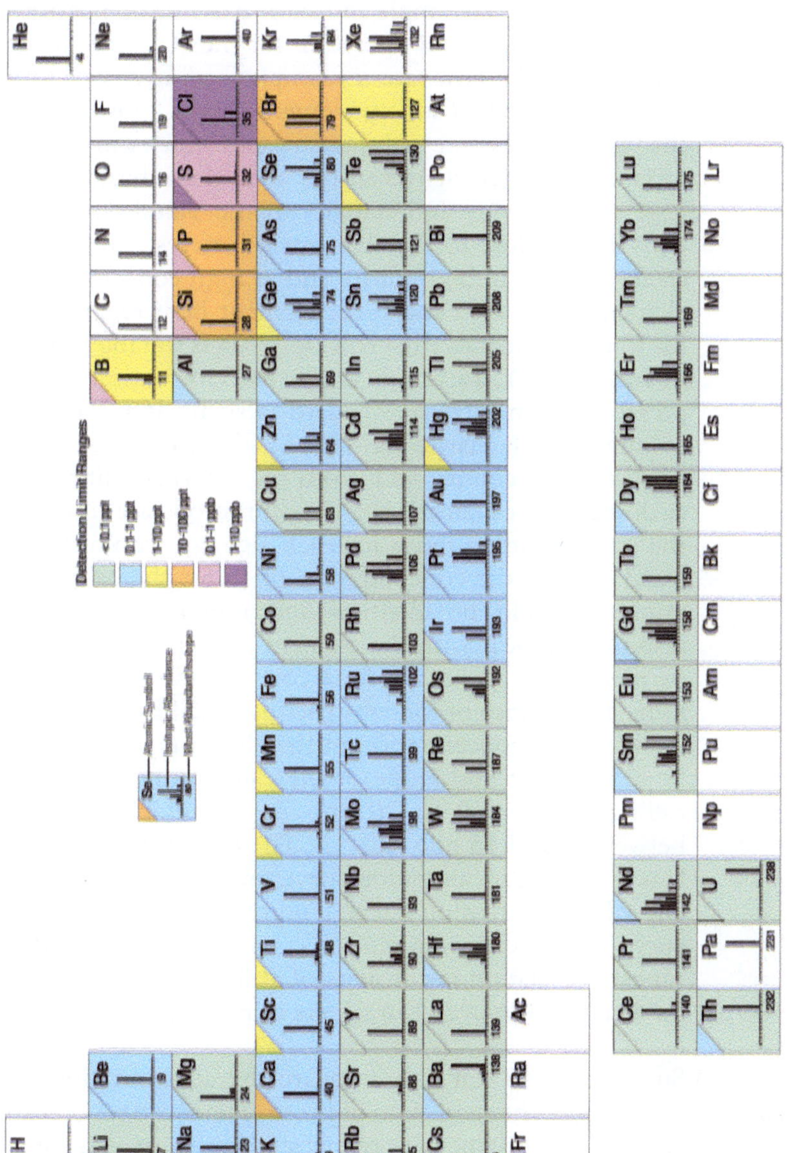

Fig. 6. ICP–MS periodic table. For each element the stable isotopes are shown. The colors represent the approximate limits of detection for a Perkin Elmer DRC ICP–MS. Elements in white cannot be determined by ICP–MS due to the high ionization potential or low abundance.

1.4. Internal Standard Addition and Calibration

The accurate determination of the concentration of all elements depends on matching the sample and standard solutions used for calibration. The determination of each individual concentration is usually done by measuring the counts of the m/z signal for a particular element followed by interpolation from a plot of counts versus element concentration from an external calibration curve. The counts for the sample and standards must be performed under identical conditions of operation for the ICP–MS. Since there are variations in the flow rates of sample feeding as well as Ar flows, it is recommended to add an element that is known to be absent in the sample as the internal standard and which can correct for those variations. Other characteristics that the internal standard must have are solubility and stability in the sample matrix, an atomic mass that is close to the elements of interest, and being commercially available in high purity and at high concentration. For light elements such as Li and B the element Be is recommended, while for transition metals (first row only) Ga is suitable, while other elements such as Y, Ta, and In can be used for heavier elements. In the presence of an internal standard (IS) the ICP–MS measures the counts for each analyte and IS for each sample, calculates the ratios of analyte/IS counts, and, from the external calibration curve for each element obtained, the concentrations are interpolated from the corresponding ratios for each element within the sample analyzed. Ideally the concentration of IS for the calibration and the samples should be identical and uniform. Since the detector response is linear with the element concentration over five orders of magnitude, generally four concentration points and a blank sample are sufficient for constructing a calibration line. It is recommended that concentration of the elements in the samples should be around the mid range of the calibration line. For example, a calibration from 0 to 20 ppb Zn is suitable to quantitate Zn between 5 and 15 ppb (μg/L). The operator can also introduce a bias in the linear calibration curve towards the low end or the high end of the calibration curve. Figure 7 shows a typical calibration curve for 18 elements plus ^{71}Ga as internal standard at 50 ppb. Linear correlation coefficients of at least 0.99 should be expected for a calibration over four orders of magnitude.

1.5. Preparation for ICP–MS Analysis of Plant Tissues

Since ICP–MS can only measure the concentration of elements, the original chemical state of those elements has no bearing on the analysis. For instance, inorganic sulfur present as sulfate is indistinguishable from sulfur derived from cysteine or thiamine. The sample preparation must destroy the all the sample tissues, unless a separation is performed before digestion to separate contributions from different organelles within cells. The more thorough the digestion of the sample, the more reliable the results can be obtained, as complete mineralization eliminates most of the sample matrix contaminants. Sample matrix contaminants can cause either

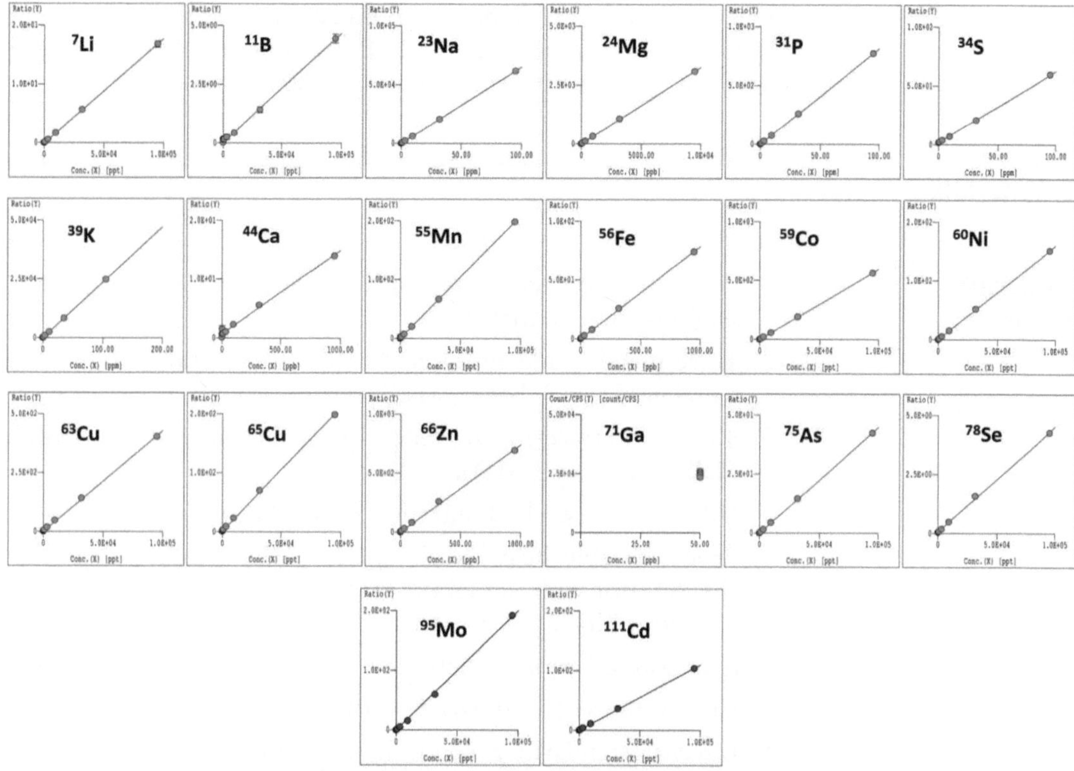

Fig. 7. Linear calibration curves for 18 elements plus Ga IS and a second (^{65}Cu) isotope for copper. The standard concentrations span about four orders of magnitude. Values are background equivalent concentrations (BECs), detection limits (DL), and correlation coefficients (*r*) for each element. Units for the *X*-axis are parts per million (ppm) for Na, P, S, and K, parts per billion (ppb) for Mg, Ca, Fe, Zn, and Ga, and parts per trillion (ppt) for all other 11 isotopes.

suppression or enhancement of one of more element signals, since the elements could be chelated by components of the plant tissue in question. One of the most frequently used protocols for analysis of plant tissues by ICP–MS involves the wet ashing technique. The tissues are destroyed by treatment with strong inorganic acid treatment such as nitric acid, hydrochloric acid, or sulfuric acid. The former is usually preferred since it can be purchased at a very high level of purity with minimal trace metal contamination, it introduces very few polyatomic species (the only elements it contains are hydrogen, nitrogen and oxygen), and it is also a mild oxidizer—a bonus when analyzing volatile elements such as sulfur and selenium. However, not all elements (Mo, W, Br, Cl, and I, for example) are readily soluble in nitric acid; so addition of a small percentage of hydrochloric acid (2%, v/v) is often recommended to keep those elements in solution. Hydrogen peroxide is also added to ensure oxidation of sulfur and selenium, which can be volatile in the corresponding reduced states ($S^{=}$ and $Se^{=}$). The internal standard is usually added at the end of the digestion step.

Final dilutions of the samples should contain less than 5% (v/v) of HNO_3 before loading onto the ICP–MS autosampler, since higher levels can damage the optical lenses of the instrument.

2. Materials

2.1. Sample Preparation

1. Distilled-deionized water (ddw), if necessary analyzed by a semiquantitative method.
2. Trace-metal grade nitric acid: 70% (w/v; 100%, v/v) trace-metal grade (J.T. Baker, Mallinckrodt or EMD).
3. 30% (w/v) trace-metal grade H_2O_2 solution (Fisher Scientific, Mallinckrodt or EMD).
4. 100 mL acid-washed digestion tubes (borosilicate glass).
5. Programmable heat block digestor (SPC Science DigiPrep). If a digestor with digital programming is not available, a standard heat block can be used.
6. Acid-washed glass funnels.
7. Falcon tubes (15 mL sized) with screw caps.
8. Internal standard: a solution of 1,000 ppm gallium in 2% HNO_3. In all instances the digestions must be performed inside a hood, preferentially one with an exhaust constructed with polypropylene or other acid-resistant material.
9. Commercial standards for each element to be analyzed or a single custom-made mixture of elements.
10. Calibrated pipettors.
11. Microcentrifuge tubes.
12. Falcon (50 mL) tubes.
13. Holding racks for 15 and 50 mL tubes (see Note 1).

2.2. Quality Control

Include at least one NIST certified reference material (CRM) along with the samples.

2.3. Calibration Curve

1. Individual element standards at concentrations at or above 1,000 ppm (see Note 2), or

   Mixture of standards purchased from an elemental analysis supplier at concentrations close to those expected in the plant tissues (see Note 3).
2. Microcentrifuge tubes.
3. 15-mL falcon tubes.
4. Trace-metal grade nitric acid (70%, w/v).
5. Calibrated pipettors.

6. Vortex unit.
7. Microcentrifuge.
8. A blank solution consisting of 50 ppm gallium in 2% (v/v) nitric acid.

2.4. Other Types of Samples (Yeast, Bacteria, Mammalian Cells)

1. Trace-metal grade nitric acid.
2. Trace-metal grade 30% (v/v) hydrogen peroxide (10 M).
3. Lyophilizer or speed Vac centrifuge.
4. Microcentrifuge tubes.
5. 15-mL falcon tubes.
6. Calibrated pipettors.
7. Vortex unit.
8. Microcentrifuge.
9. Small oven or heat block capable of reaching (85°C).

2.5. Instrument Operating Conditions

The following are the parameters that are normally used for elemental analysis of plants using an Agilent 7500cx ICP–MS: power, 1,500 W; RF matching, 1.8 V; sample depth, 8 mm; Ar carrier flow, 0.90–1.10 L/min; Ar makeup flow, 0.15–0.2 L/min; peristaltic pump, 0.1 rps; sample flow rate, 70 μL/min; He gas flow (collision cell), 5.0 mL/min; integration time per peak, 0.1 s for all elements except Se, 0.5 s; replicate readings, three per isotope peak. It is recommended that 2% (v/v) nitric acid be used as carrier for the injection of the samples and for rinsing the probe of the autosampler. Although an autosampler is not required, it is highly recommended for obtaining precise element concentrations and for convenience during long runs. A six-port injection valve with a sample loop of the desired volume can be used for loading the samples into the nebulizer, push the carrier solution, and rinse the sample probe and tubing. My setup includes an injection valve with Teflon tubing and PEEK valves, all from Elemental Scientific Inc. (Omaha, NE).

3. Methods

3.1. Experimental Sample Preparation

1. Weigh approximately 0.25 g of oven-dried tissue into labeled 100 mL digestion tubes. (Depending on the nature of samples, smaller or larger amounts can be used and final resuspension volume can be adjusted.) The digestion tubes must be made out of borosilicate glass.
2. In a hood, add 3 mL of concentrated HNO_3, rinsing down the tube sides during the addition. Predigest samples overnight by leaving in hood with no heat.

3. Make sure that the samples are fully submerged in the liquid. Place digestion tubes in rack, then put funnels on each tube. Lower onto the digestion block. Heat at 100°C (standby setting) for 1.5–2 h. Make sure that the brown gas has begun to diminish before moving to the next step.
4. Remove samples from the block and let cool for 5 min.
5. Add 2 mL 30% H_2O_2 slowly to each sample. Allow frothing to settle. Place tubes back on block.
6. Heat for 30 min at 100°C, then at 125°C (program 2) for 1.5 h.
7. Repeat steps 4–6 if you have large samples that still have material in tube. Note that soil-grown grass samples will have silicate crystals at this stage that will never dissolve.
8. Lift rack and let cool for a few minutes, then remove funnels. Increase block temperature to 150°C (program 3), heat for 1 h. Check for dry tubes, and remove if they are dry.
9. Heat at 165°C (program 4) for 1 h. This should bring the samples to dryness. If sample is still wet, heat a little longer.
10. Remove dry samples one by one from the block. When they are cool, add 1% HNO_3 spiked with 50 ppb gallium. For ICP–MS analysis, add 10–15 mL, for other applications adjust volume as needed.
11. Vortex vigorously.
12. Transfer to labeled screw-top 15 mL tubes. (Crystals will settle to the bottom.) The samples at this point are stable for an indefinite time before ICP–MS analysis (see Note 4).

3.2. Quality Control

1. Add a National Institute of Standards and Technology (NIST) standard sample along with samples. For peach leaves use about 0.1 g., for wheat flour use 0.25 g.
2. Include a blank sample with each run.

3.3. Calibration Protocol

The most convenient method for the construction of a calibration curve is to mix standards of all the elements that are going to be measured into a single working standard mix. Since almost all biological elements are soluble in 2% (v/v) nitric acid, precipitation should not be a concern. One cannot overemphasize the need to have standards from at least two different vendors for the purpose of validation. If cost is an issue, a mixture of standards from one vendor and a mixture prepared from available laboratory salts should be used. The in-house mixture of standards can be periodically sent for analysis to an external laboratory.

Figure 7 shows a typical calibration curve obtained with a He flow in the collision cell (5 mL/min) for 18 common elements.

The counts are plotted as ratios over the counts for 50 ppb Ga, which is constant in all the standards. The ratio of intercept/slope corresponds to the background equivalent concentration (BEC). A related parameter is the detection limit (DL), which is expressed as three times the standard deviation of the blank counts ratio. The more stable the reading of the blank, the lower the detection limit. As a general rule both BECs and DLs should be of similar order, and the practical limit of quantitation (LOQ) for an element is equivalent to the sum of the BEC plus ten times the DL (2).

The standard operating procedure for the preparation of the dilutions is as follows:

1. Prepare a stock of 50 ppm gallium by dilution of 20 μL of the 1,000 ppm standard with 980 μL of ddw. Prepare a blank solution consisting of 2% (v/v) nitric acid and 50 ppb Ga by mixing 1 mL of trace-metal grade nitric acid, 50 μL of 50 ppm Ga and qualification standard (QS) with water to 50 mL (see Note 5).
2. Mix 999 μL of the stock mixture of element standards with 1 μL of 50 ppm Ga. This solution will have 999 ppm Na, P, S, K, and so on, plus 50 ppb Ga.
3. Mix 90 μL of the element mix/Ga stock with 810 μL of the blank solution into a 1.5 mL microcentrifuge tube. Label the tube S-8. Dispense 810 μL of the blank solution onto six separate microcentrifuge tubes labeled S-6 to S-1 (see Note 6).
4. Mix 30 μL of the element mix/Ga with 870 μL of the blank solution into another 1.5 mL microcentrifuge tube. Label this tube S-7. Mix standards S-8 and S-7 by inversion or vortexing.
5. Remove 90 μL from each S-8 and S-7 standards and add them to the 810 μL of blank solution in the tubes labeled S-6 and S-5, respectively. Cap and mix S-6 and S-5 thoroughly.
6. Remove 90 μL from each S-6 and S-5 and deliver them into the 810 μL of blank solution in tubes labeled S-4 and S-3, respectively. Cap and mix S-4 and S-3.
7. Remove 90 μL from each S-4 and S-3 and deliver them into the tubes containing 810 μL of blank solution and labeled S-2 and S-1. Cap and mix S-2 and S-1 (see Note 7).
8. Prepare a qualification of calibration standard (QCS) either by twofold dilution of S-8 or by 20-fold dilution of the stock mixture prepared in step 2. This calibration standard is to be read with the samples to ensure the validity of the calibration, especially if a large set of samples are going to be run (see Note 8).

3.4. Other Types of Samples (Yeast, Bacteria, Mammalian Cells)

Yeast cells are best analyzed by dissolution of the cell pellet into concentrated nitric acid followed by oxidation and dilution.

1. Collect about 10^6 cells by centrifugation into a microcentrifuge tube. If necessary, wash the pellet with either ddw or TE

buffer (10 mM Tris-HCl pH 8.0, 1 mM EDTA). Ensure that the washes do not add any metal contamination. Include an empty microcentrifuge tube as a mock sample.

2. Dry the pellets using lyophilization or a speed vacuum centrifuge.
3. Resuspend the cell pellet into 500 μL of concentrated nitric acid that was spiked with 50 ppb Ga.
4. Incubate the suspended samples at room temperature for several hours (can be overnight) or at 85°C for 1–2 h or until the pellet is completely dissolved. Cool down to room temperature.
5. If desired, add also 50 μL of 30% (w/v) metal-grade H_2O_2, also spiked with 50 ppb Ga. At this point the samples can be incubated overnight at room temperature.
6. Dilute the samples at by at least 20-fold so the final nitric acid concentration is 5% (v/v) with 50 ppb Ga in 1% HNO_3 (see Note 8).

4. Notes

1. A more convenient and speedy method involves the use of a programmable acid-resistant microwave oven. The MARS model produced by CEM (Matthews, NC, USA) is quite suitable for this. Digestions can be carried out in sealed containers at 160°C for as little as 30 min or until all solids are dissolved.
2. Most laboratories have stock solutions at concentrations of 1,000 mg/L (1,000 ppm) or more of each element which are purchased from companies specializing in metal analysis (SPEX Certiprep, Inorganic Ventures, Thermo-Fisher and Sigma-Aldrich). Those standards have a shelf life of about 1 year from the point at which they are opened. These solutions are provided with a certificate of analysis and their concentrations are traceable to standards at the National Institute of Standards and Technology (NIST).
3. Standards in which the elements are premixed and certified are also available at a higher cost. I normally use a standard mix that contains 1,000 ppm of Na, P, S, and K, 100 ppm of Mg, 10 ppm of Ca, Fe, and Zn, and 1 ppm of each Li, B, Mn, Co, Ni, Cu, As, Se, Mo, and Cd.
4. Always analyze a blank/mock sample in parallel. The perfect validation of the analytical method is the complete analysis of a certified reference material. That is, a specific plant tissue sample that has previously been analyzed by several laboratories and has been found to contain levels of elements within a cer-

tain range, generally using ICP–MS or ICP–optical emission spectroscopy (OES).

5. It is strongly recommended that enough blank is prepared so that there is enough volume for both the solutions for the calibration curve as well as dilutions of the samples (see item 12 of Subheading 2.1 on plant tissue preparation).
6. Prepare not more than 1.5 mL of each standard at a time since only about 150 μL of each standard are used per injection. It is impossible to prevent the slow release of metal ions from the plastic from the tubes (typically Cu and Zn) even if the microcentrifuge tubes have been acid washed. Also, glassware contains high levels of Li, B, Si, and other elements, which can slowly leach in the presence of strong acids. Small amounts of standard solutions do not need to be kept for more than a few days, a time during which leaching can be disregarded. Some elements (halogens) tend to be volatile, even if all containers are kept closed at all times. Storing the standards frozen or at 4°C does not slow down the leaching and may actually cause some elements to precipitate.
7. Plasticware: although most types of plastic are resistant to strong acids such as hydrochloric acid and nitric acid, the cleanest material is by far Teflon. However for regular experiments the cost of Teflon is prohibitive to most research laboratories. The next choice as far as contamination is LDPE (low-density polyethylene), which is quite inexpensive and relatively free of plasticizers, which add metals as part of the process of incorporation into the plastic during manufacturing. Also avoid any containers that have been stained. The addition of dyes to tubes, vials, and pipette tips significantly adds traces of metals that increase the background levels.
8. Although 1% (v/v) nitric acid is quite appropriate as the final concentration, the stock of nitric acid tends to accumulate metals from the storage bottles and plasticware (microcentrifuge tubes, pipette tips). In such case, lower nitric acid concentrations work as well; however at least 0.1% (w/v) is necessary to maintain the IS and other elements in solution.

Acknowledgment

The work described here has been supported by NIH award RR017675.

References

1. Baxter I (2009) Ionomics: studying the social network of mineral nutrients. Curr Opin Plant Biol 12:381–386
2. Salt DE (2004) Update on plant ionomics. Plant Physiol 136:2451–2456
3. Salt DE, Baxter I, Lahner B (2008) Ionomics and the study of the plant ionome. Annu Rev Plant Biol 59:709–733
4. Fleet JC, Replogle R, Salt DE (2011) Systems genetics of mineral metabolism. J Nutr 141:520–525
5. Becker JS, Becker JS (2010) Imaging of metals, metalloids, and non-metals by laser ablation inductively coupled plasma mass spectrometry (LA-ICP-MS) in biological tissues. Methods Mol Biol 656:51–82
6. Ahrends R, Pieper S, Neumann B, Scheler C, Linscheid MW (2009) Metal-coded affinity tag labeling: a demonstration of analytical robustness and suitability for biological applications. Anal Chem 81:2176–2184
7. Baxter IR, Vitek O, Lahner B, Muthukumar B, Borghi M, Morrissey J, Guerinot ML, Salt DE (2008) The leaf ionome as a multivariable system to detect a plant's physiological status. Proc Natl Acad Sci USA 105: 12081–12086
8. Buescher E, Achberger T, Amusan I, Giannini A, Ochsenfeld C, Rus A, Lahner B, Hoekenga O, Yakubova E, Harper JF, Guerinot ML, Zhang M, Salt DE, Baxter I (2011) Natural genetic variation in selected populations of *Arabidopsis thaliana* is associated with ionomic differences. PLoS One 5:e11081
9. Beauchemin D (2002) Inductively coupled plasma mass spectrometry. Anal Chem 74: 2873–2894
10. Perkin-Elmer, I. (2011) Thirty minute guide to ICP-MS. www.perkinelmer.com
11. Taylor HE (2001) Inductively coupled plasma mass-spectrometry, practices and techniques. Academic, New York
12. Thomas R (2000) Practical guide to ICP-MS: a tutorial for beginners, 2nd edn. Academic, New York
13. Houk RS (1986) Mass spectrometry of inductively coupled plasmas. Anal Chem 58:97A–105A
14. Dams RFJ, Goossens J, Loens L (1996) Spectral and non-spectral interferences in inductively coupled plasma mass spectrometry. Mikrochim Acta 119:277–286
15. Tan SH, Horlick G (1986) Background spectral features in inductively coupled plasma/mass spectrometry. Appl Spectrosc 40:445–460
16. Vaughn MA, Horlick G (1986) Oxide, hydroxide and doubly charged species in inductively coupled plasma/mass spectrometry. Appl Spectrosc 40:434–445

Chapter 15

The Plant Volatilome: Methods of Analysis

Carlo Bicchi and Massimo Maffei

Abstract

Analysis of plant volatile organic compounds (VOCs) and essential oils (EOs, collectively called the plant volatilome) is an invaluable technique in plant biology, as it provides the qualitative and quantitative composition of bioactive compounds. From a physiological standpoint, the plant volatilome is involved in some critical processes, namely plant–plant interactions, the signaling between symbiotic organisms, the attraction of pollinating insects, a range of biological activities in mammals, and as an endless source of novel drugs and drug leads. This chapter analyses and discusses the most advanced methods of analysis of the plant volatilome.

Key words: Volatile organic compounds (VOCs), Essential oils, Gas chromatography, GC detectors, Electronic nose, Nanomaterials

1. Introduction

The plant volatilome is defined as the complex blend of essential oils (EOs) and volatile organic compounds (VOCs) fed by different biosynthetic pathways and produced by plants, constitutively and/or after induction, as a defense strategy against biotic and abiotic stress (1). VOCs are released from leaves, flowers, and fruits into the atmosphere and from roots into the soil, while EOs are produced by specialized secretory tissues. VOCs also attract pollinators, seed dispersers, and other beneficial animals and microorganisms, and serve as signals in plant–plant communication (2). The plant volatilome is also involved in signaling between symbiotic organisms. The plant volatilome has wide agricultural and industrial applications: from the search for sustainable methods for pest control to the valuable production of flavors, fragrances, and phytochemicals (1).

Volatilomics, the study of the volatilome, includes the qualitative and quantitative analysis of EOs and VOCs emitted by plants.

Jennifer Normanly (ed.), *High-Throughput Phenotyping in Plants: Methods and Protocols*, Methods in Molecular Biology, vol. 918, DOI 10.1007/978-1-61779-995-2_15, © Springer Science+Business Media, LLC 2012

Improvements in analytical techniques and molecular and biochemical methods have made EOs and VOCs one of the best-studied groups of plant secondary metabolites (1, 3, 4). Volatilomics has reached threshold levels allowing the characterization and quantification of volatiles with limits resembling the detection ability of molecular sensors present in living organisms (3, 5, 6). Gas chromatography (GC) as such or in combination with mass spectrometry (GC–MS) is the method of choice for volatilomics. GC and GC–MS nowadays evolve toward the speeding up of separation and detection, while keeping or improving separation and producing reliable qualitative and quantitative results (3, 4). Miniaturization of GC and GC–MS instruments has reached amazing levels allowing full portability of highly sophisticated and sensitive VOC analyzers. Integrated instrumental multichannel solutions combine micro-GC modules to single quadrupole mass detectors, being perfect for environmental monitoring performed by mobile laboratories. Another cutting-edge technology is proton transfer reaction-mass spectrometry (PTR-MS), which is a combination of a PTR drift tube and a quadrupole mass spectrometer. So far, several VOCs have been analyzed by PTR-MS (5, 6).

Other types of sensors have been developed for monitoring low concentration of VOCs: electrochemical sensor, optical sensor (chemiluminescence method and Fabry–Perot interferometer), metal oxide semiconductor field effect transistors with Pt-gate (Pt MOSFET), conducting polymer sensors, and piezoelectric sensors (7, 8). All of these technologies are grouped under the umbrella of the so-called "electronic nose," which is based on the analysis using a semi-selective electronic sensor array, resulting in recognition of volatile odor patterns (8). Finally, nanotechnology has also been applied to the volatilome science. Porous anodic alumina (PAA) has been used to adsorb VOCs dissolved in liquids and present in a gas phase (9).

2. Materials

EOs and VOCs can be extracted and analyzed from both fresh and dried plant materials. When using fresh material, particular attention must be paid on the health status of plants. Plants must not show necrotic areas (visible as brown spots on leaves or other tissues) and tissues must be selected at the same developmental stage if comparative analyses are needed. Since the content of water may vary, it is a good practice to use some of the fresh material to calculate the dry matter percentage (after drying at 105°C from 2 to 8 h, depending on tissue/organ). When evaluating the EO composition of new species, it is important to collect several individuals from different populations in order to minimize the phenotypic

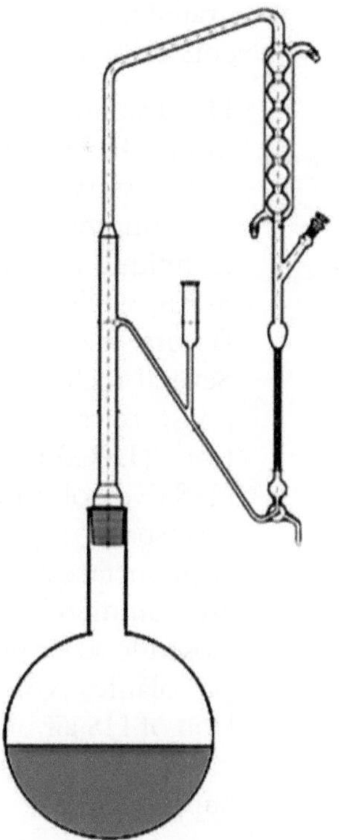

Fig. 1. A typical Clevenger circulatory hydrodistillation apparatus reported in the European Pharmacopoeia (12). The plant material is placed inside the glass containers along with water. Application of a heat source allows water to boil and extract EOs which are condensed along with water by the upper condensing system. The different density of EOs with respect to water allow separation and collection.

plasticity effects on terpenoid synthesis, depending on plant responses to environmental conditions. When running VOC analyses from living plants it must be remembered that rooted plants in pots respond differently than cuttings, and that soil in pots may contain microorganisms that can produce VOCs (10).

2.1. Steam Distillation and Hydrodistillation of EOs

1. An EO is classically obtained using equipment based on the circulatory distillation approach introduced by Clevenger (11) and described in several pharmacopoeias. Only the products obtained by hydrodistillation or steam distillation can be called EOs (with the sole exception of cold-pressed citrus oils, see Note 1). Figure 1 shows a diagram of the apparatus reported in the European Pharmacopoeia (12).

2.2. Headspace Sampling by Direct Sampling

Since hydrodistillation is time-consuming and cannot be combined online with analysis, solvent-free (or solvent-less) sample preparation techniques have been developed for the direct VOC capturing by head space (HS) collection (13). Here, analytes are isolated from a matrix without using a liquid solvent, allowing fully automatic analysis systems in which sample preparation and analysis are integrated in a single step. Adsorbed VOC are then desorbed using solvents or by thermal desorption.

1. HS sampling operates in either the static (S-HS) or the dynamic mode (D-HS) (see Note 2), but an additional approach, high concentration capacity headspace techniques (HCC-HS), acts as a bridge between S-HS and D-HS (14). The HCC-HS technique is simple, fast, easy to automate, and as reliable as S-HS, while at the same time showing analyte concentration factors that are very often comparable to those of D-HS. Several techniques based on this approach such as: HS-solid-phase microextraction (HS-SPME), in-tube sorptive extraction (INCAT, HS-SPDE), headspace sorptive extraction (HSSE), solid-phase aroma concentrate extraction (SPACE), headspace liquid-phase microextraction (HS-LPME), and large surface area HCC-HS sampling (MESI, MME, HS-STE) are commercially available (13). Among the several methods, closed-loop stripping systems have broad utility for the collection of volatiles: volatiles are collected during continuous circulation of HS air inside closed chambers (see Note 3) in which air circulation pumps are connected to sorpting columns or coated supports (15).
2. Gas-tight syringe: direct HS is possible by removing an aliquot of the headspace with a gas-tight syringe (see Note 4) and injecting it directly into the GC. The process can be automated with commercial HS autosamplers (e.g., Gerstel MultiPurpose Sampler, MPS). However, direct HS sampling requires a sufficiently high concentration of VOCs in the HS to provide suitable quantities in the sample taken for GC analysis. Thus, the method is limited by the need for satisfactory sensitivity (15).

2.3. HS Sampling with Porous Polymers

1. Several adsorbing systems are available for HS analysis of VOCs such as silica gel, activated charcoal, anasorb 747, and carboxens. The most commonly used in plant volatilome analyses are porous polymers. Some of them are more suitable for solvent desorption: Chromosorb 106, Amberlite XAD-4, Porapak Q, and Hayesep D are some examples (16). With thermal desorption the most commonly used sorbents are Tenax TA and Carbotrap (Scientific Instrument Services, Inc.). Tenax TA (see Note 5) is a macroporous, semicrystalline polymer manufactured from diphenyl-*p*-phenylene oxide (DPPO).
2. Clean glass container (the volume depending on the tissue/organ size or the amount of dry material).
3. Tenax TA fluxing system (see Note 5).
4. Glass column (the size depending on the amount of the tissues to be extracted, see Note 5).
5. Calibrated Internal Standards (see Note 6).

2.4. HS Sampling with SPME

SPME is a technique in which sampling and pre-concentration of analytes are combined into a single step, and then the analytes adsorbed on a fiber are directly transferred into a standard GC (17). The fiber possesses a coating of a sorbent material (typically polydimethylsiloxane, Carboxen, or Carbotrap). The use of SPME (see Note 7) has several advantages such as lightweight and compactness, no need for reagents for desorption and high sensitivity, linear response, and generally independent of humidity (except at very high humidities). However it must be taken into consideration that analytes of low volatility do not reach partition equilibrium quickly, storage stability is poor, samples are not time-integrated, and gas standards are required for calibration (see Note 6, (18)).

2.5. HS Sampling by Stir Bar Sorptive Extraction and HSSE

Another approach for sample enrichment is the use of stir bars coated with the sorbent polydimethylsiloxane (PDMS), referred to as stir bar sorptive extraction (SBSE). This system is ideal for detection of VOCs both in liquid phases and in gas phases. Stir bars with a length of 10 and 40 mm coated with 55 and 219 μL of PDMS liquid phase, respectively, are used (see Note 8). Depending on the sample volume and the stirring speed, typical stirring times for equilibration are between 30 and 60 min. Detection limits using mass selective detection are in the low ng/L range for a wide selection of analytes. A comparison between SBSE and other methods shows that SBSE has higher sensitivity (19, 20). Headspace sorptive extraction (HSSE) uses SBSE to detect airborne VOCs. Here, sampling is performed by suspending the coated stir-bar in the headspace vial and the polymer is in static contact with the vapor phase of a solid or liquid matrix. One of the drawbacks is related to the fact that the coated stir bar cannot be directly desorbed in a simple split/splitless injection port of a gas chromatograph. Hence the analyte has to be back extracted into a suitable solvent, which adds an additional step to the overall analytical method, or a specially designed Thermal Desorption Unit (TDU, Gerstel GmbH & Co. KG, see below) needs to be used (20). Figure 2 shows the application of different HS detection systems.

2.6. Thermal Desorption

SPME, SBSE, HSSE, and porous polymers involve the use of a TDU followed by GC to recover the accumulated analytes. Thermal desorption (TD) is performed at temperatures in the 150–300°C range. TD requires the use of a special unit on the GC, the TDU, which consists of two programmable temperature vaporizers (PTVs). The first PTV is heated in order to desorb the solutes from the coatings, the second is kept cool (temperatures in the −150°C and 40°C range) in order to cryofocus the desorbed analytes before entering the GC (16).

2.7. Liquid Desorption

Liquid desorption (LD) is an alternative to TD when a TDU unit coupled with GC is not available. LD is also used when thermally

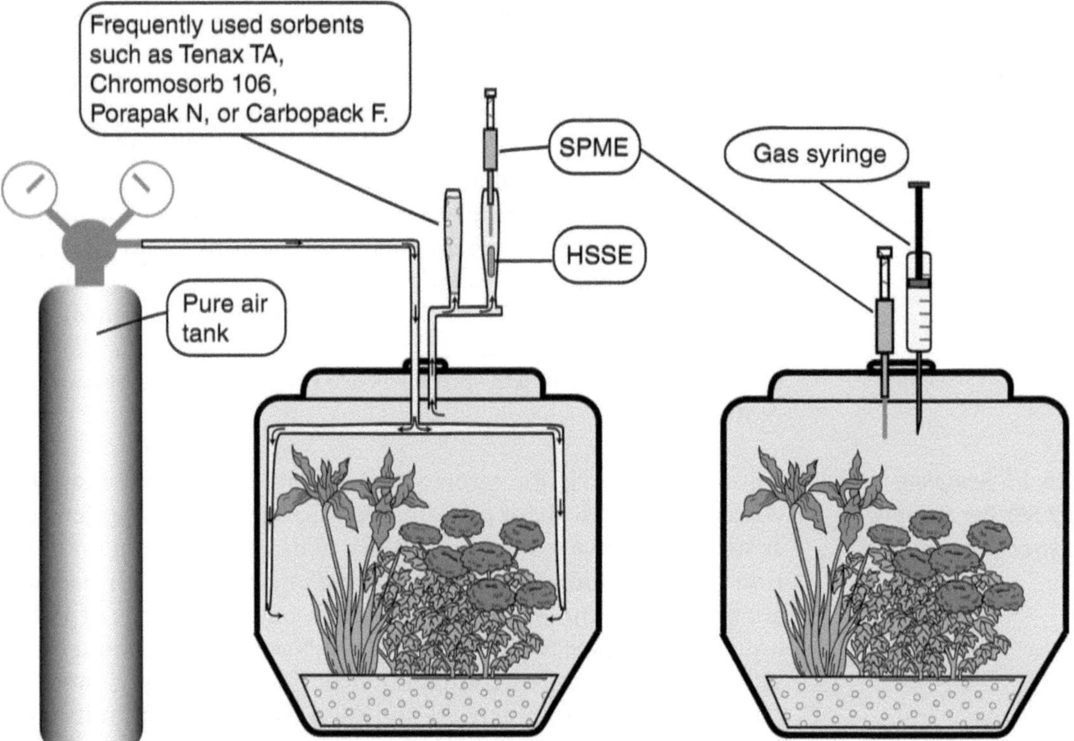

Fig. 2. Main techniques used for the detection of HS-VOCs. A tank of pure air is connected to a sealed jar where plants in pots or plant cutting are placed. Clean air circulates inside the jar and moves plant VOCs, which are forced to pass through different sorpting systems. Alternatively, plant VOCs can be analyzed by direct SPME/HSSE analysis (in this case it is called "static sampling") or by a gas-tight syringe.

labile solutes are analyzed, when the separation is carried out using liquid chromatography (LC) or capillary electrophoresis (CE). During LD, SBSE/HSSE, SPME, and other sorbting systems are immersed or eluted with a solvent for the chemical desorption of the extracted solutes (see Note 9). The recovery of the analytes during LD is strongly determined by the K_{ow} partition coefficient (see Note 10).

2.8. Gas Chromatography

GC uses short bore columns (length: 10–60 m, inner diameter 0.10 mm stationary phase film thickness 0.10–0.40 μm) and modern GC instrumentation are provided with automatic injectors, electronic flow control, oven with high precision temperature and temperature rate controls, detectors with high sensitivity, electronic stability (low signal-to-noise ratio), and frequency of signal acquisition (21). The method translation approach introduced by Klee and Blumberg (21) makes it possible to find the optimal separation/speed trade-off for a conventional GC method and to derive (fast) GC conditions from it automatically (22). The most used GC stationary phases in the volatilome analysis are apolar polydimethylsiloxane as such (OV-1, DB-1, HP-1, Mega 1, Se-30, etc.) or with 5% of phenyl groups (DB-5, HP-5, SE-52, Mega 5, etc.) and moderately polar polyethylene glycols (DB-Wax, HP-wax, CW-20 M, Megawax, etc.).

2.9. Detectors for GC (Flame Ionization Detector/Thermal-Conductivity Detector)

Detection in GC can be obtained by conventional systems or by mass spectrometry (MS). GC detectors can be classified as universal or selective. Universal detectors measure all (or nearly all) components in a mixture, although their quantitative response may vary with analyte structures. Selective detectors only respond to components with a specific structural characteristic in their molecules. A GC detector should be sensitive enough to detect solute amounts in the 10^{-6} and 10^{-9} g range, linear over a wide concentration range, and stable (23, 24).

Volatilome analysis mainly requires universal detectors. The most used of them is the flame ionization detector (FID) where solute molecules eluting from the GC columns are combusted by a hydrogen-rich flame between two electrodes producing small charged species. The existing current varies depending on the analyte concentration. FID is sensitive (minimum detectable amounts in the order of 10^{-12} g/s for hydrocarbons), stable, and linear with solute concentrations up to six orders of magnitude. It can operate at very high acquisition frequencies (up to 500 Hz) that makes it suitable for fast and ultra-fast GC (Fast-GC, UFM-GC, GC × GC, etc. (23, 24)).

The thermal-conductivity detector (TCD) (or hot-wire detector) measures the difference in thermal conductivity between pure carrier gas and carrier gas plus a solute eluting from the GC column. The column effluent flows through a thermostated cavity containing a resistor element, heated by a constant electric current. The flowing gas changes the thermal resistor conductivity inducing a variation in element temperature and, as a consequence, in electric resistance. Reference is a parallel cell, through which pure carrier gas flows. Compared to FID, TCD is nondestructive but its sensitivity is much lower (minimum detectable amounts in the order of 10^{-8} g/s for hydrocarbons (23, 24)).

2.10. Mass Selective Detectors/GC–MS

MS as a detector system overcomes the main limit of GC, i.e., the limited information on the chemical nature of the separated analytes because it provides information specific for each separated component, i.e., its spectrum (6–8). MS operates on the ionization and fragmentation of a compound, followed by the physical separation of charged fragments and their detection. The resulting mass spectrum (ion intensity vs. mass/charge, m/z) is highly informative of the analyte original structure. In GC–MS ionization is mainly produced by electron impact (EI) or chemical ionization (CI), while ion separation is obtained by mass analyzers operating on different principles. The mass analyzers used most often in volatilome analysis are quadrupole (qMS), time of flight (TOF-MS), and ion trap (IT-MS) analyzers, although the most popular GC–MS detectors for routine analysis are MS with single quadrupole analyzers due to their reliability, performance, and acceptable cost (7, 8).

Currently available MS instrumentation can be grouped according to the information they provide, i.e., low-resolution instruments (qMS, IT-MS, and high speed TOF-MS) give nominal molecular weights and fragmentation of an analyte, while high-resolution instruments (high-resolution TOF-MS and hybrids, see below) give the precise elemental composition of nominal masses. Further structural information and data can be obtained by combining more than one analyzer, which enables the analyst to operate in tandem MS–MS (MS^n) mode. These combinations range by the well-known triple quadrupole and IT^n to the so-called "hybrid masses," where in general a quadrupole or an IT analyzer is combined with a TOF-MS with orthogonal acceleration and array detectors originally developed for LC–MS but recently also applied to GC–MS (25).

2.11. Qualitative Analysis

Component identification can be achieved simply by GC results or, more reliably, by combining GC and MS data. The identification of a component separated by GC is generally carried out through its retention time or retention index. Retention indices (*I*s) were first introduced by Kováts (see Note 10, (26)) for isothermal analysis and then by Van den Dool (27, 28) for temperature programmed analysis (linear retention indices, I^Ts). I^T "measures" the physicochemical interaction of an analyte with the chromatographic system adopted (or in the case of GC, with a given stationary phase (29)). I^Ts are determined vs. an homologous series of linear hydrocarbons or fatty acids methyl or ethyl esters by running all analyses under highly standardized GC conditions.

Another method for compound identification of GC data in programmed temperature analysis is retention time locking (RTL, (30)): this approach standardizes the analyte retention behavior but operates without index calculation.

2.12. Identification of Compounds Analyzed by GC–MS

Elaboration of GC–MS data exploits new powerful software programs, exhaustive and specialized libraries, and collections of spectra. Recent tools such as MS spectra deconvolution software of co-eluting peaks (31), and interactive combination of chromatographic linear retention indices (I^Ts) and MS data also concur to give highly effective component identification (31, 32).

The coupling of GC with MS dramatically increases the reliability of identification because the two parameters are highly orthogonal, being based on completely different principles. Linear retention indices (I^Ts) are very useful for analyte identification because volatilome samples often consist of complex groups of components with very similar chemical and physical characteristics (e.g., mono- and sesquiterpenoids, homologous series, alcohols, and aldehydes) showing very similar mass spectra (29). The most recent mass spectrum library software packages include retention index information to facilitate peak assignment and, in some cases, enable these two parameters to be used "interactively" (31, 33, 34). RTL too can be combined with an additional mass spectrum library with a commercially available RTL-based software package (Flavfid), thus exploiting retention time-locked GC–FID and GC–MS data in the identification process (32).

2.13. Quantitative Analysis

The high variability of data obtained from the investigated matrix composition makes it hard to indicate a universal approach to quantitatively evaluate the volatilome composition. The most widely used approaches are: (a) relative percentage abundance, (b) internal standard normalized percentage abundance, and (c) "absolute" or true quantitation of one or more components (target analysis) with or without a validated method. FID is the detector of choice for quantitation, whereas MS cannot be used for normalized percentage abundance as a detector for quantitation, because the analyte structure conditions not only the formation of ions but also their abundances (35, 36). Relative percentage abundance can correctly be used only to evaluate relative component ratios within the same sample.

Internal standard normalized percentage abundance is the ideal approach when a group of samples is compared, as often occurs for volatilome studies. Raw data must first be corrected vs. analyte response factors to the detector, then normalized vs. an internal standard (or at least vs. an external standard if an automatic injector is available). Percentage abundance must be calculated vs. the sum of the areas of a fixed number of selected components taken as markers, better if common to all the samples investigated. Detector response factor must be used because, depending on the compound structure, FID gives responses to some compounds that can differ. In general, a response factor representative of each class or subclass of compounds (hydrocarbons, aldehydes, alcohols, esters, etc.) calculated on a model compound for each class to overcome the lack of pure standards should be determined and applied. The true quantitation of one or more marker components is obtained from its chromatographic area in selective ion monitoring (SIM) mode with MS normalized vs. an internal (or external) standard and calculated via a calibration curve constructed from amounts of pure standard of the marker(s) investigated in the operative concentration range (37, 38).

2.14. Multidimensional GC

The complexity of the volatilome often requires the adoption of multidimensional techniques. In 1987, Giddings (39) defined multidimensional separation as: "…an orthogonal two-column separation, with complete transfer of solute from the separation system 1 (column 1) to the separation system 2 (column 2), such that the separation performance from each system (column) is preserved." Two multidimensional approaches are at present available: GC–GC and GC×GC.

In heart-cut 2D (GC–GC) dedicated time-programmable interfaces transfer automatically and online one or a very few components (peaks) eluting from the first column (first dimension, 1D) for a limited time fraction of the whole chromatographic run, to a second column (second dimension, 2D) coated with a different stationary phase (40, 41). The main limit of heart-cut 2D is that only a small number of fractions possibly eluting at relatively different temperatures can be transferred from 1D to 2D within a single GC run, in particular when a second oven is not available for the independent heating of the second column.

In two-dimensional comprehensive GC (GC×GC) each component eluting from the first column is online and automatically trapped, refocused, and reinjected into the second column by a modulator—a thermal or valve-based focusing device—in a fixed time (4–8 s), which is also the time allowed for the analysis in the second dimension (42). GC×GC overcomes the limitation of GC–GC, because each peak eluting from the first column is online transferred and analyzed with the second column. GC×GC was extensively adopted for the analysis of ultra-high complexity samples consisting of thousands of components, such as those in the mineral oil industry (43), and it is now successfully applied in volatilome analysis, where very often samples contain hundreds of components, many of them decisive to define the biological phenomenon under investigation.

2.15. Portable GC and GC–MS

Micro-GC and micro-GC–MS (SRA Instruments, Cernusco sul Naviglio, MI, Italy). Some key features include: start-up of 30 min, analysis time 1–2 min, and continuous automatic cycle and acquisition. These analyzers can accommodate from 1 to 4 analytical channels equipped with injector, column, and detector, as well as solenoid valves for the automatic selection of the sampling lines. The use of micro-GC is particularly suitable for atmospheric VOCs determination and field studies on the plant volatilome, with concentration limits of few ppb for each single component (see Note 11).

2.16. Lightweight Transportable GC

Miniaturized GC instrument: zNose® (Electronic Sensor Technology, Newbury Park, CA, USA), combines fast trapping of VOCs with fast online GC analysis. The detector installed on this instrument is a highly sensitive surface acoustic wave (SAW) quartz microbalance detector, in which VOC analytes are condensed on the surface of an oscillating crystal. The high sensitivity of the SAW detector (in the ppbv range) allows reduced times for volatile sampling on a Tenax TA trap (20–40 s). The instrument is also equipped with a short (1 or 5 m) capillary column (DB-5) of length, and following rapid thermal desorption compounds are separated and analyzed by the SAW detector. While the short column reduces the resolution of VOCs with similar retention times (such as many components of the volatilome), there are several reports of successful application of the zNose® (3, 44).

2.17. Proton Transfer Reaction-Mass Spectrometry

The instrument, which is commercially available from Ionicon Analytik GmbH, Austria, principally allows fast detection of most VOCs in combination with low detection limits (10–100 pptv). PTR-MS can replace GC–MS in many applications of VOC analysis and achieve a response time measured in seconds per compound. PTR-MS uses "soft" chemical ionization of VOC molecules by reaction with hydroxonium ions (H_3O^+) produced by an external ion source (see Note 12, (15)).

2.18. Electronic Nose and Detection of the Plant Volatilome

In the human olfactory system, olfactory neurons produce electrical stimuli generated by the specific interaction of odors with the appropriate receptors, and a single olfactory neuron responds to several different odors while a pattern recognition process produced by multiple olfactory neurons identifies and classifies odors. Similarly, electronic noses are based on analysis using a semi-selective electronic sensor array, resulting in recognition of volatile odor patterns (8). The use of metal oxide thin films sensors is desirable for low power consumption, small dimension of the sample, small response times; moreover, their selectivity could be increased by placing them into an array. Compactness and low cost will also be advantageous for multisite monitoring (45). Conducting polymers are widely used as sensor elements in electronic noses due to their ability to adjust their conductivity in response to organic compounds. Sensors using polymer-coated crystals can detect volatile molecules adsorbed on the surface by measuring the frequency change upon mass increase, whereas optical sensors have been widely used as chemical sensors in many areas as their output signal can be precisely measured and defined (11, 12).

2.19. Other VOC Adsorbing Nanostructured Materials

PAA (see Note 13) has a huge number of nanoscale holes the cell diameter (pore to pore) of which can be controlled from about 10 nm when grown in the anodization solutions of oxalic and/or sulfuric or phosphoric acids (see Note 14). PAA has been used for decades as protection and hard coatings or adhesive layers and as structured supports for the elimination of VOCs (46). Recently, a multistep anodization and leaching process was employed to produce three-dimensional nanometer scale structured alumina plates (see Note 15), used to adsorb VOCs dissolved in liquids and present in a gas phase. After exposure to VOCs, PAA was analyzed by GC–MS after cryo-desorption through a

TDU. A direct comparison between PAA and SPME or SBSE showed that PAA is a suitable and inexpensive material for the adsorption of VOCs with adsorbing properties comparable to the more expensive SPME and SBSE (9). Several types of experiments can be performed by using PAA such as: static headspace (SH), dynamic headspace (DH), and liquid extraction (LE).

1. Aluminum foils (99.99% purity, 0.1 mm thickness; cut in plates of 40 mm × 20 mm).
2. Sandpapers of different grains and diamond paste on velvets.
3. Electropolishing unit.
4. Perchloric acid:ethanol 1:4.
5. Anodization solution: 0.4 M phosphoric acid.
6. Unit to keep anodization solution at 0°C under a constant current of 1.2 mA/cm^2 for 10–90 min
7. PAA.
8. Chemical etching solution: 0.5 M CrO_3 and 0.5 M H_3PO_4.
9. Soxhlet apparatus.
10. Acetone.

3. Methods

3.1. Steam Distillation

1. Weigh fresh or dry material before distillation.
2. Distill the plant material for at least 1 h, depending on tissue/organ type, with the apparatus shown in Fig. 1.
3. Collect the EO floating on the water layer.
4. Calculate the volume in order to express it as a function of the dry or fresh weight.
5. Store EO in a screw-capped clean glass vial at 3–5°C in the dark (see Note 3).
6. Analyze by GC (see Subheadings 2.8 and 2.9).
7. Identify molecules by GC–MS (see Subheadings 2.8 and 2.10).

3.2. HS Sampling by Direct Sampling

1. Place the fresh or dry material into a sealed glass vial.
2. Allow for VOCs to saturate the vial.
3. Use a heating system (40–50°C) to increase the release of VOCs from tissues.
4. Use a gas-tight syringe and extract a volume of air (1–3 mL).
5. Inject air into a GC injector.

3.3. HS Sampling with Porous Polymers

1. Place the fresh or dry material in a clean glass container (see Note 3).
2. Apply a calibrated amount of one or two Internal Standards to the Tenax TA system for semiquantitative analysis (see Note 6).
3. Connect the column containing Tenax TA to the glass container as shown in Fig. 2.
4. Flux through the system for 2–24 h, depending on the experiment.
5. Remove the column and desorpt HS-VOCs by either solvent or thermal desorption (see Note 5).

3.4. HS Sampling with SPME

1. Place fresh or dry material into a sealed vial (see Note 3).
2. Prepare a water bath to warm the vial during extraction.
3. Insert the needle of the SPME device and extract the fiber according to the manufacturer instructions.
4. Expose the fiber for 1–4 h.
5. Retract the fiber and inject into a GC injector.

3.5. HS Sampling by SBSE and HSSE

3.5.1. Stir Bar Sorptive Extraction

1. Put in contact the polymer-coated stir-bar with the solutes by immersion into a headspace vial that contains the liquid sample.
2. Stir under controlled physical conditions.
3. After extraction (from a few minutes to few hours, depending on concentration), remove SBSE and rinse with distilled water in order to remove salts, sugars, proteins, or other sample components.
4. Dip SBSE on a clean paper tissue to remove water.
5. Submit SBSE to desorption.

3.5.2. Head Space Solvent Extraction

1. Put in contact the polymer-coated stir-bar with the vapor phase of a solid or liquid matrix by suspending the coated stir-bar in the headspace vial.
2. After HSSE rinse the coated stir-bar with distilled water and dip it on a clean paper tissue.
3. Submit HSSE to desorption.

3.6. Thermal Desorption

1. Insert the SBSE/HSSE into the glass liner provided with the TDU.
2. Insert the liner inside the TDU.
3. Program the PTVs temperature (see Note 16).
4. Desorb stir bars and analyze by GC or GC–MS.
5. In case of SPME, use the TDU adapter for SPME and follow steps 3 and 4 of this subheading.

3.7. Liquid Desorption

3.7.1. For SBSE/HSSE and SPME

1. Use a minimum stripping solvent volume (see Note 17) to guarantee the complete immersion of the coated supports for 1–5 min.
2. Use mechanical shaking, variable temperature, or sonication to accelerate LD.
3. Recover solutes, concentrate (when needed), and inject into the GC or GC–MS.

3.7.2. For Other Sorpting Systems

1. Elute the column containing Tenax TA or other sorpting systems with a small volume (0.3–1 mL) of the appropriate solvent.
2. Collect eluate into a small cone-shaped vial (see Note 3).
3. Inject into the GC or GC–MS.

3.8. Gas Chromatography

1. Set analysis conditions: injector, detector and oven temperatures and temperature program, split ratio and flow conditions, and MS conditions and tuning when combined with GC (see Note 18 for reference to articles in which various classes or types of VOC and EO have been analyzed).
2. Upload a suitable sample volume (in general 1 μL) into the micro-syringe manually or automatically.
3. Select the injection mode (split/splitless, see Note 19).
4. Inject the sample into the GC system and start GC temperature program.
5. Run GC analysis under rigorously controlled conditions (see Note 20).
6. After analysis, store the acquired data and proceed to their elaboration (qualitative and quantitative analysis) by using GC or GC–MS data station software and databases.

3.9. Detectors for GC (FID/TCD/MSD/TOF)

1. Select and switch on the detector and apply operative conditions.
2. Wait for system stability.
3. Check for detector response repeatability, linearity range, and sensitivity with a reference mixture.
4. Run sample analysis.

3.10. Mass Selective Detectors/GC–MS

1. Tune GC–MS system performance by using manufacturer's instructions.
2. Apply the selected analysis method (see Notes 18–20).
3. Apply a reference tune to the MS system to obtain reproducible spectra.
4. Run sample analysis.
5. Continue with data elaboration (see Subheadings 3.11 and 3.12).

3.11. Qualitative Analysis

1. Set analysis conditions: i.e., injector, detector and oven temperatures and temperature program, split ratio and flow conditions, and MS conditions and tuning when combined with GC.
2. Inject references, a homologous series of linear hydrocarbons, fatty acid methyl or ethyl esters and sample(s) under the same conditions (see Subheading 2.11).
3. After analysis, store the acquired data and prepare the reference table by assigning to each homologue a value corresponding to the number of carbon atoms multiplied by 100 (e.g., decane equal to 1,000).
4. Calculate I^{T}s of each sample component versus the reference table through the Van den Dool algorithm: $I^{T} = 100\,N + 100n\,((t_{R}A - t_{R}N)/(t_{R}N + n - t_{R}N))$.
5. Identify each analyte by matching with the I^{T}s library.

3.12. Identification of Compounds Analyzed by GC–MS

1. Fix an acceptation value for the unknown mass spectrum/library spectrum matching (see Subheading 2.12).
2. Select the top of the investigated peak with the cursor in the total TIC chromatogram.
3. View the mass spectrum.
4. Match manually or automatically mass spectrum and retention index with those stored in the database.

3.13. Quantitative Analysis

1. Select a suitable internal standard for the set of samples under investigation.
2. Analyze all samples in the set.
3. Select the markers characterizing the samples in the set.
4. Determine class or subclass detector response factors.
5. Select (automatically) marker and internal standard areas.
6. Normalize automatically marker areas versus internal standard and detector response factor.
7. Calculate (automatically) relative percent abundance of each marker in each sample.
8. Compare manually or automatically the normalized data within the set.

3.14. Multidimensional GC

1. Select column combination (see Note 21).
2. Select chromatographic conditions (see Notes 18–20).
3. Select modulation time and flow rate (see Note 22).
4. Run sample analysis.
5. Run data elaboration (see Note 23).

3.15. Portable GC and GC–MS

1. Set internal pump sampling system.
2. Set cycle length: typically 6 min for the complete cycle: (thermal desorption at 200°C) sampling, thermal desorption, injection, cleaning, analysis, and cooling.
3. Adjust sampling flow from 100 to 500 mL/min.
4. Adjust sampling temperature from 30°C up to 100°C.
5. Adjust desorption temperature up to 250°C.
6. Set sample injection heater transfer line up to 150°C.
7. Adjust temperature of transfer line to the micro-GC up to 150°C.
8. Couple the micro-GC channel to MSD.
9. Simultaneously acquire micro-TCD and MS chromatograms.
10. Use total ion acquisition mode for a sensitivity to ppm level.
11. Use SIM mode for a sensitivity to ppb level.

3.16. Lightweight Transportable GC

1. Connect the terminal air inlet of the zNose® to a glass jar containing fresh plant organs.
2. Allow clean filtered air to flux inside the jar.
3. Run the zNose® software by setting cycles of analyses.
4. Identify peaks based on previously tested pure standards (see Note 6).

3.17. Proton Transfer Reaction-Mass Spectrometry

1. Carefully chose the ion source to avoid contamination with "impurity" ions (see Note 12)
2. Inject analytes into the drift tube (see Note 24).
3. Set the operating pressure (see Note 25).
4. Adjust the mass flow controller (MFC, see Note 26).
5. Analyze VOCs (see Note 27).

3.18. Electronic Nose and Detection of the Plant Volatilome

1. Prepare a thin film metal oxide gas sensors by sputtering using rheotaxial growth and thermal oxidation (RGTO) technique (see Note 28).
2. Use a glass plant growth chamber for in vivo tests by using plants and/or other organisms.
3. Use an array of VOCs sensors (see Note 29).
4. Detect VOCs after a given time, depending on organisms.

3.19. Other VOC Adsorbing Nanostructured Materials

The following procedure is one of the most used methods for PAA use as a VOC adsorbing material.

1. Before all experiments, PAA is cleaned with a Soxhlet apparatus by using acetone.

2. Mechanically polish aluminum foils (see Note 13) using sandpapers of different grains and diamond paste on velvets and then electropolish (1.5 A for 30 min in a solution of perchloric acid:ethanol 1:4) in order to reduce surface micro-roughness (see Note 15).
3. Anodization of PAA is performed in a beaker filled with a solution of 0.4 M phosphoric acid and the solution is kept at a temperature of 0°C under a constant current of 1.2 mA/cm^2 for 10–90 min (see Note 14).
4. The chemical etching is carried out by dipping the PAA in a mixture of 0.5 M CrO_3 and 0.5 M H_3PO_4 at 60°C for 3 h. The electrochemical process leads to membranes with suitable features that enhance adsorption of VOCs. For SH sampling, a 200 mL beaker containing peppermint essential oil is placed in a sealed glass container.
5. A platinum grid is placed at 9 cm from the upper edge and is used to hold PAA.
6. The extraction lasts from 1 to 3 h.
7. PAA is desorpted by using a TDU connected to a cryofocusing system, which uses liquid CO_2 or liquid N_2 as the cooling agent.

3.20. Concluding Remarks

Gas chromatography is the technique of election to study the complexity of the plant volatilome and to obtain unequivocal information on sample composition. Thanks to their separative power, GC techniques have played a fundamental role to detect (new) odorous compounds (often present as a trace) and elucidate their structure, and directly contributed to the development of sensory analysis, e.g., electronic noses and nanostructured adsorbing materials. Dynamic and static analyses of the plant volatilome are of paramount importance to evaluate both the ecological function of these direct defense and indirect defense signaling molecules and the economic value in the flavor and fragrance industry.

4. Notes

1. A typical distillation apparatus can be assembled from different glassware. A nice example on how to build a distiller for aromatic plants can be found at http://heartmagic.com/eoinstructions.html. Alternatively glassware for distillation are available at SCHOTT.
2. Static sampling refers to sorption on solid phases at still air. Dynamic sampling implies the flux of air through the sample.

3. Closed chambers are usually represented by glass containers of different size. These can be simple screw-capped vials, where the cap can contain a septum and the shape of the vial can be regular or conical. Large glass container (the volume depending on the tissue/organ size or the amount of material) are usually desiccators or custom-made glass domes. Suppliers such as Sigma-Aldrich offer a variety of vials whereas glass vacuum desiccators can be purchased from several glassware dealers.
4. Different kinds of gas-tight syringes can be purchased at the Hamilton, CH-7402 Bonaduz, GR, Switzerland.
5. Tenax TA can be purchased at the Sigma-Aldrich. Typical fluxing systems for Tenax is the following: (a) a vacuum pump that extracts air from the glass container, (b) a GC-grade air generator that blows air in the glass container, (c) a pump that uses ambient air, (d) compressed air from a lab compressor. With options (c) and (d), it is useful to clean air with charcoal filters. When using glass column (the size depending on the amount of the tissues to be extracted), plug one end with clean rock wool and fill it with Tenax TA. Then plug the other end with rock wool. Breakthrough volume data for hydrocarbons, alcohols, glycols, alkenes, acetates, acids, aldehydes, ketones, halogens, amines, aromatics, and terpenoids, expressed in liters per gram of resin, are available at http://www.sisweb.com/index/referenc/tenaxta.htm.
6. Refer to (ref. 47) for internal and external standard calibration.
7. Further detailed information on the theory, optimization, and different types of fiber adsorbents are available from Supelco (Bellefonte, PA, USA).
8. SBSE (Gerstel GmbH & Co. KG). The 10-mm stir bars are best suited for stirring sample volumes from 10 up to 50 mL, whereas 40-mm stir bars are more ideal for sample volumes up to 250 mL.
9. TA is rarely used with solvent desorption, as it has low capacity for volatile compounds, and is incompatible with many solvent systems (see ref. 16).
10. The octanol/water partition coefficient (K_{ow}) is defined as the ratio of a chemical's concentration in the octanol phase to its concentration in the aqueous phase of a two-phase octanol/water system.

    The value of Kováts index can be calculated by the equation:

$$I = 100 \times \left[n + (N - n) \frac{t'_{r(\text{unknown})} - t'_{r(n)}}{t'_{r(N)} - t'_{r(n)}} \right]$$

where *I* is the Kovats retention index; *n*, the number of carbon atoms in the smaller *n*-alkane; *N*, the number of carbon atoms in the larger *n*-alkane; t_r, the retention time.

11. This unit will increase the detection limit up to several orders of magnitude.
12. A hollow-cathode discharge through water vapor to generate H_3O^+ with high purity, this remains the most commonly employed ion source in PTR-MS. Primary H_3O^+ ions enter a drift tube that is flushed continuously with ambient air and undergo nonreactive collisions with any of the common components in air (N_2, O_2, Ar, CO_2) (5).
13. Aluminum foils (99.99% purity, purchased by Sigma-Aldrich) of 0.1 mm thickness can be cut in plates of 40 mm × 20 mm for a better use in TDU.
14. An oxide film can be grown on aluminum by an electrochemical process called anodizing. There are process conditions which promote growth of a thin, dense, barrier oxide of uniform thickness. The thickness of this layer and its properties vary greatly depending on the metal.
15. Polishing is done with sandpapers of different grains and diamond paste on velvets. Anodization is performed in a beaker filled with a solution of 0.4 M phosphoric acid and the solution shall be kept at a temperature of 0°C under a constant current of 1.2 mA/cm^2 for 10–90 min. The best chemical etching solution is 0.5 M CrO_3 and 0.5 M H_3PO_4.
16. Inject 1 μL (or more); injection conditions: initial temperature: 30–50°C; final temperature: operative injection temperature (250°C or more); injector temperature rate: 100–700°C/min.
17. Check that solvents are compatible with the polymer (acetonitrile, methanol, or mixtures with water or aqueous buffers are the most common desorption solvents).
18. References for articles in which EO and VOC have been analyzed (see refs. 3, 13–15, 29, 36, 37, 44).
19. Split or splitless? Splitless injection enables the transfer of all injected sample into the column and it is mainly used when specific trace components have to be detected or quantified. Split injection affords to transfer to the column sample amounts compatible with its loadability, thanks to a split valve at the base of the hot injector that divides the flowstream from the injector between column and outside in a fixable ratio.
20. GC analysis under rigorously controlled conditions assures repeatability of the chromatographic process and therefore requires columns with the same characteristics, in combination with flowrate, split ratio, injection temperature, oven temperature program.

21. In selecting the column combination for multidimensional GC, the column set must be selected with the highest orthogonality, i.e., stationary phases must have selectivities as different as possible, e.g., apolar/polar.
22. In GC × GC, analytes eluting from the first dimension column (1D) are trapped, focused, and then rapidly injected, as a narrow band of few milliseconds, in the second dimension column (2D). This process is actuated by a *modulator*, a thermal or valve-based focusing system. Each single modulator cycle takes a fixed time (4–8 s) and each fraction, injected online into the second column must be analyzed in a time equal to that of the successive modulation.
23. For an example refer to (ref. 48).
24. In an attempt to confine the reaction between H_3O^+ and the analyte gases to the drift tube, some research groups have employed a Venturi-type inlet to minimize backstreaming of gases into the source drift region.
25. The typical operating pressure in the tube is in the region of 2 mbar, and the electric field strength (E) is generally near 60 V/cm.
26. A potential disadvantage of MFCs is that some VOCs can linger on the stainless steel surfaces in the MFC interior, causing memory effects in time-resolved measurements. To solve the problem, use a gas line constructed of materials that are much less "sticky" than stainless steel, namely, Teflon and perfluoroalkoxy (PFA) polymer.
27. A small fraction (typically 1%) of the primary H_3O^+ ions transfers its protons to VOCs which are present as trace gases in air and which have proton affinities higher than that of water. VOCs become protonated and protonation does not generally cause a fragmentation of the ionized molecule. The ions are transported through the drift tube by a homogeneous electric field. The time response is determined by the time the air spends in the drift tube (<0.2 s). Primary and product ions enter a small intermediate chamber where air is pumped out. The ions are extracted into the detection chamber where their masses are analyzed using a quadrupole mass spectrometer.
28. Different oxides can be used like SnO_2 doped with Ag, Au, and Mo, mixed Sn and In oxide and WO_3.
29. Array of sensors: (1) SnO_2–Au operated at 400°C; (2) SnO_2–Ag operated at 400°C; (3) WO_3 operated at 400°C; (4) SnO_2–Mo operated at 400°C; (5) mixed Sn and In oxide operated at 475°C; (6) In_2O_3 operated at 475°C.

References

1. Maffei ME, Gertsch J, Appendino G (2011) Plant volatiles: production, function and pharmacology. Nat Prod Rep 28:1359–1380
2. Maffei ME (2010) Sites of synthesis, biochemistry and functional role of plant volatiles. S Afr J Bot 76:612–631
3. Rubiolo P, Liberto E, Sgorbini B, Russo R, Veuthey JL, Bicchi C (2008) Fast-GC-conventional quadrupole mass spectrometry in essential oil analysis. J Sep Sci 31:1074–1084
4. Maštovská K, Lehotay SJ (2003) Practical approaches to fast gas chromatography-mass spectrometry. J Chromatogr A 1000:153–180
5. Blake RS, Monks PS, Ellis AM (2009) Proton-transfer reaction mass spectrometry. Chem Rev 109:861–896
6. de Gouw J, Warneke C (2007) Measurements of volatile organic compounds in the earths atmosphere using proton-transfer-reaction mass spectrometry. Mass Spectrom Rev 26:223–257
7. Baratto C, Faglia G, Pardo M, Vezzoli M, Boarino L, Maffei M, Bossi S, Sberveglieri G (2005) Monitoring plants health in greenhouse for space missions. Sens Actuators B Chem 108:278–284
8. Oh EH, Song HS, Park TH (2011) Recent advances in electronic and bioelectronic noses and their biomedical applications. Enz Microb Technol 48:427–437
9. Mombello D, Li Pira N, Belforte L, Perlo P, Innocenti G, Bossi S, Maffei ME (2009) Porous anodic alumina for the adsorption of volatile organic compounds. Sens Actuators B Chem 137:76–82
10. Del Giudice L, Massardo DR, Pontieri P, Bertea CM, Mombello D, Carata E, Tredici SM, Tala A, Mucciarelli M, Groudeva VI, De Stefano M, Vigliotta G, Maffei ME, Alifano P (2008) The microbial community of Vetiver root and its involvement into essential oil biogenesis. Environ Microbiol 10:2824–2841
11. Clevenger JF (1928) Apparatus for the determination of volatile oil. J Am Pharm Assoc 17:346–349
12. EDQM (2008) European Pharmacopoeia. Strasbourg
13. Rubiolo P, Sgorbini B, Liberto E, Cordero C, Bicchi C (2010) Essential oils and volatiles: sample preparation and analysis. A review. Flav Fragr J 25:282–290
14. Bicchi C (2004) Special issue: analysis of flavors and fragrances. J Chromatogr Sci 42:401
15. Tholl D, Boland W, Hansel A, Loreto F, Rose USR, Schnitzler JP (2006) Practical approaches to plant volatile analysis. Plant J 45:540–560
16. Harper M (2000) Sorbent trapping of volatile organic compounds from air. J Chromatogr A 885:129–151
17. Chai M, Pawliszyn J (1995) Analysis of environmental air samples by solid-phase microextraction and gas-chromatography ion-trap mass-spectrometry. Environ Sci Technol 29:693–701
18. Elke K, Jermann E, Begerow J, Dunemann L (1998) Determination of benzene, toluene, ethylbenzene and xylenes in indoor air at environmental levels using diffusive samplers in combination with headspace solid-phase microextraction and high-resolution gas chromatography-flame ionization detection. J Chromatogr A 826:191–200
19. Baltussen E, Sandra P, David F, Cramers C (1999) Stir bar sorptive extraction (SBSE), a novel extraction technique for aqueous samples: theory and principles. J Microcol Sep 11:737–747
20. Prieto A, Basauri O, Rodil R, Usobiaga A, Fernandez LA, Etxebarria N, Zuloaga O (2010) Stir-bar sorptive extraction: a view on method optimisation, novel applications, limitations and potential solutions. J Chromatogr A 1217:2642–2666
21. Klee MS, Blumberg LM (2002) Theoretical and practical aspects of fast gas chromatography and method translation. J Chromatogr Sci 40:234–247
22. Agilent (2011). http://www.chem.agilent.com
23. Grob RL, Barry EF (2004) Modern practice of gas chromatography. Wiley, Hoboken
24. McNair H, Miller JM (2009) Basic gas chromatography. Wiley, Hoboken
25. Portoles T, Sancho JV, Hernandez F, Newton A, Hancock P (2010) Potential of atmospheric pressure chemical ionization source in GC-QTOF MS for pesticide residue analysis. J Mass Spectrom 45:926–936
26. Kováts E (1958) Gas-chromatographische charakterisierung organischer verbindungen. Teil 1: retentions indices aliphatischer halogenide, alkohole, aldehyde und ketone. Helv Chim Acta 41:1915–1932
27. van den Dool H (1974) Standardisation of gas chromatographic analysis of essential oils. Ph.D. Thesis, Groningen
28. van den Dool H, Kratz PD (1963) A generalization of the retention index system including linear temperature programmed gas-liquid partition chromatography. J Chromatogr 11:463–471
29. d'Acampora Zellner B, Bicchi C, Dugo P, Rubiolo P, Dugo G, Mondello L (2008) Linear

retention indices in gas chromatographic analysis: a review. Flav Fragr J 23:297–314

30. Blumberg LM, Klee MS (1998) Method translation and retention time locking in partition GC. Anal Chem 70:3828–3839
31. AMDIS - version 2.65. (2007) http://chemdata.nist.gov/mass-spc/amdis
32. Anon. (2004) Flavors RTL databases for GC-FID and GC/MS. Agilent Technologies, Santa Clara
33. FFNSC (2011) MS Library ver. 1.3. Chromaleont, Messina
34. NIST/EPA/NIH (2011) NIST 05 – mass spectral library. National Institute of Standards and Technology, Gaithersburg
35. Cicchetti E, Merle P, Chaintreau A (2008) Quantitation in gas chromatography: usual practices and performances of a response factor database. Flav Fragr J 23:450–459
36. Bicchi C, Liberto E, Matteodo M, Sgorbini B, Mondello L, Zellner BD, Costa R, Rubiolo P (2008) Quantitative analysis of essential oils: a complex task. Flav Fragr J 23:382–391
37. Cordero C, Bicchi C, Joulain D, Rubiolo P (2007) Identification, quantitation and method validation for the analysis of suspected allergens in fragrances by comprehensive two-dimensional gas chromatography coupled with quadrupole mass spectrometry and with flame ionization detection. J Chromatogr A 1150: 37–49
38. Chaintreau A, Joulain D, Marin C, Schmidt CO, Vey M (2003) GC-MS quantitation of fragrance compounds suspected to cause skin reactions. 1. J Agric Food Chem 51:6398–6403
39. Giddings JC (1987) Concepts and comparisons in multidimensional separation. J High Res Chromatogr 10:319–323
40. Deans DR (1968) A new technique in heart-cutting in gas chromatography. Chromatographia 1:18–21
41. Mondello L, Catalfamo M, Dugo C, Dugo P (1998) Multidimensional tandem capillary gas chromatography system for the analysis of real complex samples. Part I. Development of a fully automated tandem gas chromatography system. J Chromatogr Sci 36:201–209
42. Liu ZY, Phillips JB (1991) Comprehensive 2-dimensional gas-chromatography using an on-column thermal modulator interface. J Chromatogr Sci 29:227–231
43. Adahchour M, Beens J, Brinkman U (2008) Recent developments in the application of comprehensive two-dimensional gas chromatography. J Chromatogr A 1186:67–108
44. Bicchi C, Cagliero C, Rubiolo P (2011) New trends in the analysis of the volatile fraction of matrices of vegetable origin: a short overview. A review. Flav Fragr J 26:321–325
45. Roeck F, Barsan N, Weimar U (2008) Electronic nose: current status and future trends. Chem Rev 108:705–725
46. Sanz O, Echave FJ, Sanchez M, Monzon A, Montes M (2008) Aluminium foams as structured supports for volatile organic compounds (VOCs) oxidation. Appl Catal A Gen 340: 125–132
47. de Oliveira EC, Muller EI, Abad F, Dallarosa J, Adriano C (2010) Internal standard versus external standard calibration: an uncertainty case study of a liquid chromatography analysis. Quimica Nova 33:984–987
48. Reichenbach SE, Tian X, Cordero C, Tao Q (2012) Features for non-targeted cross-sample analysis with comprehensive two-dimensional chromatography. J Chromatogr A 1226:140–148

Chapter 16

High-Throughput Monitoring of Plant Nuclear DNA Contents Via Flow Cytometry

David W. Galbraith and Georgina M. Lambert

Abstract

Interest in measuring the nuclear holoploid genome sizes of higher plants reflects not just the status of the nucleus as a defining characteristic of eukaryotic organisms. Higher plants also attract interest in that they display an unusually large range of genome sizes, current measurements indicating an almost 2,500-fold difference between the smallest and the largest. Scientists would like to learn more about the significance of nuclear genome sizes, in terms of molecular and cytological mechanisms regulating the interaction of the nucleus with the cytoplasm and regulating the observed increases and decreases of genome sizes observed within and across families, genera, and species. We would like to understand their adaptive significance through charting their distribution within populations and ecosystems. Further, since genome size values are only available for a small minority of the ~650,000 species of angiosperms (known and yet undiscovered), we would like to systematically survey plant genome sizes globally before their extinction as a consequence of anthropogenic change. Flow cytometry is accepted as the method of choice for genome size measurements, these measurements being based on fluorescent staining of the nuclear DNA. Flow cytometry offers exceptional ease of use, accompanied by high accuracy and reproducibility and low cost. This chapter provides a general discussion of flow cytometric methods for measuring plant genome sizes, and detailed methods for carrying out these analyses.

Key words: Plants, Genome sizing, Nucleus, Flow cytometry

1. Introduction

1.1. Genome Size Measurements in Multicellular Eukaryotes

The nuclear holoploid genome size (C-value) as reflected by its measured content of DNA is the quintessential cellular descriptor of eukaryotic organisms. Higher plants are remarkable in that they display an unusually large range of genome sizes (1), the smallest currently identified value (2) being that of the carnivorous plant *Genlisea margaretae* (a 1 C value of 0.0648 pg or 63.4 Mbp), and the largest value (3) that of *Paris japonica* (152.23 pg;

Jennifer Normanly (ed.), *High-Throughput Phenotyping in Plants: Methods and Protocols*, Methods in Molecular Biology, vol. 918, DOI 10.1007/978-1-61779-995-2_16, © Springer Science+Business Media, LLC 2012

148.8 Gbp), a 2,349-fold difference. Genome size measurements traditionally employed quantitative microscopy based on Feulgen staining (4, 5), but although some activity continues (6), microscopy has largely been replaced by flow cytometry based on fluorescent staining of the nuclear DNA (7). Of the 7,058 prime entries listed today in the Kew C-value database (http://data.kew.org/cvalues/), 3,320 were produced using flow cytometry, and of the 10,277 total entries in the database, flow cytometry was used in 4,945 of these. Tellingly, ~85 % of the 1,860 first estimates recently listed (4) were estimated with flow cytometry. The popularity of flow cytometry reflects its ease of use, excellent accuracy and dynamic range, high sample throughput rates, low cost per measurement, and general applicability across plant families, genera, and species (4, 8). Given the large numbers of angiosperm species for which nuclear DNA contents are not known (9), flow cytometry represents a cost-effective way to obtain this important information.

Flow cytometric analysis requires the use of samples comprising objects in suspension, which pass within a liquid stream through the focus of an intense light source. The light-scatter and fluorescence signals produced by these objects provide quantitative information about their biological properties (10). The specialized organs of multicellular eukaryotes are not typically found in the form of natural single-cell suspensions, and this is true of flowering plants at most growth stages. For plants, the multicellular sporophyte, with its characteristic organs (leaves, roots, flowers, etc.), contains tissues comprising complex interspersions of different cell types, the multicellular form in plants being a consequence of the sharing by daughter cells of a connected cell wall following cell division within meristematic zones (11).

In order to employ flow cytometry to analyze plants, it is necessary to devise methods to produce single-cell suspensions. This can be done by dissolving the cellulosic cell wall under hyperosmotic conditions, which releases individual cells in the form of protoplasts. However, for analysis of nuclear genome sizes, an alternative approach, involving the flow analysis of cellular homogenates, rather than of protoplasts, is much simpler, both conceptually and empirically. Plant tissues and organs can be converted into cell-free homogenates by hand-chopping using a razor blade, and the constituents of the homogenates are then suitable for flow analysis. This approach can be generalized to all multicellular eukaryotes. The speed of sample processing, which can be done on ice and also avoids enzymatic incubation steps, minimizes unwanted changes associated with cellular perturbation. Although preparation of homogenates is a simple way of handling samples, flow cytometers are not typically used for the analysis of homogenates, in which the objects of interest are an extreme minority of the objects in the sample, and the samples themselves contain very high concentrations of objects detectable by the flow cytometer. This situation typically causes confusion when first encountered, since flow cytometry usually involves analysis of cell suspensions in which the objects of interest comprise the vast majority of the objects detectable by the instrumentation.

For the measurement of nuclear genome sizes, consideration must also be given to the selection of the fluorochrome to be employed for these measurements. Although a number of fluorochromes specifically bind to DNA, including DAPI, the Hoechst family of fluorochromes, mithramcyin and the other chromomycins, and Draq5, many of these exhibit pronounced base-pair preferences, and therefore provide fluorescence signals that are systematically biased according to the AT:GC ratio within the genome of interest (10). Propidium iodide and ethidium bromide, which intercalate the stacked base-pairs within the helix, are agnostic to base composition, and therefore are preferred for absolute genome size measurements, although treatment with ribonuclease is required to eliminate fluorescence contributions from RNA (10) (see Note 1).

This chapter provides details as to how to employ flow cytometry to handle the complications presented by crude homogenates and, at the same time, specifically to measure plant nuclear genome sizes. We provide details and helpful tips for successful operation of two representative cytometric instruments, the Accuri C6 and the Becton Dickinson LSR II. Our experience has involved working with numerous instruments, including the Coulter EPICS and Elite series, the Cytomation MoFlo, the Becton Dickinson FACScan, Aria, and LSR II, the BD Accuri C6, the Life Technologies Attune, and the SONY iCyt Eclipse. The described methods are applicable to these instruments, and to eukaryotic organisms in general, although specific operational details will vary.

1.2. Applying Flow Cytometry for Analyzing Minority Subpopulations in Homogenates

A flow cytometer operates through the detection of light-scatter and fluorescence signals from objects entering the area of illumination. Since more objects scatter light than produce fluorescence, flow cytometers typically use light-scatter signals to trigger the cycle of detection at the point that this object is illuminated (10). In the measurement of nuclear DNA contents within plant tissue homogenates using DNA-specific fluorochromes, the torrent of light-scatter signals produced by subcellular debris may obscure detection of the objects of interest, not only since they may comprise <2 % of the detected objects, but also a consequence of a general overwhelming of the detection and display hardware and software. To combat issues of background noise, all flow cytometers have discriminators (basically a background offset adjustment) that can be used to exclude signals derived from debris (10). The utility of a discriminator assumes that the signals from noise are lower than those from the objects of interest (in our case, nuclei).

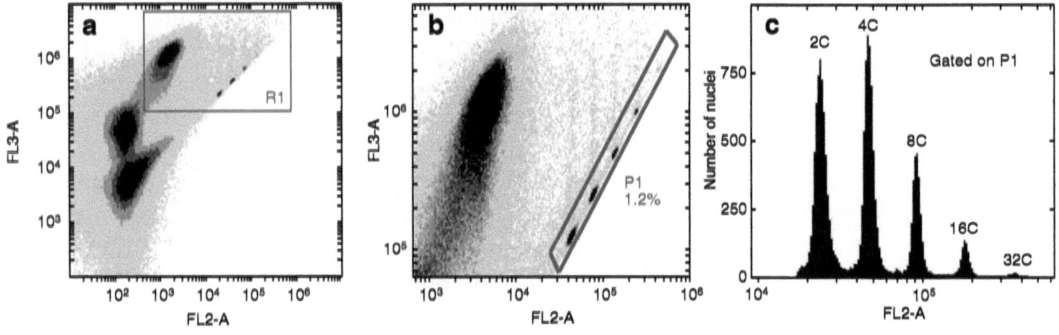

Fig. 1. Flow cytometric analysis of homogenates prepared from *Arabidopsis thaliana* (thale cress) leaf (**a–c**) and root (**d–f**) tissues. (**a**) Biparametric contour plot of FL2-A versus FL3-A fluorescence emission. (**b**) Enlargement of the square region-of-interest (R1) containing the nuclei. The nuclei are enclosed by polygonal region P1. (**c**) Uniparametric histogram of F2-A fluorescence, gated on region P1 of panel (**b**). *Abbreviations*: 2C, 4C, 8C, 16C, 32C designate C-values for the individual peaks (from (8)).

It also presupposes availability of knowledge of the light-scattering properties of the nuclei which, for novel samples, evidently is not available. Further complications are associated with the intrinsic autofluorescence of subcellular organelles other than nuclei, for example chloroplasts, but these can typically be ameliorated by a combination of selecting the appropriate optical filters for detection, and utilizing the spectral characteristics of DNA-fluorochrome complexes as a means to differentiate nuclei and debris.

1.3. Typical Results

Figure 1 illustrates typical profiles obtained from use of the Accuri C6 after preparing homogenates from leaf tissues of *Arabidopsis thaliana*. The general population of fluorescent objects contained in this homogenate (Fig. 1a) includes a series of clusters, having fluorescence emission properties that are correlated between the two channels. Since these channels specify fixed regions of fluorescence emission on the shorter and longer wavelength sides of the peak of the PI-DNA emission spectrum, the amounts detected by the two channels are in fixed proportion, a property that is extremely useful for the identification of the nuclei. This is particularly obvious for plants that display endoreduplication, such as Arabidopsis. Region R1, displayed as a region-of-interest through software-based enlargement of the two dimensional density plot (Fig. 1b), is then used for placing a gate (Region P1) that surrounds the endoreduplicated nuclei. Region P1 is then used to gate the fluorescent signals displayed in a conventional uniparametric histogram (Fig. 1c), in which the individual populations of 2 C, 4 C, 8 C, and 16 C nuclei are clearly distinguished. DNA contents can be assigned to the peaks through determination of their peak positions. Figure 2 illustrates a similar analysis for a plant species (*P. sativum*) that does not exhibit somatic endoreduplication. The helpful distribution of correlated PI fluorescence is still perceptible within the biparametric density plot (Fig. 2a) although

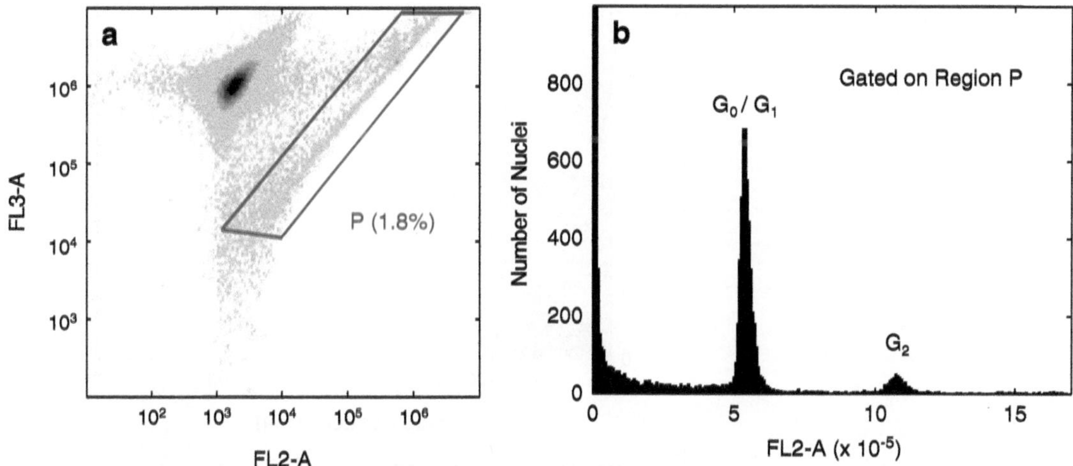

Fig. 2. Flow cytometric analysis of homogenates prepared from *Pisum sativum* (pea) seedling tissue. (**a**) Biparametric contour plot of FL2-A (585/40 nm) versus FL3-A (>670 nm) fluorescence emission. (**b**) Uniparametric histogram of FL2-A fluorescence, gated on region P1 of panel (**a**) (from (8)).

the cluster representing the nuclei is less immediately obvious. Nonetheless, positioning of a polygonal gate around this correlated fluorescence distribution provides ready identification of G_0/ G_1 and G_2 nuclei (Fig. 2b). In this latter case, the histogram is displayed using a linear abscissa and illustrates excellent coefficients of variability (CVs) for the peaks. For measurements of plant genome sizes, obtaining CVs lower than 4 % is desirable, and may require systematic modification to the chopping procedure and/or the chopping buffers (see below).

Figure 3 provides a comparable analysis of PI-RNAse-treated homogenates prepared from two plant species using the LSR II. In this case, examination of the biparametric density plot of side scatter versus propidium iodide fluorescence allows identification of the relative positions of the nuclei and the debris. Gating to exclude debris (the nuclei in this case represent ~3.9 % of the total particles detected) provides good quality uniparametric distributions (Fig. 3b, c).

Figure 4 illustrates the important point that nuclear DNA contents linearly scale according to propidium iodide fluorescence, across an extraordinarily wide range of values, most probably including all samples that will be obtained from the angiosperms (8, 9). For measurement of the nuclear DNA contents of "unknown" species, one therefore need only include, as an internal control, a sample of tissue of known genome size, which is then co-chopped with the unknown. The relative peak positions are then used to define the amount of nuclear DNA of the unknown sample. Including multiple standards, spanning as much of the dynamic range as is available for the instrument and as required for the species of interest allows ready confirmation

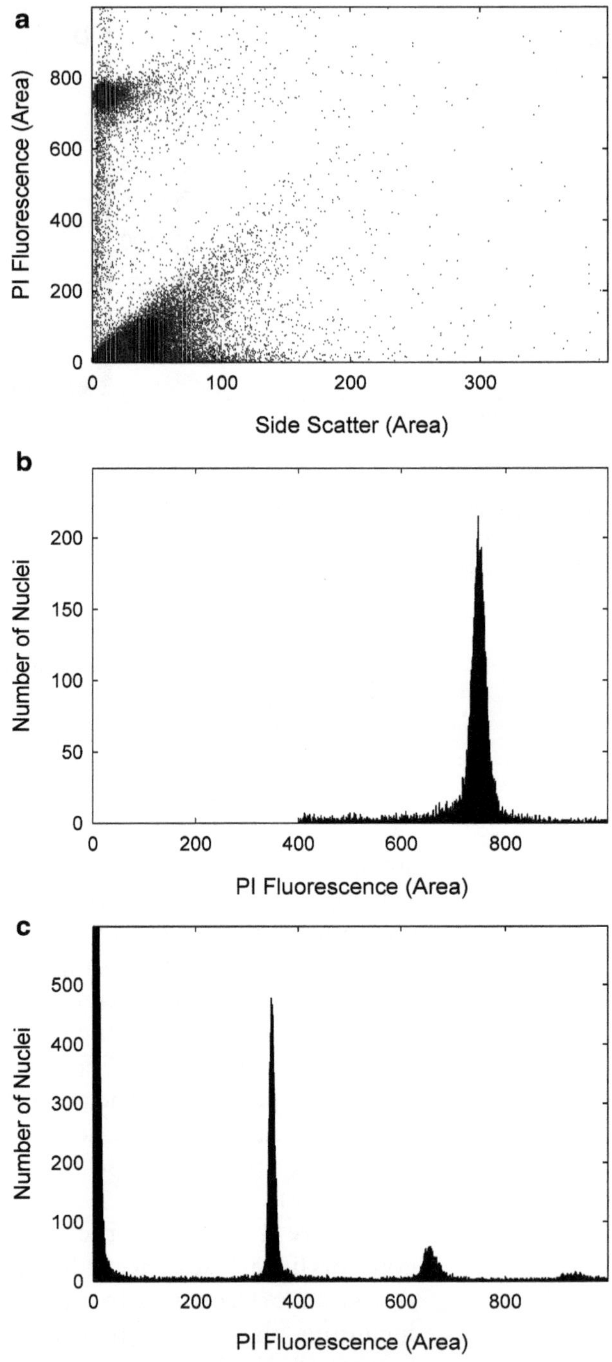

Fig. 3. (**a**) Flow cytometric analysis of homogenates prepared from *N. tabacum* leaf tissue: a biparametric contour plot of side scatter versus propidium iodide fluorescence emission. (**b**) Uniparametric histogram of propidium iodide fluorescence, gated to remove all objects having a value <400 within panel (**a**). (**c**) Uniparametric flow cytometric analysis of homogenates prepared from *Pisum sativum* (pea) seedling tissue.

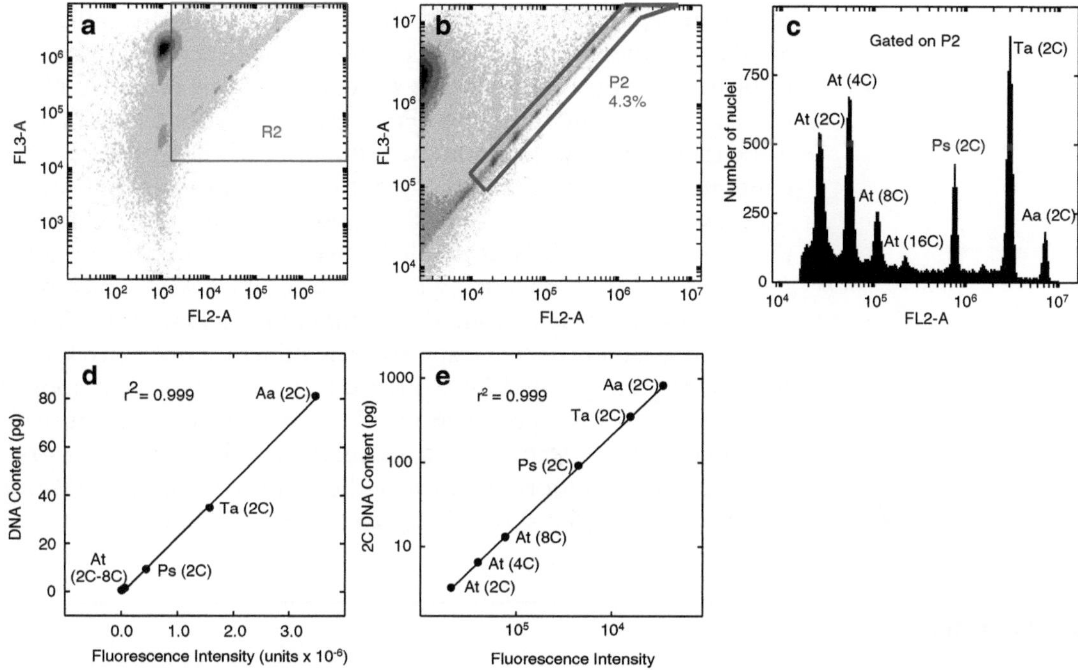

Fig. 4. Simultaneous analysis of mixtures of plant homogenates, illustrating the dynamic range and linearity of measurements that can be achieved. (**a**) Biparametric contour plot of FL2-A versus FL3-A fluorescence emission. (**b**) Enlargement of the square region-of-interest (R2) containing the nuclei. The nuclei are enclosed by polygonal region P2. (**c**) Uniparametric histogram of F2-A fluorescence, gated on region P2 of panel (**b**). (**d**) Plot of DNA content versus the mean fluorescence values of the 2 C peak positions for the four species. (**e**) As for panel (**d**) except employing a log scale, allowing identification of the six DNA content peaks. *At Arabidopsis thaliana*, *Ps Pisum sativum*, *Ta Triticum aestivum*, *Aa Alstromoeria aurea* (from (8)).

of the linearity of the measurements and the overall reliability of the standards that are being employed.

Issues associated with defining and using plant standards depend in part on the types of applications for which the measurements are being made (12). For identification of plants within different ploidy classes (haploid, diploid, triploid, tetraploid, etc.), the measurements need only be capable of recognizing obvious differences in genome size (1X, 2X, 3X, and so on). In this case, since sample throughput is probably the most important consideration, internal standards become redundant, and even as external standards need only be used infrequently for monitoring day-to-day instrument performance. Similar arguments govern the need for sample replication—the fact that obvious differences in nuclear DNA content are easy to accurately detect implies that the frequency of false-negative or false-positive results becomes vanishingly low, and sample replication can be reduced to a minimum. In contrast, for the detection of subtle differences in nuclear DNA content, for example to recognize aneuploidy or to confirm euploidy, within plants regenerating from tissue or organ culture, this absolutely requires the presence of an internal standard, and demands sufficient replications to attain the desired statistical significance. Similar arguments apply to measurements made across populations of plants, whether representing agricultural crops or native plants in their natural environments. For some applications, for example in genomics and biotechnology, estimation of the absolute DNA content, either in mass units (pg/nucleus) or base-pair equivalents, is necessary. However, avoiding an obsession with the accuracy of these measurements is counseled; given the rapid and ongoing decreases in DNA sequencing costs, the accuracy with which plant genome sizes can be measured becomes much less significant than an ability to rapidly generate approximate values for unknown species that subsequently could be sequenced. In other words, it appears more important to recognize whether a species has an unusually large genome, for example *P. japonica*, than to fret as to the reasons for differences in reported values for the DNA contents of *Raphanus sativus* (1.02–1.26 pg) or *Vicia faba* (25.59–27.00 pg).

2. Materials

Always wear laboratory coats, disposable gloves, and protective eyewear. All chemicals are reagent grade, unless otherwise noted. The following section lists specialty chemicals, kits, and equipment for the described applications and methods.

2.1. Isolation of Nuclei

1. Chopping buffer (7): 45 mM $MgCl_2$, 30 mM sodium citrate, 20 mM 3-(N-morpholino) propane sulfonic acid (MOPS), pH 7.0 adjusted with NaOH. This should be filter-sterilized and stored as 50 mL aliquots at −20 °C. Once thawed, the buffer may be kept at 4 °C for up to 1 week.
2. Propidium iodide (PI) stock solution: 1 mg/mL prepared in deionized water (diH_2O). This should be stored as 1 mL aliquots at −20 °C until the day of use.
3. 10 mg/mL solution of DNAse-free RNAse A.
4. Similar fluorescent particles (SPHERO™ Rainbow Calibration Particles) are also available from Spherotech (Lake Forest, IL).
5. Single-edged razor blades (VWR, West Chester, PA; catalog number 55411–050).
6. Other filter porosities are available, and may be used for larger or smaller nuclei.
7. Plastic petri dishes diameter: 60 mm.
8. Prechilled metal or ceramic plate embedded in an ice filled tray.

3. Methods

3.1. Preparation of Plant Homogenates for Analysis of Nuclei

1. Excise plant materials (organs or tissues) and, if necessary, wash using diH_2O. Transfer to a plastic petri dish (60 mm × 15 mm). Perform the remaining procedures on ice, and preferably in a walk-in cold room. It is convenient to place the petri dish on a prechilled metal plate embedded in an ice filled tray (see Notes 2–6). Notes 7–10 discuss sampling of seeds, herbarium species, high-throughput sampling, and eukaryotic species other than plants.
2. Add chopping buffer (2 mL chopping buffer and 0.5 g fresh weight of the sampled tissue is appropriate for this size of petri dish). Chop tissues using a new razor blade for 2–3 min. Discard the razor blade after one use.
3. Filter the homogenate through a 30 μm CellTrics® disposable filter.
4. Add a 0.5 mL aliquot to a labeled tube containing 2.5 μL of a 10 mg/mL solution of DNAse-free RNAse A. Incubate on ice for 10 min. Add PI to a final concentration of 50 μg/mL.
5. Incubate the stained samples, on ice and in darkness, for 20 min prior to flow cytometric analysis.

3.2. Analysis of Genome Sizes Through Flow Cytometry

Switch on the flow cytometer and establish that the instrument is operating according to the manufacturer's specifications. The following methods are specific to (a) the Accuri C6 and (b) the BD LSR II.

3.2.1. Using the Accuri C6

1. Flow cytometry. The Accuri C6 instrument is switched on, and allowed to warm up for 15 min prior to starting analyses. Waste and sheath fluid bottles are checked during this time, and are emptied or refilled, as required. The CFlow software is started. Performance validation of the Accuri C6 flow cytometer is done using six- and eight-peak fluorescent bead mixtures (Spherotech) according to the manufacturer's instructions (CFlowUser Guide pp. 6–7). Finally, the Plant DNA Analysis Template is opened. This template comprises the following histograms: FSC-A versus SSC-A, FL1-A versus FL2-A, FL3-A versus FL2-A, and a univariate histogram of FL2A. The fluorescence signals (pulse area measurements) are screened by the following filter configurations: (a) FL-1: 530/14 nm band-pass filter, (b) FL-2: a 585/20 nm band-pass filter, and (c) FL-3: a 670 nm long-pass filter. Analysis of the homogenates is based on light-scatter and fluorescence signals produced from 20 mW laser illumination at 488 nm.

2. Plant homogenate samples are introduced to the SIP, and the CFlow RUN tab is clicked to initiate data acquisition. The threshold levels are adjusted manually to prevent detection of irrelevant debris found in plant homogenates, which typically have FALS signal values lower than those of nuclei. The default threshold values on the C6 are set to 80,000, although it should be noted that this setting may exclude very small nuclei from analysis. The threshold is set at 10,000 for FALS, with a secondary threshold of 1,000 for FL-2. These values should be adjusted during data acquisition, observing the position of the nuclei on the bivariate dot plots, with the general goal of retaining all nuclei are on scale, but with the lowest possible amount of debris appearing in these plots. The flow cytometer is routinely operated using the "Slow Flow Rate" setting (14 μL sample/minute), meaning that data acquisition for a single sample will occupy 3–5 min.
3. The PI-stained nuclei appear on the bivariate plots as a region of dots clustered around a diagonal line. This region contains two populations of nuclei for non-endoreduplicated species, representing 2 C (G_1/G_0) and 4 C (G_2) nuclei; additional populations are seen for tissues exhibiting somatic endoreduplication. A polygonal gate is then drawn to enclose the nuclei, and univariate histograms of PI fluorescence (FL2-A) are accumulated, based on this gate. The zoom tool is used to locate and isolate the peaks appearing within this univariate histogram. Regions of identification are then placed across the lowest peaks to export values representing peak positions and CVs. These data are used to calculate the DNA contents of the nuclei (see Note 11).

3.2.2. Using the BD LSRII

1. The instrument is powered on, and allowed to warm up. During this time the sheath tanks should be emptied and refilled as required. The BD DIVA software is opened, and a histogram panel is created that comprises three bivariate dot plots (PI area versus log PI height, PI area versus 90 ° light scatter, and log PI height versus log FITC or chlorophyll, depending on whether the plant samples contain other fluorophores such as transgenically expressed GFP or pigments within chloroplasts), and one linear histogram (PI area). The fluorescence signals in the histogram panel are defined by the following filter combinations: (a) FITC: 525/50 nm band-pass filter, (b) PI: 660/20 nm band-pass filter, and (c) chlorophyll: a 685 nm long-pass filter. For the measurements of nuclear DNA content illustrated here, the samples are illuminated by a 532 nm laser operating at 150 mW.
2. Photomultiplier voltages are adjusted within a range of 350–450 V such that the population of nuclei is visible on the dot plot. Thresholding is applied to the PI fluorescence signal,

and its magnitude is adjusted such that the positions of the nuclei on the bivariate dot plots are on scale with the least amount of debris appearing in these plots. A polygonal or amorphous gate is placed around the population of nuclei identified within the bivariate dot plots, and the univariate histogram is generated based on nuclei falling within that gate. The flow cytometer is routinely operated at the Slow Flow Rate setting (12 μL sample/min), or at a data acquisition rate of 200–500 events per second.

3. The FCS3.0 data files produced by the LSR II are extracted into CSV format using FCSExtract 1.02 (http://research.stowers-institute.org/efg/ScientificSoftware/Utility/FCSExtract/), and histograms created using SigmaPlot 11 (Systat Software, Inc., San Jose, CA).

3.3. Data Analysis

Nuclear DNA contents are calculated based on the linear relationship that exists between the amount of fluorescence produced by PI-stained nuclei and their DNA contents. To provide a calibration curve between these two values, standards are employed (see refs. 13–17 for standards and discussion of multiple issues). Table 1 provides values for standards from a combination of these sources.

4. Notes

1. *Standardization*. Determination of nuclear DNA contents, as described above, allows calculation of absolute genome sizes, in base-pair equivalents, using a simple mathematical relationship (see refs. 4, 15). Assignment of a particular peak as representing the 2 C DNA content, and further making the connection to an unambiguous ploidy level can be surprisingly complicated. See refs. 15, 16 for a detailed discussion of these issues.
2. *Throughput*. The standard chopping procedure takes under 5 min to complete, implying that a single individual can generate a pipeline that could process as many as 12–15 samples per hour. Efficiencies are improved by involving more individuals at the homogenization steps, since the staining and filtration steps can be readily automated, and since many flow cytometers have carousel-type multiple sampling features; third-party sample stations exist for flow cytometers that do not have these features.
3. *Statistical issues*. As for any type of quantitative measurement applied to biological samples, care must be taken to provide sufficient replications to satisfy statistical concerns. Care must be taken to define the particular null hypothesis, and then sufficient samples be taken, and these must be of the correct

Table 1
Plant reference standards recommended for estimation of nuclear DNA contents

Plant species and cultivar/variety	2 C DNA content (pg)
Arabidopsis thaliana L. (Heynh.) "Columbia"[a]	0.32
Raphanus sativus L. "Saxa"[b]	1.11
Sorghum bicolor L. "Pioneer 8695"[c]	1.74
Solanum lycopersicum L. "Stupické polní rané"[b]	1.96
Glycine max Merr. "Polanka"[b]	2.50
Medicago sativa L. "Cimarron"[a]	3.50
Zea mays L. "CE-777"[b]	5.43
Pisum sativum L. "Ctirad"[b]	9.09
Pisum sativum L. "Minerva Maple"[c]	9.56
Hordeum vulgare cv. Sultan[c]	11.12
Secale cereale L. "Dankovské"[b]	16.19
Vicia faba L. "GSO11"[c]	26.66
V. faba L. "Inovec"[b]	26.90
Triticum aestivum L. "line 812"[a]	34.65
Allium cepa L. "Ailsa Craig"[c]	33.55
Allium cepa L. "Alice"[b]	34.89

[a](8)
[b]Seeds may be obtained free of charge by contacting dolezel@ueb.cas.cz (14)
[c](13)

provenance to address the question(s) of interest. When considering samples taken from a single tissue, the examples of the sources of variation that require analysis can include: machine variation (day-to-day), technical variation associated with the staining method, and biological variation associated with (a) leaf position, (b) position, within the leaf, (c) different plants, (d) different growing conditions (position in the field, within the greenhouse), (e) geographical location, and (f) anything else that can be varied and that might be relevant.

4. *Low yields of nuclei.* Occasionally, low yields of intact nuclei from the chopping procedure may be encountered. To optimize yield, the chopping procedure should aim to efficiently break open as many cells as possible without damaging the nuclei. Low yields are generally due to the source tissue being in poor condition; ideal samples comprise fresh, young tissues that are well hydrated. Living samples but that are from old or

dried-out tissues as contrasted to specimens deliberately dried under controlled conditions (see Note 9) do not generally provide good quality histograms. Grasses are often problematic, most likely due to structural considerations relating to the cell wall composition. Scaling up the sample size and homogenization volume may be necessary to obtain sufficient nuclei for analysis. In terms of sample collection, optimization has to be done on an individual basis. For example, the tissues and organs of some species can be refrigerated in damp paper towels for several days prior to analysis. The tissues of other species are cold-sensitive and collapse on refrigeration. Freezing samples prior to analysis is not recommended under any circumstance.

5. *Secondary compounds and various mucilaginous materials.* Secondary products are of concern if they are released during the chopping procedure and subsequently interfere with the DNA-fluorochrome binding and/or the fluorescence emission. The adverse effects of secondary products are potentiated by changes in the redox state of the tissue as a consequence of homogenization—the generally reduced state of the cytoplasm is replaced by a generally oxidized state of the homogenate. These effects can be partially controlled using internal standards that are co-chopped with the species of interest, and can also be ameliorated by the inclusion of components in the chopping buffer that either absorb these secondary products, or maintain a generally reducing state, or both. Additives that sequester secondary products include polymers such as polyethylene glycol and polyvinylpyrrolidone. Antioxidants include mercaptoethanol, dithiothreitol, and ascorbic acid. In all cases, an empirical approach is recommended, and comparisons to the fluorescence yield of ethanol-fixed, intact tissue samples (using qualitative observation under the fluorescence microscope) can provide a useful estimation as to whether the amounts of fluorescence produced from isolated nuclei are similar to those in intact tissues.

   Other species are problematic for flow cytometric analysis for different reasons, notably lactiferous species containing latex of various composition, and species which release mucilaginous materials during homogenization. In all cases, careful observation of the samples during chopping is recommended; any changes in homogenate color and consistency are causes for concern. Alternatives can include selecting different tissues or organs for analysis. Additional suggestions for sample preparation under recalcitrant situations are found in ref. (14).

6. *Poor CVs.* A reasonable expectation is that CVs for the G_0/G_1 peaks should be lower than 5 % since, under optimal conditions and for selected species, values as low as <1 % may be obtained. It can be argued that CVs >5 %, in themselves, may not degrade

measurement of the peak position itself. Poorer CVs can be due simply to excessive chopping, or to the use of a dull razor blade. Sources for razor blades are numerous (including VWR, Fisher Scientific, and the Personna America Safety Razor Company). Blades should have an edge thickness of 0.22 mm or finer. Double-edged shaving blades are the best for this application.

7. *Sampling seeds.* We have shown that dry seeds can be employed for nuclear DNA content measurements (see ref. (18)). Care is required to distinguish between diploid embryo tissues and seed storage tissues having other ploidy levels (for example, the triploid endosperm).
8. *Sampling herbarium specimens.* A number of laboratories have described the use of dried plant samples, including archival herbarium specimens, for genome size measurements (see additional information and a thorough discussion, see ref. (19)).
9. *High-throughput sample processing.* Given the large numbers of samples generated by plant biotechnology companies, there is considerable interest in automating the homogenization procedure. Bead-beating within 96-well plates can be used to produce *Brassica napus* homogenates suitable for staining and flow analysis, and this approach greatly increases sample throughput (see ref. (20)). Optimization of the homogenization process will be required for different species and tissue types.
10. *Processing eukaryotic species other than plants.* The concept of homogenization followed by flow cytometric analysis of nuclei has been shown to work for insects (see ref. (21)) as well as mammalian tissues and organs (Galbraith and Lambert, unpublished).
11. *Unstable (drifting) peak positions.* Drifts in peak positions can reflect an insufficient period of staining prior to analysis, or degradation in the sample over time. It is a simple matter to systematically measure the peak positions at various times after staining, and this should be done for all new samples. Our recommendation of 20 min is a reasonable starting point.

Acknowledgments

Part of the development of the methods described in this chapter involved support from the National Science Foundation Plant Genome Research Program.

References

1. Gregory TR (2005) The C-value enigma in plants and animals: a review of parallels and an appeal for partnership. Ann Bot 95:133–146
2. Greilhuber J, Borsch T, Müller K, Worberg A, Porembski S, Barthlott W (2006) Smallest angiosperm genomes found in Lentibulariaceae, with chromosomes of bacterial size. Plant Biol 8:770–777
3. Pellicer J, Fay MF, Leitch IJ (2010) The largest eukaryotic genome of them all? Bot J Linn Soc 164:10–15
4. Bennett MD, Leitch IJ (2011) Nuclear DNA amounts in angiosperms: targets, trends and tomorrow. Ann Bot 107:467–590
5. Michaelson MJ, Price HJ, Ellison JR, Johnston JS (1991) Comparison of plant DNA contents determined by Feulgen microspectrophotometry and laser flow cytometry. Am J Bot 78:183–188
6. Praça-Fontes MM, Carvalho CR, Clarindo WR (2011) C-value reassessment of plant standards: an image cytometry approach. Plant Cell Rep. doi:10.1007/s00299-011-1135-6
7. Galbraith DW, Harkins KR, Maddox JR, Ayres NM, Sharma DP, Firoozabady E (1983) Rapid flow cytometric analysis of the cell cycle in intact plant tissues. Science 220:1049–1051
8. Galbraith DW (2009) Simultaneous flow cytometric quantification of plant nuclear DNA contents over the full range of described angiosperm 2 C values. Cytometry 75A:692–698
9. Galbraith DW, Bennetzen JL, Kellogg EA, Pires JC, Soltis PS (2011) The genomes of all angiosperms: a call for a coordinated global census. J Bot. doi:10.1155/2011/646198
10. Shapiro HM (2003) Practical flow cytometry, 4th edn. Wiley, New York
11. Bell PR, Helmsley AR (2000) Green plants: their origin and diversity. Cambridge University Press, Cambridge
12. Galbraith DW, Bartos J, Dolezel J (2005) Flow cytometry and cell sorting in plant biotechnology. In: Sklar LA (ed) Flow cytometry in biotechnology. Oxford University Press, New York
13. Johnston JS, Bennett MD, Rayburn AL, Galbraith DW, Price HJ (1999) Reference standards for determination of DNA content of plant nuclei. Am J Bot 86:609–613
14. Dolezel J, Greilhuber J, Suda J (2007) Estimation of nuclear DNA content in plants using flow cytometry. Nat Protoc 2:2233–2244
15. Greilhuber J, Dolezel J (2009) 2 C or not 2 C: a closer look at cell nuclei and their DNA content. Chromosoma 118:391–400
16. Dolezel J, Greilhuber J (2010) Nuclear genome size: are we getting closer? Cytometry 77A:635–642
17. Praça-Fontes MM, Carvalho CR, Clarindo WR, Cruz CD (2011) Revisiting the DNA C-values of the genome size-standards used in plant flow cytometry to choose the "best primary standards". Plant Cell Rep 30:1183–1191
18. Sliwinska E, Pisarczyk I, Pawlik A, Galbraith DW (2009) Measuring genome size of desert plants using dry seeds. Botany 87:127–135
19. Bainard JD, Husband BC, Baldwin SJ, Fazekas AJ, Gregory TR, Newmaster SG, Kron P (2011) The effects of rapid desiccation on estimates of plant genome size. Chromosome Res 19:825–842
20. Cousin A, Heel K, Cowling WA, Nelson MN (2009) An efficient high-throughput flow cytometric method for estimating DNA ploidy level in plants. Cytometry 75A:1015–1019
21. Brown JK, Lambert GM, Ghanim M, Czosnek H, Galbraith DW (2005) Nuclear DNA content of the whitefly *Bemisia tabaci* (Aleyrodidae: Hemiptera) estimated by flow cytometry. Bull Entomol Res 95:309–312

References

1. [illegible] (2005) The C-value paradox, [illegible] genes and genomes: a review of [illegible] Ann Bot [illegible]
2. [illegible] J, [illegible] (2003) [illegible] nuclear genome size and [illegible] [illegible]
3. [illegible] (2010) The largest [illegible] genome of [illegible] [illegible]
4. [illegible] (2011) [illegible] [illegible] 1017–1030
5. [illegible] J (2011) [illegible] [illegible] DNA [illegible] plant [illegible]
6. [illegible]
7. Galbraith DW, [illegible] (2002) [illegible]
8. [illegible]
9. Galbraith DW, [illegible] (2011) The [illegible] [illegible]
10. Shapiro HM (2003) Practical flow cytometry, 4th edn. Wiley, New York
11. [illegible] (2006) [illegible] [illegible]
12. Galbraith DW, Harkins KR, Maddox JM, Ayres NM, Sharma DP, Firoozabady E (1983) Rapid flow cytometric analysis of the cell cycle in intact plant tissues. Science 220:1049–1051
13. [illegible] (1998) [illegible] [illegible]
14. [illegible] (2005) Estimation of nuclear DNA content in plants using flow cytometry. Nat Protoc 2:2233–2244
15. [illegible] (2003) [illegible] [illegible]
16. [illegible] (2002) [illegible] [illegible]
17. [illegible]
18. [illegible] (2011) [illegible] [illegible]
19. [illegible] (2008) [illegible] [illegible]
20. [illegible] (2007) [illegible] [illegible]

Chapter 17

Transient RNAi Assay in 96-Well Plate Format Facilitates High-Throughput Gene Function Studies in Planta

Shu-Zon Wu and Magdalena Bezanilla

Abstract

Here we describe a rapid high-throughput method for performing RNA interference (RNAi) in moss, in which phenotyping is performed within 1 week after transformation. The moss *Physcomitrella patens* is a great plant model system for reverse genetic studies due to its amenability to homologous recombination as well as RNAi. Our lab has developed a rapid RNAi assay to screen for growth phenotypes in moss protonemal tissue. Here we describe how we have recently further facilitated this assay by modifying the PEG-mediated transformation protocol allowing for transformations to be carried out in a semiautomated fashion in a 96-well plate format.

Key words: *Physcomitrella patens*, RNAi, PEG-mediated transformation, Protoplast, Tip growth, Gene function

1. Introduction

Gene-knockdown mediated by RNAi is a widely used reverse-genetic approach to study gene function. Genes of interest are specifically silenced and the resulting phenotypes are analyzed in order to decipher gene function. Genome-wide RNAi screens are able to phenotypically assay thousands of genes at a time. In animal systems these approaches have been applied to whole organisms in *Caenorhabditis elegans* (1) and to cell-based assays in *Drosophila* (2, 3) and in human (4, 5) cells. These studies have proven to be very informative, uncovering new genes involved in important cellular processes and signaling pathways (4). To date, similar methods are not available for plants.

The moss *Physcomitrella patens* is a great model system for studying plant gene function using reverse-genetic approaches (6). Mosses are nonvascular, multicellular terrestrial plants with simple developmental structures and relatively few cell types.

Jennifer Normanly (ed.), *High-Throughput Phenotyping in Plants: Methods and Protocols*, Methods in Molecular Biology, vol. 918, DOI 10.1007/978-1-61779-995-2_17, © Springer Science+Business Media, LLC 2012

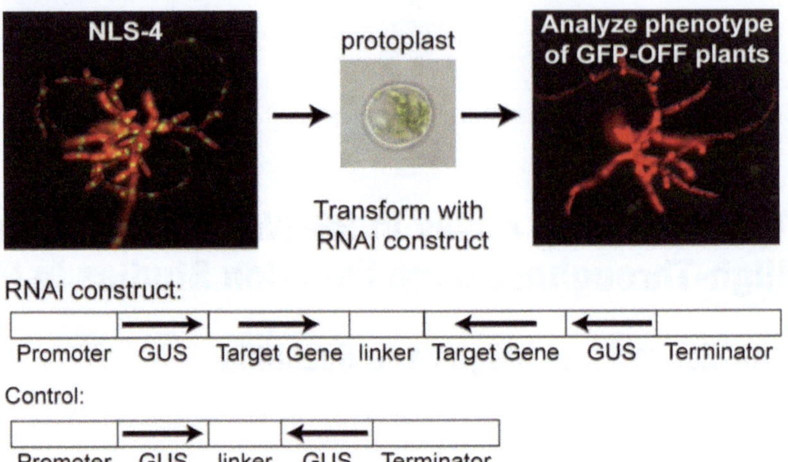

Fig. 1. Transient RNAi assay. An RNAi assay performed in the NLS-4 line, which expresses a nuclear-localized GFP:GUS protein. Protoplasts isolated from NLS-4 are transformed with control and RNAi constructs. One week after transformation, phenotypes of regenerated plants lacking nuclear GFP signal are imaged. Control plasmid silences only nuclear-localized GFP:GUS, while the RNAi construct simultaneously silences the nuclear-localized GFP:GUS and the targeted gene.

Experimentally, the haploid tissues can be propagated vegetatively by blending in water and cultured on medium agar plates overlaid with a layer of cellophane. The haploid nature of the tissues makes genetic manipulation of the organism and the interpretation of the phenotypes straightforward. Moss protoplasts can be transformed and regenerated into a new plant, allowing us to genetically manipulate the organism at the single cell level and study the resulting phenotypes in a whole plant. Additionally, RNAi is quite effective in *P. patens* (7, 8), providing a great tool for rapid gene silencing and phenotypic characterization. Since moss shares many common processes with vascular plants, the data sets generated from studies in moss are potentially useful for many plant systems.

Gene silencing induced by RNAi in moss is done by transforming moss protoplasts with a DNA construct that expresses inverted-repeat sequences of the targeted gene. The RNAi assay is performed in a stable transgenic line expressing a nuclear-localized GFP–GUS fusion protein, NLS-4 (8). The RNAi construct contains a fragment of the target gene fused to 400 bp of GUS in an inverted repeat orientation. The inverted repeats are separated by a ~400 bp linker. Expression of the RNAi construct is driven by the maize ubiquitin promoter, which is a strong constitutive promoter in moss (8). Once introduced into cells and transcribed, the mRNA will pair and form double-stranded RNA molecules, which then triggers RNAi-mediated gene silencing of the targeted cDNA and the GFP:GUS fusion protein. The effectiveness of gene silencing is determined by the disappearance of the nuclear GFP:GUS protein. Only transgenic plants lacking nuclear GFP are analyzed (Fig. 1).

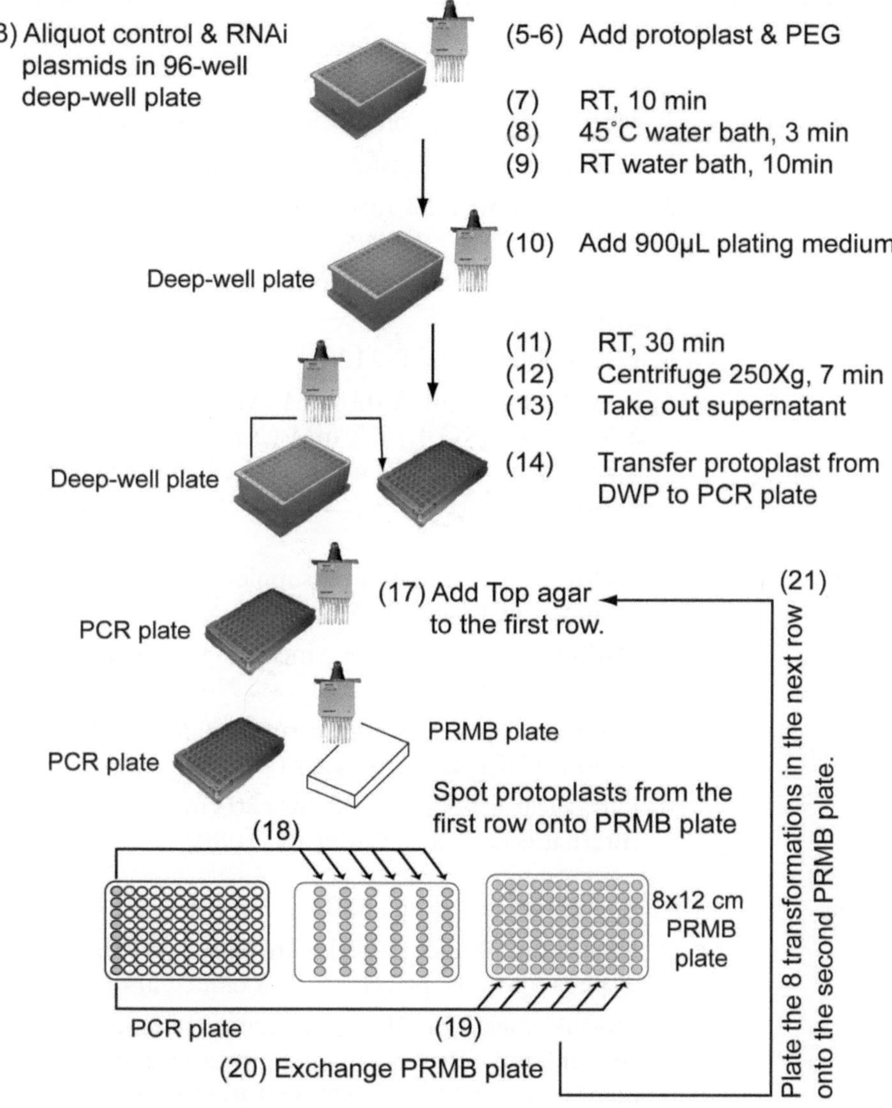

Fig. 2. Transformation procedures outlined in Subheading 3.2. Numbers in parentheses indicate steps in Subheading 3.2.

Our lab has successfully used this approach to silence individual genes as well as gene families (9–11).

Here we describe our modified method for performing RNAi in 96-well plates (Fig. 2). This method not only increases the number of genes that can be handled at a time, but also reduces the demand on reagents. Pipetting steps are performed by an Eppendorf EpMotion 5070 to increase the number of samples that can be handled and to ensure the consistency between each transformation. All parameters described here for liquid-handling are optimized for EpMotion 5070.

2. Materials

2.1. Reagents

1. Moss tissue: 7-day-old NLS-4 tissue (see Note 1).
2. Moss protoplast transformation reagents (see Note 2).
3. 3 M solution: 9.1% mannitol, 15 μM $MgCl_2$, and 0.1% MES pH 5.6.
4. PEG solution: 8.5% mannitol, 100 mM $Ca(NO_3)_2$, 10 mM Tris–HCl pH 8.0.
5. PEG 8000 (Sigma P-2139).
6. $PpNH_4$ medium: 1.03 mM $MgSO_4$, 1.86 mM KH_2PO_4, 3.3 mM $Ca(NO_3)_2$, 2.7 mM $(NH_4)_2$-tartrate, 45 μM $FeSO_4$, 9.93 μM H_3BO_3, 220 nM $CuSO_4$, 1.966 μM $MnCl_2$, 231 nM $CoCl_2$, 191 nM $ZnSO_4$, 169 nM KI, 103 nM Na_2MoO_4 (see Note 3).
7. PRMB: $PpNH_4$ medium supplemented with 6% mannitol, 10 mM $CaCl_2$, and 0.8% agar (see Note 4).
8. Plating medium: $PpNH_4$ medium supplemented with 8.5% mannitol and 10 mM $CaCl_2$ (see Note 4).
9. Top agar: $PpNH_4$ medium supplemented with 6% mannitol, 10 mM $CaCl_2$ and 1.6% agar (see Note 4).
10. Ultra clear cellophane, 28 cm × 50 cm roll (Research Products International Corp., Order no. 1080).

2.2. Equipment

1. Eppendorf centrifuge 5804 and rotor A-2-DWP.
2. EpMotion 5070 equipped with 1-mL and 50-μL eight-channel dispensing tools, placed inside a tissue culture hood.
3. Module racks for 30-mL reservoirs connected to a heating element and a thermocouple to measure temperature. The element is connected to a temperature-control unit that measures the temperature of the block (Fig. 3).
4. Thermoadapter for 96-well PCR plate connected to a heating element and a thermocouple to measure temperature. The element is connected to a temperature-control unit that measures the temperature of the block (Fig. 3).
5. Deepwell plate 96/1,000 μL (Eppendorf Cat no. 951032603).
6. Deepwell Mat 96 (1.2 mL) (Eppendorf Cat no. 951030121).
7. Twin-tec PCR plate 96, skirted (Eppendorf Cat no. 951020401).
8. Reservoir Rack (Eppendorf Cat no. 960002148).
9. EpMotion reservoirs 30 mL, 100 mL (Eppendorf Cat no. 960051009).
10. Ep T.I.P.S. Motion 40–1,000 μL (Eppendorf Cat no. 960050088).

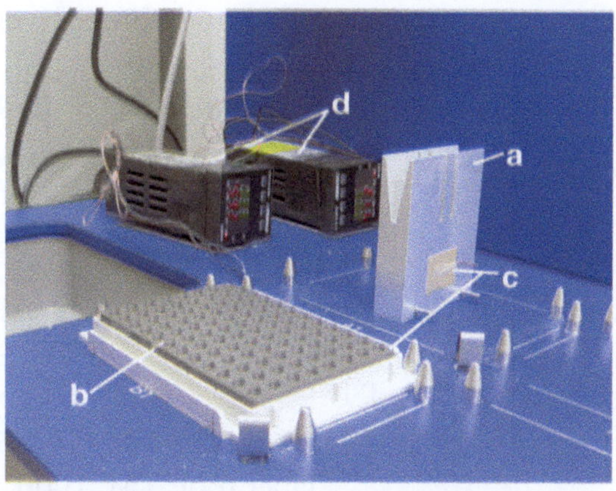

Fig. 3. Heating system for top agar. Module-rack for 30-mL reservoir (**a**) and thermoadaptor for 96-well PCR plate (**b**) are attached to heating elements and thermocouples (**c**), which are each connected to a separate temperature-control unit (**d**). Heating elements are on the side of the module-rack and beneath the thermoadaptor.

11. Ep T.I.P.S. Motion 1–50 μL (Eppendorf Cat no. 960050002).
12. One well dish non-treated sterile with Lid, 127.8 mm × 85.5 mm (nunc 267060).

3. Methods

3.1. Preparation Before Transformation

1. The design and generation of RNAi constructs has been described (7, 8). Make RNAi constructs and prepare RNAi and control plasmids using a commercially available mini-prep kit (see Note 5).
2. Autoclave deep-well plates, deep-well mat, PCR plates, and reservoirs.
3. Prepare PRMB plates: Pour 40 mL of PRMB in each 8 cm × 12 cm 1-well dish and let the agar solidify (see Note 6).
4. Cut cellophane into 8 cm × 12 cm rectangular shape. Autoclave cellophane in a glass petri-dish containing water.
5. In the tissue culture hood, carefully overlay the surface of each PRMB plate with a piece of wet cellophane. Make sure there are no wrinkles or air bubbles. Let the plates dry in the tissue culture hood till there is no visible water on the surface of the cellophane.

3.2. Transformation

1. Prepare PEG: weigh out 4 g of PEG 8000 into 50 mL conical tube, melt in microwave, watch carefully while melting. In a sterile hood, add 10 mL of PEG solution into the same conical tube, mix well by vortexing. This solution can be used after 2 h or can be stored at −20 °C for long-term storage.

2. Transformation procedures are outlined in Fig. 2.
3. Aliquot plasmid DNA into a 96-well deep well plate. Each well holds an independent transformation (see Note 7).
4. Generate protoplasts as previously described (12). Resuspend protoplasts in 3 M solution at a density of 2×10^6 cells/mL.

On EpMotion5070 (see Note 8):

5. Put protoplasts in a 30 mL reservoir, aliquot 50 μL into each well in the 96-well deep-well plate with the 1-mL eight-channel pipette.

   EpMotion Command: Reagent Transfer

   Pipet tool: TM_1000_8

   Filter tips: no

   Volume: 50 μL

   Transfer type: pipette

   Source: Tubs_1 (30 mL reservoir)

   Destination: DWP 96_1

   Liquid type: glycerol

   Dosing parameter:

   – Speed aspiration: 10,000 mm/s
   – Speed dispense: 20,000 mm/s
   – Delay blow: 150 ms
   – Speed blow: 66,000 mm/s
   – Movement blow: 90% of max movement
   – Prewetting: 0 cycles

   Change tips before each aspiration

   Change tips after 0 aspirations

   Mix before no

   Mix after no

   Special aspirate from bottom

6. Put the PEG solution in a 30 mL reservoir, aliquot 50 μL into each well in the 96-well deep-well plate with the 1-mL eight-channel pipette. Mix by pipetting up and down.

   EpMotion Command: Reagent Transfer

   Pipet tool: TM_1000_8

   Filter tips: no

   Volume: 50 μL

   Transfer type: pipette

   Source: Tubs_1 (30 mL reservoir)

Destination: DWP 96_1

Liquid type: glycerol

Dosing parameter:

- Speed aspiration: 5,000 mm/s
- Speed dispense: 110,000 mm/s
- Delay blow: 400 ms
- Speed blow: 100,000 mm/s
- Movement blow: 90% of max movement
- Prewetting: 0 cycles

Change tips before each aspiration

Change tips after 0 aspirations

Mix before no

Mix after yes

- No. of cycles: 3
- Speed: 5 mm/s
- Volume: 80 μL
- Fixed height: yes
- Asp./Disp.: 1,000/3,000 mm

7. Put the deep-well mat on the deep-well plate. Let the plate stand at room temperature for 10 min.

Out of the hood: make sure the lid on the deep-well plate is tightly sealed.

8. Incubate the deep-well plate in a 45 °C water bath for 3 min (see Note 9).
9. Incubate the deep-well plate in a room temperature water bath for 10 min (see Note 9).

On EpMotion5070:

10. Put the plating medium supplemented with 10 mM $CaCl_2$ in a 100-mL reservoir. Aliquot 900 μL into each well in the 96-well deep-well plate with the 1-mL eight-channel pipette. Mix by pipetting up and down.

    EpMotion Command: Reagent Transfer

    Pipet tool: TM_1000_8

    Filter tips: no

    Volume: 900 μL

    Transfer type: pipette

    Source: Tubs_1 (100 mL reservoir)

    Destination: DWP 96_1

    Liquid type: protein

Dosing parameter:

- Speed aspiration: 7,700 mm/s
- Speed dispense: 7,700 mm/s
- Delay blow: 0 ms
- Speed blow: 15,400 mm/s
- Movement blow: 0% of max movement
- Prewetting: 0 cycles

Change tips before each aspiration

Change tips after 0 aspirations

Mix before no

Mix after yes

- No. of cycles: 2
- Speed: 6 mm/s
- Volume: 800 μL
- Fixed height: no
- Asp./Disp.: 0/0 mm

11. Let the plate stand at room temperature for 30 min.

Out of the hood: make sure the lid on the deep-well plate is tightly sealed.

12. Spin down protoplasts at 250 × *g* for 7 min.

On EpMotion 5070:

13. Take out 900 μL of the supernatant from each well in the 96-well deep-well plate with the 1-mL eight-channel pipette.

    EpMotion Command: Pool to One Destination

    Pipet tool: TM_1000_8

    Filter tips: no

    Volume: 900 μL

    Transfer type: pipette

    Source: DWP 96_1

    Destination: Tubs_1 (100 mL reservoir)

    Liquid type: Speed_xs

    Dosing parameter: default

    Change tips before each aspiration

    Change tips after 0 aspirations

    Mix before no

    Mix after no

14. Resuspend protoplasts and transfer 75 μL to a 96-well PCR plate with the 1-mL eight-channel pipette.

    EpMotion Command: Sample Transfer

Pipet tool: TM_1000_8

Filter tips: no

Volume: 75 μL

Transfer type: pipette

Source: DWP 96_1

Destination: PCR 96_1

Liquid type: glycerol

Dosing parameter:

- Speed aspiration: 20,000 mm/s
- Speed dispense: 10,000 mm/s
- Delay blow: 70 ms
- Speed blow: 66,000 mm/s
- Movement blow: 90% of max movement
- Prewetting: 0 cycles

Change tips before aspirating a new sample

Change tips after 0 aspirations

Mix before yes

- No. of cycles: 5
- Speed: 5 mm/s
- Volume: 70 μL
- Fixed height: no
- Asp./Disp.: 2,000/5,000 mm

Mix after no

Special aspirate from bottom

15. Place the 96-well PCR plate containing the protoplasts on a thermoadapter connected to a temperature-controlling unit. Keep the PCR plate at 33 °C.
16. Put top agar in a 30-mL reservoir. Place the reservoir on a holder connected to the temperature-control unit. Keep the top agar at 50 °C.
17. Aliquot 45 μL of top agar into each well in the 96-well PCR plate with the 50-μL eight-channel pipette. Mix by pipetting up and down.

    EpMotion Command: Reagent Transfer

    Pipet tool: TM_50_8

    Filter tips: no

    Volume: 45 μL

    Transfer type: pipette

    Source: Tubs_1

Destination: PCR 96_1

Liquid type: glycerol

Dosing parameter: default

Change tips keep tips, do not change tips

Change tips after 0 aspirations

Mix before no

Mix after yes

- No. of cycles: 5
- Speed: 10 mm/s
- Volume: 35 μL
- Fixed height: no
- Asp./Disp.: 2,000/5,000 mm

18. Dispense the protoplast–agar mixture in 6 μL spots in a 96-well pattern on a PRMB plate overlaid with cellophane. Plate the even numbered columns (see Fig. 2, Note 10).

EpMotion Command: Sample Transfer

Pipet tool: TM_50_8

Filter tips: no

Volume: 6 μL

Transfer type: multidispense

Source: PCR 96_1

Destination: PCR 96_2 (PRMB plate)

Liquid type: glycerol

Dosing parameter:

- Speed aspiration: 15,400 mm/s
- Speed dispense: 20,000 mm/s
- Delay blow: 0 ms
- Speed blow: 20,000 mm/s
- Movement blow: 90% of max movement
- Prewetting: 0 cycles

Change tips before each aspiration

Change tips after three aspirations

Mix before no

Mix after no

Special aspirate from bottom

19. Repeat step 17, plate the odd-numbered columns (see Fig. 2, Note 10).
20. Replace the PRMB plate with a new PRMB plate.
21. Repeat steps 16–18 for each column (see Note 11).

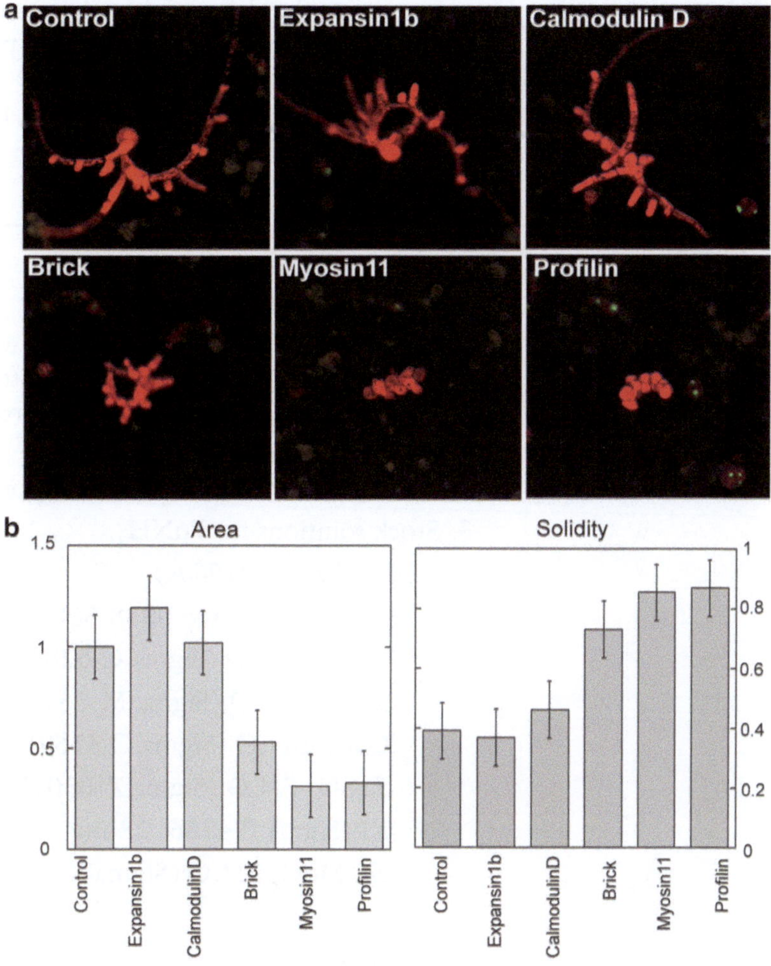

Fig. 4. (**a**) Representative images of 1-week-old plants transformed with control and RNAi constructs visualized by epifluorescence stereomicroscopy. The *red color* is the chlorophyll autofluorescence. The absence of nuclear GFP indicates they are undergoing active gene silencing. (**b**) Quantification of RNAi-induced phenotypes (control, $n=151$; expansin 1b, $n=34$; calmodulin D, $n=54$; brick, $n=96$; myosin 11, $n=69$; profilin, $n=87$). *Error bars* indicate SE. *Y*-axes represent area normalized to the control (*left*) and solidity (*right*). Solidity is a ratio of area to convex hull area.

22. Regenerate protoplasts on PRMB for 4 days.
23. Four days after transformation, transfer protoplasts by lifting the cellophane and moving them to $PpNH_4$ plates containing antibiotic selection (see Notes 12 and 13).
24. Let the plants grown on $PpNH_4$ for another 3 days.
25. One week after transformation, image plants lacking expression of GFP in the nucleus for phenotypic analysis (see Notes 14 and 15).
26. Examples of RNAi phenotypes are shown in Fig. 4. For each RNAi construct, at least 25 plants were imaged for growth by

measuring the plant area based on chlorophyll autofluorescence. The degree of polarization of the plants was estimated using solidity. Solidity is the ratio of the area over the convex hull area, which is the smallest convex polygon that contains the shape under analysis (see Note 16).

4. Notes

1. Moss tissue should be propagated weekly. Tissue is ground in water with a homogenizer and plated on $PpNH_4$ plates overlaid with cellophane (Fig. 5). Growth chamber condition: 25 °C, 16 h light and 8 h dark.
2. All reagents should be autoclaved or filter-sterilized.
3. Stock solutions for $PpNH_4$:

   Micro Elements (1000×):

   H_3BO_3 (Amresco 0588-500 G) 614 mg
   $CuSO_4 \cdot 5H_2O$ (Sigma C-6283) 55 mg
   $MnCl_2 \cdot 4H_2O$ (Sigma M-3643) 389 mg
   $CoCl_2 \cdot 6H_2O$ (Sigma C-3169) 55 mg
   $ZnSO_4 \cdot 7H_2O$ (Sigma Z-0501) 55 mg
   KI (Sigma P-4286) 28 mg
   $Na_2MoO_4 \cdot 2H_2O$ (Sigma S-6646) 25 mg

Fig. 5. Propagation of moss tissue. (**a**) One-week-old moss tissue growing on a $PpNH_4$ plate overlaid with cellophane. (**b**) Using a metal spatula, gently scrape moss tissue off the plate. (**c**) Put harvested moss tissue in sterile water in a culture tube. (**d**) Grind moss tissue with a tissue homogenizer and then plate onto a new $PpNH_4$ plate.

Measure out the chemicals on an analytical balance.

Add H_2O to 1 L. Autoclave and store at 4 °C.

500× stock solutions (make separately)

127 g/L $MgSO_4 \cdot 7H_2O$ (Sigma 230391-500 G)

125 g/L KH_2PO_4

400 g/L $CaNO_3 \cdot 4H_2O$ (Sigma 237124-500 G)

250 g/L di-ammonium tartrate (Sigma A4767-500 G)

Autoclave and store at 4 °C

To make 1× $PpNH_4$ medium, fill a beaker with distilled water to 3.5 L, add 8 mL of each 500× stock solution, 4 mL of micro elements, and 50 mg of $FeSO_4 \cdot 7H_2O$ (Amresco 0387-500 G). Bring the volume to 4 L. For agar plates add 7 g of agar per liter of media. Autoclave. Media can be used immediately or stored at room temperature.

4. PRMB, plating medium, and top agar should be made without $CaCl_2 \cdot 2H_2O$ and autoclaved. Before use, add 20 mL of 500 mM $CaCl_2$ to each liter of medium, shake well.
5. Each transformation requires 2–3 μg of plasmid DNA. Carry out mini-preps from 5 mL of overnight bacterial culture and elute in 40 μL. One mini-prep is generally enough for three transformations.
6. Make sure the surface of the agar plate is level and the height of the agar medium is uniform for each plate.
7. The volume of DNA should be less than 15 μL.
8. All methods described here are based on EpMotion 5070 program settings. All parameters are optimized to handle protoplasts and other transformation reagents.
9. The deep-well plates float on water easily.
10. In the EpMotion 5070 method, choose a 96-well PCR plate to represent the rectangular PRMB plate. This will allow the machine to pipette onto the plate in a 96-well pattern. Set the liquid volume of the PCR plate as 8 μL, so when the pipette dispenses protoplasts onto the PRMB plate, the tip will be at the surface of the plate.
11. Eight transformations in one column are plated on one PRMB plate. Each transformation is plated in a row of 12 spots (see Fig. 2).
12. Suggested antibiotic concentration: hygromycin (15 μg/mL), kannamycin (30 μg/mL), zeocin (50–100 μg/mL).
13. RNAi constructs generated from pUGGi and the control plasmid pUGi carry the hygromycin resistance gene.

14. With good transformation efficiency, each spot should have at least ten transformants (antibiotic-resistant plants). The number of plants lacking nuclear GFP varies and depends on gene that is being silenced.
15. Plants are imaged with a stereo fluorescence microscope (Leica MZ16FA). Images are acquired with a color camera (Leica DFC 300FX) using the GFP2 filter (excitation 480/40 nm, emission 400 long pass) for chlorophyll and GFP.
16. Image-acquisition can be facilitated by an automatic stage on the microscope. Image analysis can be performed/automated with macros developed for ImageJ as described in (see ref. 9).

Acknowledgments

We thank Lawrence Winship for building the temperature control systems. We thank Paula Franco and Luis Vidali for initiating the project and David O'Donnell for performing pilot screens. We are grateful to members of the Bezanilla, Baskin and Hepler labs for helpful suggestions and recommendations. This work was supported by a fellowship from the David and Lucille Packard Foundation to (MB) and the University of Massachusetts – Amherst Plant Biology Graduate Program (SW).

References

1. Kamath RS et al (2003) Systematic functional analysis of the *Caenorhabditis elegans* genome using RNAi. Nature 421:231–237
2. Goshima G et al (2007) Genes required for mitotic spindle assembly in *Drosophila* S2 cells. Science 316:417–421
3. Wheeler DB et al (2004) RNAi living-cell microarrays for loss-of-function screens in *Drosophila melanogaster* cells. Nat Meth 1:127–132
4. Boutros M, Ahringer J (2008) The art and design of genetic screens: RNA interference. Nat Rev Genet 9:554–566
5. Collinet C et al (2010) Systems survey of endocytosis by multiparametric image analysis. Nature 464:243–249
6. Cove D et al (2006) Mosses as model systems for the study of metabolism and development. Annu Rev Plant Biol 57:497–520
7. Bezanilla M et al (2003) RNA interference in the moss *Physcomitrella patens*. Plant Physiol 133:470–474
8. Bezanilla M et al (2005) An RNAi system in *Physcomitrella patens* with an internal marker for silencing allows for rapid identification of loss of function phenotypes. Plant Biol 7:251–257
9. Vidali L et al (2009) Rapid formin-mediated actin-filament elongation is essential for polarized plant cell growth. Proc Natl Acad Sci USA 106:13341–13346
10. Vidali L et al (2007) Profilin is essential for tip growth in the moss *Physcomitrella patens*. Plant Cell 19:3705–3722
11. Augustine RC et al (2008) Actin depolymerizing factor is essential for viability in plants, and its phosphoregulation is important for tip growth. Plant J 54:863–875
12. Schaefer D et al (1991) Stable transformation of the moss *Physcomitrella patens*. Mol Gen Genet 226:418–424

Chapter 18

A High-Throughput Biological Conversion Assay for Determining Lignocellulosic Quality

Scott J. Lee, Thomas A. Warnick, Susan B. Leschine, and Samuel P. Hazen

Abstract

Lignocellulosic biomass is a source of low cost polysaccharides that some microbes can deconstruct and convert into liquid transportation fuel. Feedstocks vary in their ease of use depending on their source and handing. Estimating conversion amenability is useful to determine the effects of biomass pretreatment and genetic potential for the purposes of energy crop breeding and genetics. Here we describe a small-scale high-throughput assay that measures ethanol production from a culture of plant biomass and the ethanologen *Clostridium phytofermentans*.

Key words: Plant biomass, Feedstock quality, Lignocellulosic biological conversion assay, Ethanol yield, *Clostridium phytofermentans*

1. Introduction

Fossil fuels are a finite resource and the practice of burning them to meet our energy needs has resulted in the highest atmospheric CO_2 concentrations in over 800,000 years (1, 2). A potential alternative is renewable energy derived from sources not irreversibly exhausted by their use. In 2010, the US produced more than 13 billion gallons (46 billion L) of ethanol mostly by fermentation of starch milled predominantly from corn grain (DOE-Energy Information Administration, 2011). The Energy Independence and Security Act of 2007 mandated that 36 billion gallons (136 billion L) of biofuels be blended into gasoline, diesel, and jet fuel by 2022. This legislation capped the amount of biofuels derived from corn starch at 15 billion gallons (60.6 billion L) and specified that 21 billion gallons (79.5 billion L) be derived from lignocellulosic

Jennifer Normanly (ed.), *High-Throughput Phenotyping in Plants: Methods and Protocols*, Methods in Molecular Biology, vol. 918, DOI 10.1007/978-1-61779-995-2_18, © Springer Science+Business Media, LLC 2012

sources such as crop or forest residues and dedicated energy crops such as energy sorghum, switchgrass, and shrub willow (3). A joint study by the DOE and USDA determined that there is an annual supply of greater than one billion tons of non-grain biomass available from forestry and agricultural resources to support a renewable biofuels industry, a great deal more than enough to meet the US government's target of 30% displacement of current US gasoline by 2030 (4). In order to better reach this goal and maintain a plant feedstock derived biofuel industry that is carbon neutral or negative, energy crop yield per unit area will need to be maximized under low input conditions and the biomass should be amenable to facile deconstruction.

A key bottleneck in cellulosic biofuels production, which limits its cost-effectiveness, is the initial breakdown of biomass into simple sugars (5). A central issue of importance for reducing processing costs is decreasing the recalcitrance of lignocellulosic biomass, which can be accomplished by several methods of chemical or physical pretreatment (6). There is also extensive genetic variation among and within plant species for feedstock properties (e.g., (7–10)). Beyond the expectation of high biomass yield with few inputs on marginal land, energy crops lack similar quality standards, a principal reason being the lack of a standing lignocellulosic biofuels industry to institute parameters. Tomatoes and beans must withstand the canning process, potatoes must chip with conventional form and color, and wheat must be a predictable ingredient for products such as pastries or leavened bread. We developed an assay to estimate the ease at which a specific feedstock can be digested and converted to ethanol using a simulated consolidated bioprocessing approach. Here we describe an approach using a well-developed microbial system, *Clostridium phytofermentans*, as a bioassay to measure feedstock digestibility and thereby determine the potential effects of altered feedstock traits and treatments. The use of *C. phytofermentans* takes into consideration specific organismal interactions, which will certainly be critical in single stage fermentation or consolidated bioprocessing. Briefly, stem tissue from completely senesced plants is pulverized in a ball mill, inoculated with *C. phytofermentans*, and allowed to grow anaerobically. Total sample requirements are as low as 20 mg, making multiple measurements of stems from very small plants such as *Brachypodium distachyon* and *Arabidopsis thaliana* possible. Supernatant ethanol concentration measured by high performance liquid chromatography (HPLC) is used as the metric to estimate feedstock quality and conversion efficiency. These results demonstrate that we are capable of measuring subtle but industrially relevant genetic diversity among energy crops and their research model species. Our novel assay permits determination of the potential effects of numerous variables in biofuel production, including the ability to measure the impact of pretreatment, conversion processes, microbial and plant genetic diversity in digestibility. In contrast to other established methods, the *C. phytofermentans* bioassay provides a direct and quantitative means of assessing feedstock quality, in terms of both digestibility and conversion.

2. Materials

2.1. Plant Biomass Washing (see Notes 1 and 2)

1. 70% (v/v) ethanol.
2. 70% (v/v) methanol.
3. Stainless steel grinding jar with one 15 mm stainless steel grinding ball per jar.
4. Retsch Mixer Mill MM400.

2.2. M6 Media Components

1. Modified Balch's trace elements (BTE) (1 L): dissolve 1.5 g nitriolotriacetic acid (NTA) in 500 mL dH_2O and adjust pH to 6.5 using 1 M KOH. Add the following salts one at a time: 3.0 g $MgSO_4{\cdot}7H_2O$, 0.5 g $MnSO_4{\cdot}4H_2O$, 1.0 g NaCl, 0.1 g $FeSO_4{\cdot}7H_2O$, 0.1 g $CoCl_2{\cdot}6H_2O$, 0.1 g $CaCl_2$, 0.1 g $ZnSO_4{\cdot}7H_2O$, 0.01 g CuSO 5 H_20, 0.01 g $AlK(SO_4)_2{\cdot}12H_2O$, 0.01 g H_3BO_3, 0.01 g $Na_2MoO_4{\cdot}2H_2O$, 0.03 g $NiSO_4{\cdot}6H_2O$, 0.02 g Na_2SeO_3, 0.02 g $Na_2WO_4{\cdot}2\ H_20$. Make up to 1 L with dH_2O. Autoclave to sterilize.
2. XT2 solution (100 mL): dissolve 0.25 g xanthine and 0.25 g thymine in 99 mL of dH_2O and 1 mL 6 N NaOH. Filter sterilize using a 0.22 μm syringe filter or autoclave.
3. AA1 solution (100 mL): dissolve 0.5 g alanine, 0.5 g arginine, 0.5 g histidine, 0.5 g isoleucine, 0.5 g leucine, 0.5 g methionine, 0.5 g proline, and 0.5 g valine in 95 mL dH_2O. Make up to 100 mL. Filter sterilize using a 0.22 μm syringe filter or autoclave.
4. Resazurin solution: resazurin sodium salt (Sigma-Aldrich) in water to 0.1% in (pH should be approximately 7.2, do not adjust; see Note 3).
5. M6 media (100 mL): dissolve 0.2 g NaH_2PO_4, 1.0 g K_2HPO_4, 0.1 g $(NH_4)_2SO_4$, 0.1 g cysteine·HCl in 92 mL dH_2O. Add 1.0 mL AA1 solution, 4.0 mL XT2 solution, 1.0 mL BTE solution. Add 0.1 mL of 0.1% resazurin solution. Transfer to round bottom flask and heat until boiling, then immediately remove from heat. Place on heated stir plate and deoxygenate under nitrogen stream until media has turned clear (see Notes 4 and 5). Utilize this media in either the 96-well or 30.0 mL tube procedures.

2.3. Culturing Components

1. CP4 vitamin mix (100 mL): dissolve 4.0 mg *p*-aminobenzoic acid, 0.1 mg biotin, 0.6 mg folinic acid, 8.0 mg nicotinamide, 0.5 mg pantethine, 0.4 mg pyridoxal HCl, 3.0 mg riboflavin, 1.0 mg thiamine in 95 mL dH_2O. Adjust pH to 7 using 1 N NaOH and dissolve. Make up to 100 mL and filter sterilize using a 0.22 μm syringe filter. Store at 4°C.
2. 10% cellobiose (w/v) solution (see Note 6).
3. 30 mL glass anaerobic culture tubes with rubber stoppers (see Note 7).
4. N_2 gas.
5. Anaerobic chamber or Hungate apparatus (see Note 4).
6. *C. phytofermentans* ISDg (ATCC 700394).
7. 2.2 mL square well autoclave safe 96-deep-well plate.
8. Adhesive foil plate seal.
9. Plate clamp (see Note 8).

2.4. HPLC

1. Screw-top HPLC vials.
2. HPLC system with Refractive Index Detector and 7.8 mm × 150 mm IC-Pak Ion Exclusion column (Waters catalog number—WAT010295).

3. Methods

3.1. Plant Biomass Washing

1. Biomass samples should be placed in plastic tube 2× the volume of the biomass to be washed. The tube should then be filled with 70% (v/v) ethanol solution and incubated in a water bath at 70°C for 30 min.
2. Centrifuge tubes for 5 min at 2,107 × *g*, pour off the supernatant, replace with fresh ethanol solution, vortex to mix, and return to the water bath for an additional 30 min.
3. Centrifuge tubes for 5 min at 2,107 × *g*, pour off the supernatant, and wash with a 70% (v/v) methanol solution as above.
4. Samples can be dried in a fume hood with the caps removed for 24 h. To accelerate drying, bake tubes in a 50°C oven overnight.
5. Using a 25.0 mL stainless steel grinding jar with one 15 mm stainless steel grinding ball per jar, homogenize samples using a Mixer Mill for 3 min at 30 Hz (see Note 1). A robotic platform can be used to facilitate high-throughput sample preparation (see Note 2).

3.2. Inoculum Preparation

1. Prepare M6 liquid media and CP4 vitamin mix according to Materials section as well as a 10% cellobiose (w/v) solution (see Notes 5 and 6).
2. Dispense 9.8 mL of M6 media into individual 30.0 mL glass anaerobic culture tubes with rubber stoppers, flush with nitrogen stream until solution becomes clear (see Note 6).
3. Cool tubes to room temperature using ice bath. Remove from nitrogen stream and seal rubber stopper tightly to ensure the media remains anaerobic while in the autoclave.
4. Autoclave, then allow tubes to cool to room temperature before using (see Notes 7 and 9).
5. *C. phytofermentans* ISDg (ATCC 700394) inoculum is prepared from frozen stock for all experiments. To prepare fresh culture, add 0.1 mL of CP4 vitamin mix and 0.1 mL of 10% cellobiose solution to one tube of M6 media.
6. Thaw a frozen culture of *C. phytofermentans* from stock and add 0.1 mL of completely thawed stock solution to freshly prepared media.
7. Incubate *C. phytofermentans* inoculum for 48 h at 30°C (see Note 10).
8. Transfer 0.1 mL of the inoculum to a fresh tube of M6 media along with 0.1 mL of CP4 vitamin mix and 0.1 mL of 10% (w/v) cellobiose solution.
9. Incubate 24 h prior to inoculating samples.

3.3. Inoculation of Samples in 96-Well Format

1. Dispense 20.0 mg of washed plant material into 2.2 mL square well autoclave safe 96-deep-well plate.
2. Add 0.98 mL of M6 media to each well using a multi-channel pipette. Seal with adhesive foil plate seal (see Note 5).
3. Before autoclaving, apply a clamp to the top of the plate to maintain the seal during sterilization (Fig. 1, see Note 8). Allow plates to cool to room temperature before inoculating.
4. Add 0.01 mL CP4 vitamin mix and 0.01 mL of prepared *C. phytofermentans* inoculum to each well utilizing a multi-channel pipet (see Notes 10 and 11).
5. Seal plate using a new adhesive foil directly over top of the perforated foil and place in incubator at 30°C for 72 h.
6. Centrifuge plates for 10 min at 2,107 × *g*.
7. Remove 1.0 mL of sample supernatant and filter into 2.0 mL screw-top HPLC vials using a 0.22 μm syringe filter.
8. Analyze samples for ethanol concentration utilizing an HPLC with refractive index (RI) detection. For example, we utilized a Waters Breeze 2 HPLC with RI detection with a 7.8 mm × 150 mm

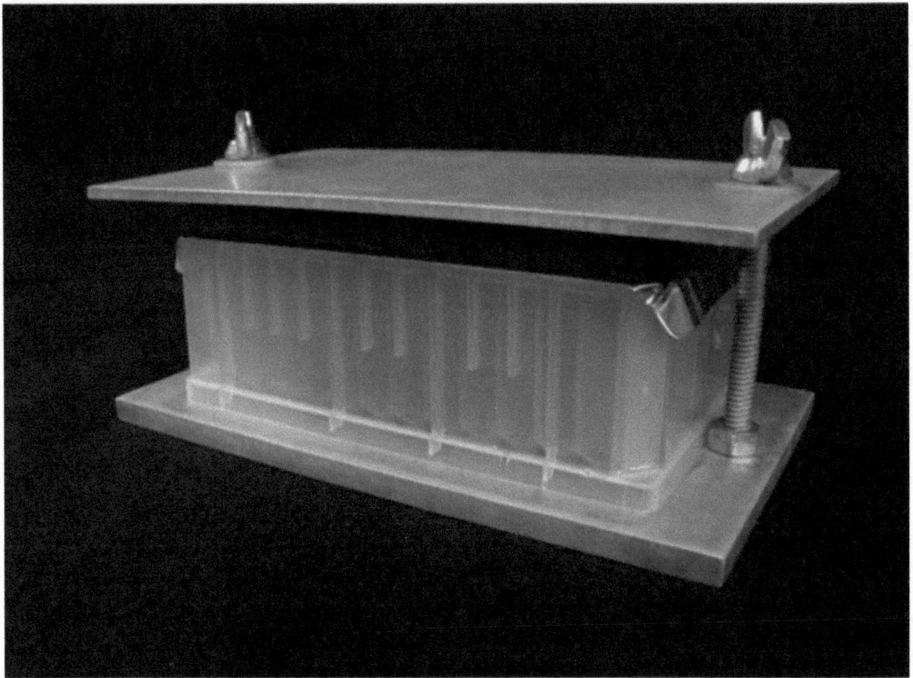

Fig. 1. 96-Well plate format for bioconversion assay. Applying pressure to the top of the plate with a clamp is necessary to maintain the foil seal during autoclaving.

IC-Pak Ion Exclusion column at 75°C with 0.01 N sulfuric acid mobile phase at a 1.0 mL/min flow rate and 5 μL injections. The RI detector was set to 30°C with a sensitivity setting of 32, 0.2 s time constant, and a 5 measurements/s sampling rate. (see Note 12).

3.4. Inoculation of Samples in 30.0 mL Glass Tube Format

1. Dispense 50.0 mg washed plant materials into 30.0 mL glass anaerobic culture tubes with rubber stoppers (see Note 4).
2. Dispense 9.8 mL of M6 media into individual 30.0 mL glass anaerobic culture tubes with rubber stoppers, flush with nitrogen stream until solution becomes clear (see Note 5).
3. Cool tubes to room temperature using ice bath. Remove from nitrogen stream and seal rubber stopper tightly to ensure media remains anaerobic while in the autoclave (see Note 7).
4. Autoclave, then allow tubes to cool to room temperature before using.
5. Transfer 0.1 mL of the prepared *C. phytofermentans* inoculum and 0.1 mL of CP5 vitamin mix, mixing gently. Seal tubes and place in incubator at 30°C for 72 h.
6. Remove 1.0 mL of sample supernatants and filter samples into 2.0 mL screw-top HPLC vials using a 0.22 μm syringe filter.
7. Analyze samples for ethanol concentration utilizing an HPLC with RI detection (see Note 12, Fig. 2).

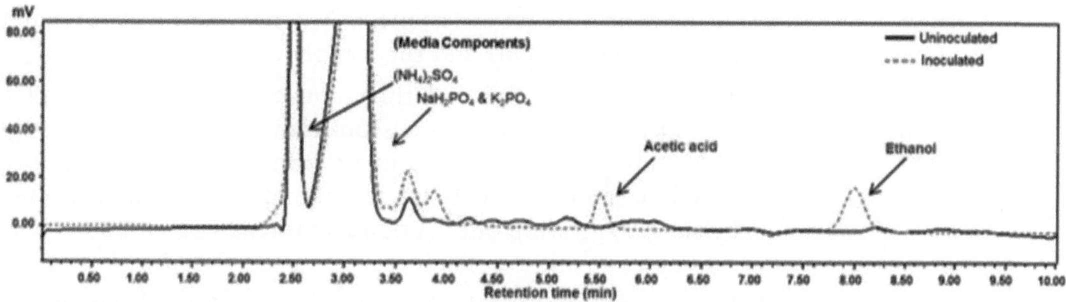

Fig. 2. *Clostridium phytofermentans* produces ethanol as a fermentation byproduct. HPLC-refractive index detection analysis of fermentation byproducts from sorghum inoculated with *C. phytofermentans*. Peaks from 2.0 to 4.0 min retention time are congruent with media additives of M6 minimal medium. Acetic acid and ethanol elute at 5.55 and 8.00 min, respectively.

4. Notes

1. Stems from woody species such as shrub willow or whole maize plants require initial processing with a cutting mill to reduce particle size before the final washing and grinding steps.
2. For higher throughput processing, Labman Automation Ltd. (Stokesley, North Yorkshire, UK) manufactures custom automated platforms to grind and dispensing weighed plant powder (see refs. (11–13)).
3. Resazurin is an indicator that will turn a solution pink in the presence of oxygen. If at any point a solution intended to be aerobic turns pink, return to nitrogen stream until the solution is clear.
4. For all bench work not done in an anaerobic chamber, we utilized a Hungate apparatus described in (see ref. (14)). Briefly, a furnace is utilized to heat copper inside of a glass tube, impure nitrogen is then passed through the tubing allowing any oxygen present to react and oxidize the copper. The deoxygenated nitrogen gas is then utilized to work anaerobically at the bench.
5. For best results keep M6 stock on a stirring hot plate just below boiling temperature and under nitrogen stream to prevent oxygenation of media.
6. The 10% cellobiose (w/v) solution serves as a simple substrate for *C. phytofermentans* cultures and can be filter sterilized using a 0.22 μm syringe filter.
7. To secure the stoppers throughout autoclaving, the tube rack should be placed inside a clamp. Briefly, two flat stainless steel plates (12.75 in. × 6.25 in. × 0.25 in.) are connected using 9 in. stainless steel bolts in each of the four corners and secured with eight 1/2 in. bolts, four 1/2 in. washers, and four metal tabs

(4 in. × 3 in.) to act as fasteners. A dense foam mat is secured between the stoppers and the top plate.

8. To maintain the foil seal throughout autoclaving, the plate should be placed inside a custom clamp. Briefly, two flat cast aluminum plates (6.5 in. × 4 in. × 0.25 in.) are connected using two 3 in. stainless steel bolts, two 0.5 in. washers, and fastened by two 1.0 in. wing nuts. A dense foam mat is secured between the top plate and the clamp.
9. Following sterilization, plates should remain inside an anaerobic chamber. Plates with plant material should be left for 24 h in anaerobic chamber to remove residual oxygen. In our experience, shrub willow biomass is considerably more recalcitrant than grasses. Add 2× biomass or culture for twice as long when working with woody species.
10. Do not add inoculum to wells that will serve as uninoculated control.
11. Media and inoculum can be added to the sealed plate by puncturing the foil above each well with the multi-channel pipette tips.
12. In order to calculate ethanol concentrations, a serial dilution of ethanol must be created and run on HPLC. Utilizing this, conduct a linear regression of the standards to create your ethanol standard curve. The linear regression of this standard curve will enable estimation of actual quantities of ethanol of your samples.

Acknowledgments

This research was supported by the Office of Science (BER) Department of Energy Grant DE-FG02-08ER64700DE to SPH. We thank Michael Keedy and Eric Bertrand (Biology Department, University of Massachusetts Amherst) for design and construction of the 96-well plate autoclave clamp.

References

1. Archer D, Eby M, Brovkin V, Ridgwell A, Cao L, Mikolajewicz U, Caldeira K, Matsumoto K, Munhoven G, Montenegro A, Tokos K (2009) Atmospheric lifetime of fossil fuel carbon dioxide. Annu Rev Earth Planet Sci 37: 117–134
2. Luthi D, Le Floch M, Bereiter B, Blunier T, Barnola J-M, Siegenthaler U, Raynaud D, Jouzel J, Fischer H, Kawamura K, Stocker TF (2008) High-resolution carbon dioxide concentration record 650,000–800,000 years before present. Nature 453:379–382
3. Ragauskas AJ, Williams CK, Davison BH, Britovsek G, Cairney J, Eckert CA, Frederick WJ Jr, Hallett JP, Leak DJ, Liotta CL, Mielenz JR, Murphy R, Templer R, Tschaplinski T (2006) The path forward for biofuels and biomaterials. Science 311:484–489
4. Somerville C (2006) The billion-ton biofuels vision. Science 312:1277

5. Lynd LR, Laser MS, Bransby D, Dale BE, Davison B, Hamilton R, Himmel M, Keller M, McMillan JD, Sheehan J, Wyman CE (2008) How biotech can transform biofuels. Nat Biotechnol 26:169–172
6. Brodeur G, Yau E, Badal K, Collier J, Ramachandran KB, Ramakrishnan S (2011) Chemical and physicochemical pretreatment of lignocellulosic biomass: a review. Enzyme Res. Article ID 787532
7. Serapiglia M, Cameron K, Stipanovic A, Smart L (2009) Analysis of biomass composition using high-resolution thermogravimetric analysis and percent bark content for the selection of shrub willow bioenergy crop varieties. Bioenergy Res 2:1–9
8. Barrière Y, Denoue D, Briand M, Simon M, Jouanin L, Durand-Tardif M (2006) Genetic variations of cell wall digestibility related traits in floral stems of *Arabidopsis thaliana* accessions as a basis for the improvement of the feeding value in maize and forage plants. Theor Appl Genet 113:163–175
9. Barriere Y, Guillet C, Goffner D, Pichon M (2003) Genetic variation and breeding strategies for improved cell wall digestibility in annual forage crops. A review. Anim Res 52:193–228
10. Vandenbrink JP, Delgado MP, Frederick JR, Feltus FA (2010) A sorghum diversity panel biofuel feedstock screen for genotypes with high hydrolysis yield potential. Ind Crop Prod 31:444–448
11. Santoro N, Cantu S, Tornqvist C-E, Falbel T, Bolivar J, Patterson S, Pauly M, Walton J (2010) A high-throughput platform for screening milligram quantities of plant biomass for lignocellulose digestibility. Bioenergy Res 3: 93–102
12. Gomez L, Whitehead C, Barakate A, Halpin C, McQueen-Mason S (2010) Automated saccharification assay for determination of digestibility in plant materials. Biotechnol Biofuels 3:23
13. Foster CE, Martin TM, Pauly M (2010) Comprehensive compositional analysis of plant cell walls (lignocellulosic biomass). Part I: lignin. J Vis Exp:e1745
14. Hungate RE (1969) A roll tube method for cultivation of strict anaerobes. Methods Microbiol 3B:117–132

Chapter 19

Carbohydrate Microarrays in Plant Science

Jonatan U. Fangel, Henriette L. Pedersen, Silvia Vidal-Melgosa, Louise I. Ahl, Armando Asuncion Salmean, Jack Egelund, Maja Gro Rydahl, Mads H. Clausen, and William G.T. Willats

Abstract

Almost all plant cells are surrounded by glycan-rich cell walls, which form much of the plant body and collectively are the largest source of biomass on earth. Plants use polysaccharides for support, defense, signaling, cell adhesion, and as energy storage, and many plant glycans are also important industrially and nutritionally. Understanding the biological roles of plant glycans and the effective exploitation of their useful properties requires a detailed understanding of their structures, occurrence, and molecular interactions. Microarray technology has revolutionized the massively high-throughput analysis of nucleotides, proteins, and increasingly carbohydrates. Using microarrays, the abundance of and interactions between hundreds and thousands of molecules can be assessed simultaneously using very small amounts of analytes. Here we show that carbohydrate microarrays are multifunctional tools for plant research and can be used to map glycan populations across large numbers of samples to screen antibodies, carbohydrate binding proteins, and carbohydrate binding modules and to investigate enzyme activities.

Key words: Carbohydrate microarrays, Glycans, High-throughput, Carbohydrate binding proteins, Carbohydrate binding modules, Polysaccharides, Oligosaccharides, Comprehensive microarray polymer profiling, Defined glycan arrays

1. Introduction

Microarray technology has revolutionized the massively high-throughput (HTP) analysis of nucleotides, proteins, and increasingly carbohydrates (1–3). Using microarrays, the abundance of and interactions between hundreds and thousands of molecules can be assessed simultaneously using very small amounts of analytes (1–3). Carbohydrate microarrays were first produced in 2002 and have been widely adopted in medical, animal, and prokaryote research—but far less so in plant research (4–7).

Jennifer Normanly (ed.), *High-Throughput Phenotyping in Plants: Methods and Protocols*, Methods in Molecular Biology, vol. 918, DOI 10.1007/978-1-61779-995-2_19, © Springer Science+Business Media, LLC 2012

This seems paradoxical when one considers the central importance of carbohydrates in plant life. Plants produce an extraordinary diversity of glycans (mono-, oligo-, and polysaccharides as well as glycoproteins), some of which can be considered amongst the most complex in nature. Almost all plant cells are surrounded by a glycan-rich cell wall. These walls form much of the plant body and are collectively the largest source of biomass on earth (8, 9). Plant cell walls display remarkable diversity and complexity and in addition to providing support, are involved in signaling and defense and are required for cell-to-cell adhesion (10). Plants also use polysaccharides for energy storage and many plant glycans are important industrially and nutritionally. Starch is the most common carbohydrate in the human diet whilst plant cell walls provide bulk materials including timber, paper, and cloth, as well as fine chemicals, food ingredients, and biofuel feedstocks (11–13). Understanding the biological roles of plant glycans, and the effective exploitation of their useful properties requires a detailed understanding of their structures, occurrence, and molecular interactions. However, glycans tend to be intractable subjects for research because unlike nucleotides, they cannot be readily sequenced, and unlike peptides and proteins, they cannot easily be produced to order. Biochemical techniques for glycan analysis including methylation analysis to determine glycosidic linkages, NMR, and other spectroscopy methods are powerful and sometimes fully quantitative, but low throughput (14). Enzyme-linked immunosorbent assays (ELISAs) are moderately HTP but do not approach the levels of multiplexed analysis achievable with microarrays. The development of rapid genome sequencing methods and improvements in protein expression techniques enable the identification and production of a large number of carbohydrate-active enzymes as well as carbohydrate binding proteins (CBPs) and carbohydrate binding modules (CBMs). There is though an ever-widening gap between our ability to identify putative carbohydrate active proteins and our ability to determine their activities (15). The need for truly HTP and versatile technology for glycan analysis is therefore becoming increasingly pressing.

The authors have shown that carbohydrate microarrays are multifunctional tools for plant research. They can be used to map glycan populations across large numbers of samples to screen antibodies, CBPs, and CBMs and to investigate enzyme activities. However, significant challenges still remain before this technology reaches its full potential in plant science. In this review we describe some of these challenges and discuss future perspectives for this technology.

1.1. The Basics of Carbohydrate Microarray Construction

All microarrays are based on the deposition or in situ synthesis of samples onto a surface, usually a membrane or slide. In the case of carbohydrate microarrays, a variety of methods have been developed to immobilize samples either covalently or non-covalently onto a variety of surfaces. These have been reviewed elsewhere (16–18), so will not be discussed in detail here but some general principles will be discussed.

There are considerable differences in the construction of arrays populated with polysaccharides and oligosaccharides (in this article oligosaccharides are considered to be glycans with degrees of polymerization between 2 and approximately 20). Plant polysaccharides can usually be immobilized non-covalently by passive adsorption onto a variety of surfaces that are routinely used for assaying biological molecules for example, nitrocellulose membranes, nitrocellulose-coated glass slides, and activated polystyrene (5, 19). This means that plant polysaccharides can normally be printed directly onto the microarray surface without any prior treatment. In contrast, oligosaccharides cannot usually be immobilized by passive adsorption and thus have to be covalently attached using one of the many chemical linker strategies described (16–18). Alternatively, oligosaccharides may be conjugated covalently to a larger molecule, usually a lipid or protein and the resulting neoglycolipid or neoglycoprotein may then be immobilized by passive adsorption onto the surfaces mentioned above (6, 7, 20, 21). Either way, strategies for the immobilization of both oligo- and polysaccharides are robust, well described, and can readily be applied to plant glycans. However, as will be discussed below, one important current bottleneck in the construction of carbohydrate microarrays for plant research is in acquiring well-defined polysaccharides and oligosaccharides with which to populate arrays.

One important technical issue surrounding carbohydrate microarray production is the choice of microarray printing robot. Most robots are designed primarily for printing nucleotides and proteins and are not optimized for carbohydrates. The printing of neoglycoproteins or neoglycolipids does not usually present difficulties because the lipid or protein moiety typically dominates the mass of the sample. Printing high molecular weight and sometimes viscous complex polysaccharides with widely varying chemical and physical properties can, however, be challenging. Two main types of microarray robot are generally available. Pin-based systems rely on either split or solid pins that dip into a solution of glycan samples that are held within a multi-well source plate. Deposition onto the slide or membrane surface is achieved when the pins touch the surface, and this inevitably tends to cause wear of the pins over time. In our experience, non-contact microarray printers, for example those that use a piezoelectric-based system for sample deposition are well suited for the production of plant glycan microarrays. Such robots are fast, accurate, do not suffer from print wear, and importantly can cope with printing glycans in harsh extraction solvents (3).

1.2. Current and Potential Uses of Carbohydrate Microarrays in Plant Research

Examples of carbohydrate microarrays in plant research are relatively scarce, but nonetheless illustrate the potential of this technology. Shipp and coworkers described the use of microarrays containing xyloglucan oligo- and polysaccharides and cellooligosaccharides as multiplexed acceptors for monitoring plant glycosyltransferase (GT) activities (22). They showed that the AtFUT1 α(1,2)fucosyltransferase could transfer (^{14}C)Fuc onto arrayed acceptors and that transfer could be detected using a phosphorimager. This work was further developed to include microarrays of donor polysaccharides that were then probed with crude plant extracts together with fluorophore-labeled acceptor oligosaccharides (23). Such research is important because it offers a potential route for the HTP screening of large numbers of GTs and may contribute to redressing the current imbalance between our ability to identify putative GTs and to determine their functions.

The authors have developed various types of carbohydrate microarray, mostly for plant cell wall research, and some of these have been described in detail previously (24–28). Although diverse in their applications, all these microarrays are based on two basic approaches. One approach that we term "extracted glycan arrays" is used primarily for monitoring the relative abundance of polysaccharides across a large set of samples, for example plant organs, tissue types, mutant populations, or developmental stages. The samples are homogenized and polysaccharides are extracted using sequential treatment with solvents that release the major cell wall polymer classes. For example, when applied to cell walls, a calcium chelator is used to remove pectins and strong base removes hemicelluloses. The extracted material is then printed as multiple microarrays, each of which is probed with monoclonal antibodies or CBMs with specificity for cell wall components. The resulting spot signals provide information about the relative abundance of the epitopes recognized across the sample set. We have used this technique, sometimes referred to as comprehensive microarray polymer profiling or "CoMPP" to analyze a wide variety of diverse plant materials (24). In one example, the fates of cell wall polysaccharides were assessed during processing of leaf material in the fungal gardens of leaf cutting ants (28). In another, we conducted a survey of cell walls in phylogenetically diverse species across the plant kingdom with the aim of gaining insight into plant cell wall evolution and diversity. This work resulted in the unexpected discovery of (1→3)(1→4)-β-D-glucan, a polysaccharide previously thought to be unique within the plant kingdom to Poales species, in the horsetail *Equisetum arvense* (29). This finding demonstrated the potential of the approach for mapping polysaccharide diversity across the plant kingdom, which has relevance for the discovery of novel bio-components and for understanding polysaccharide evolution.

An example of a typical CoMPP experiment is shown in Fig. 1. In this case the technique was used to study the cell walls of various parts of lily (*Lilium longiflorium*) flowers. Flowers were split into separate organs (Fig. 1a), which were then homogenized and extracted with 1,2-cyclohexanediaminetetraacetic acid and sodium hydroxide. Extractions were spotted as microarrays and probed with a range of monoclonal antibodies with specificities for epitopes present on cell wall polysaccharides. Mean spot signals from microarrays are presented as a heatmap (Fig. 1b), which shows the occurrence of epitopes across the organs analyzed. This kind of analysis can be applied to as little as 0.5 mg of cell wall material (alcohol insoluble residue) and takes approximately 2–3 days. CoMPP is often used synergistically with lower throughput techniques such as methylation analysis or immunocytochemistry. For example, in the case of the Lily flower study, CoMPP analysis indicated the presence of several epitopes within style material, which is of interest in relation to understanding the process of pollen tube growth down the stylar transmitting tract. Pollen tube cell walls adhere to the surface of the cells that line the lumen of the transmitting tract, and this adhesion is thought to be essential for the delivery of the sperm cell to the ovule (30). Stylar pectin has been shown to be important in this process (31) and consistent with this, immunofluorescent labeling of transverse sections through styles showed that the cell walls lining the transmitting tract contained the pectic homogalacturonan epitope recognized by mAb JIM7 (Fig. 2a). Interestingly, we also observed that the xyloglucan epitopes recognized by mAbs LM15 and the arabinogalactan epitope recognized by mAb JIM8 were also present at this location and therefore may also be worthy of further investigation with respect to possible roles in pollen tube adhesion to the transmitting tract (Fig. 2b, c, e–g). This example exemplifies how CoMPP used together with other methods can be a powerful tool for investigating plant systems.

The other main approach for plant microarray production involves the production of "defined glycan arrays" that are populated with oligosaccharides or polysaccharides of known structures. This array type is valuable for studying the binding of antibodies, CBMs or lectins and can also be used to investigate enzyme activities. In many cases the types of assays performed using defined glycan arrays are similar in principle to ELISA assays but have the advantages of requiring less reagents and of being higher throughput. The GT assays described above (22, 23) are example of this approach, and we have previously described the use of defined glycan microarrays for screening the specificities of plant cell wall directed monoclonal antibodies and CBMs (19, 32). We have also shown how defined glycan arrays can be used to investigate the activities of cell wall active enzymes (25). Our experience with defined glycan arrays suggests that they have a wide applicability in

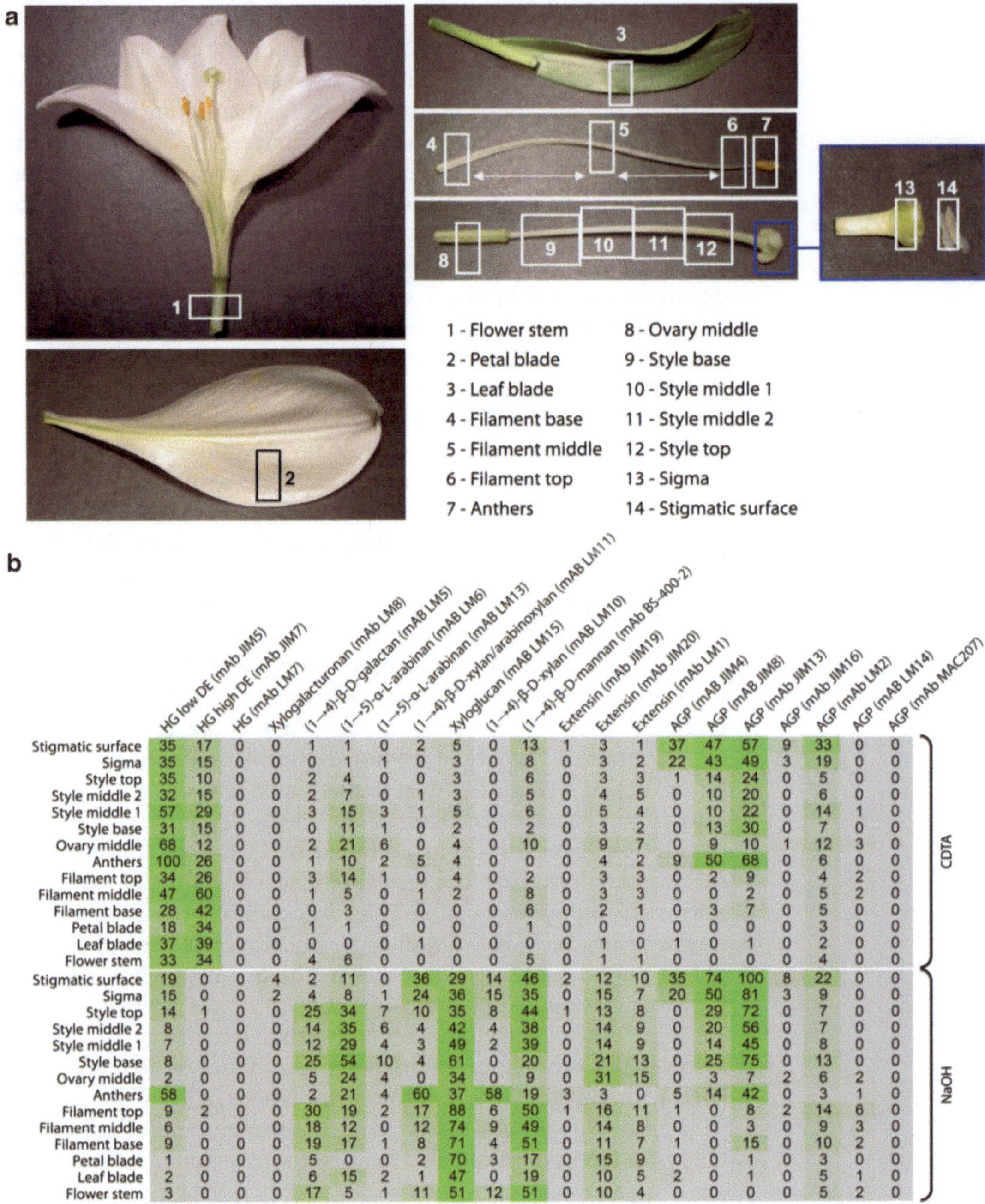

Extraction	Organ	HG low DE (mAb JIM5)	HG high DE (mAb JIM7)	HG (mAb LM7)	Xylogalacturonan (mAb LM8)	(1→4)-β-D-galactan (mAB LM5)	(1→5)-α-L-arabinan (mAB LM6)	(1→5)-α-L-arabinan (mAB LM13)	(1→4)-β-D-xylan/arabinoxylan (mAB LM11)	Xyloglucan (mAB LM15)	(1→4)-β-D-xylan (mAB LM10)	(1→4)-β-D-mannan (mAb BS-400-2)	Extensin (mAb JIM19)	Extensin (mAb JIM20)	Extensin (mAb LM1)	AGP (mAB JIM4)	AGP (mAB JIM8)	AGP (mAb JIM13)	AGP (mAb JIM16)	AGP (mAb LM2)	AGP (mAB LM14)	AGP (mAb MAC207)
CDTA	Stigmatic surface	35	17	0	0	1	1	0	2	5	0	13	1	3	1	37	47	57	9	33	0	0
CDTA	Sigma	35	15	0	0	0	1	1	0	3	0	8	0	3	2	22	43	49	3	19	0	0
CDTA	Style top	35	10	0	0	2	4	0	0	3	0	6	0	3	3	1	14	24	0	5	0	0
CDTA	Style middle 2	32	15	0	0	2	7	0	1	3	0	5	0	4	5	0	10	20	0	6	0	0
CDTA	Style middle 1	57	29	0	0	3	15	3	1	5	0	6	0	5	4	0	10	22	0	14	1	0
CDTA	Style base	31	15	0	0	0	11	1	0	2	0	2	0	3	2	0	13	30	0	7	0	0
CDTA	Ovary middle	68	12	0	0	2	21	6	0	4	0	10	0	9	7	0	9	10	1	12	3	0
CDTA	Anthers	100	26	0	0	1	10	2	5	4	0	0	0	4	2	9	50	68	0	6	0	0
CDTA	Filament top	34	26	0	0	3	14	1	0	4	0	2	0	3	3	0	2	9	0	4	2	0
CDTA	Filament middle	47	60	0	0	1	5	0	1	2	0	8	0	3	3	0	0	2	0	5	2	0
CDTA	Filament base	28	42	0	0	0	3	0	0	0	0	6	0	2	1	0	3	7	0	5	0	0
CDTA	Petal blade	18	34	0	0	1	1	0	0	0	0	1	0	0	0	0	0	0	0	3	0	0
CDTA	Leaf blade	37	39	0	0	0	0	0	1	0	0	0	0	1	0	1	0	1	0	2	0	0
CDTA	Flower stem	33	34	0	0	4	6	0	0	0	0	5	0	1	1	0	0	0	0	4	0	0
NaOH	Stigmatic surface	19	0	0	4	2	11	0	36	29	14	46	2	12	10	35	74	100	8	22	0	0
NaOH	Sigma	15	0	0	2	4	8	1	24	36	15	35	0	15	7	20	50	81	3	9	0	0
NaOH	Style top	14	1	0	0	25	34	7	10	35	8	44	1	13	8	0	29	72	0	7	0	0
NaOH	Style middle 2	8	0	0	0	14	35	6	4	42	4	38	0	14	9	0	20	56	0	7	0	0
NaOH	Style middle 1	7	0	0	0	12	29	4	3	49	2	39	0	14	9	0	14	45	0	8	0	0
NaOH	Style base	8	0	0	0	25	54	10	4	61	0	20	0	21	13	0	25	75	0	13	0	0
NaOH	Ovary middle	2	0	0	0	5	24	4	0	34	0	9	0	31	15	0	3	7	2	6	2	0
NaOH	Anthers	58	0	0	0	2	21	4	60	37	58	19	3	3	0	5	14	42	0	3	1	0
NaOH	Filament top	9	2	0	0	30	19	2	17	88	6	50	1	16	11	1	0	8	2	14	6	0
NaOH	Filament middle	6	0	0	0	18	12	0	12	74	9	49	0	14	8	0	0	3	0	9	3	0
NaOH	Filament base	9	0	0	0	19	17	1	8	71	4	51	0	11	7	1	0	15	0	10	2	0
NaOH	Petal blade	1	0	0	0	5	0	0	2	70	3	17	0	15	9	0	0	1	0	3	0	0
NaOH	Leaf blade	2	0	0	0	6	15	2	1	47	0	19	0	10	5	2	0	1	0	5	1	0
NaOH	Flower stem	3	0	0	0	17	5	1	11	51	12	51	0	10	4	0	0	0	0	5	2	0

Fig. 1. Comprehensive microarray polymer profiling (CoMPP) analysis of lily flower cell walls. (**a**) Lily flowers were divided into separate organs, which were then homogensized prior to the extraction of cell wall polysaccharides using CDTA and NaOH which predominantly release pectins and hemicelluloses, respectively. (**b**) Heatmap showing the relative abundance of cell wall glycans recognized by monoclonal antibodies (mAbs) in a range of lily flower organs. Color intensity is proportional to mean spot signal. The highest mean spot signal in the data set was assigned a value of 100 and all other signals adjusted accordingly. *HG* homogalacturonan, *DE* degree of methyl-esterification, *AGP* arabinogalactan-protein.

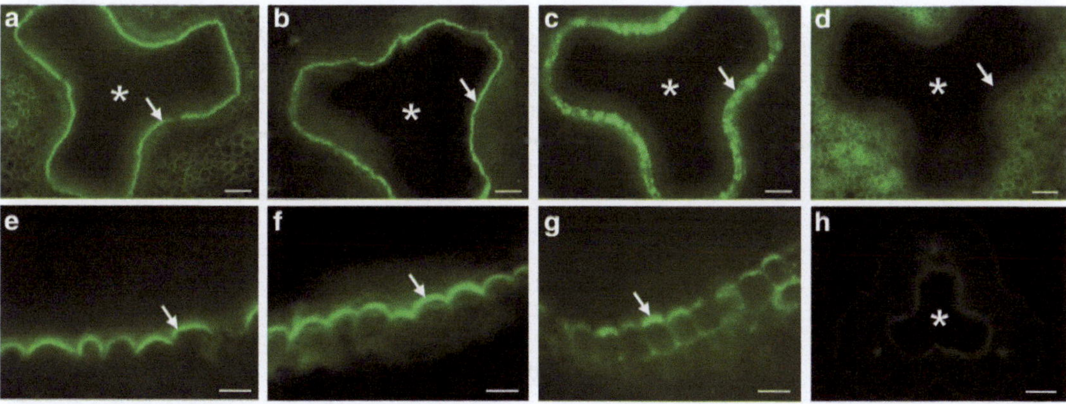

Fig. 2. Immunofluorescence labeling of transverse sections through lily styles. Micrographs showing the central portion of lily styles labeled with (**a**) mAb JIM7 (anti-pectic homogalacturonan with a high degree of methyl esterification), (**b**) mAb LM15 (anti-xyloglucan), (**c**) mAb JIM8 (anti-arabinogalactan protein), and (**d**) mAb LM2 (anti-arabinogalactan protein). (**e–g**) Higher magnification micrographs showing the cells lining the stylar transmitting tract labeled with (**e**) mAb LM15 (anti-xyloglucan), (**f**) mAb CCRC M87 (anti-xyloglucan) and mAb JIM8 (anti-arabinogalactan protein). The micrograph in (**h**) shows negative control labeling during which primary antibody was omitted. mAb binding was detected using FITC-conjugated secondary antibodies. The *asterisk* indicates the center of the lumen of the stylar transmitting tract and the *arrows* indicate the lining of the transmitting tract. Scale bars: (**a**)–(**d**) 100 μm; (**e**)–(**g**) 20 μm; (**h**) 200 μm.

plant research for investigating a range of carbohydrate-mediated interactions, as has been shown in other areas of biology. However, the development of this technology is currently limited by the lack of well-defined samples with which to populate arrays. In rare cases it is possible to obtain naturally occurring samples of relatively pure plant polysaccharides. For example, gum exuded from the bark of Acacia trees is a source of almost pure arabinogalactan protein (33). In most cases though, polysaccharides must be isolated from plant materials by the selective use of solvents followed by chromatographic separation. Oligosaccharides can be produced by fractionation of polysaccharides by appropriate hydrolase enzymes and then subsequent separation and purification. However, these processes are laborious and rarely yield oligo- and polysaccharides with purities greater than 90 %. Enzymatic synthesis is a powerful approach for oligosaccharide production (34, 35), but this approach is hampered by the fact that relatively few GTs are available for this purpose. Currently the most promising route for the production of well-defined and pure oligosaccharides for plant carbohydrate microarray production is chemical synthesis.

1.3. Chemical Synthesis of Plant Oligosaccharides

Chemical synthesis offers a solution to the chronic scarcity of plant oligosaccharides but is far from trivial. In contrast to the other major biopolymers and their oligomers, RNA/DNA and proteins/peptides, synthesis of carbohydrates is considerably more complex, since the number of permutations of even small oligosaccharides are very large due to pyranose/furanose diversity, the possibility of branching, linkage regiochemistry (1→2/1→3, etc.),

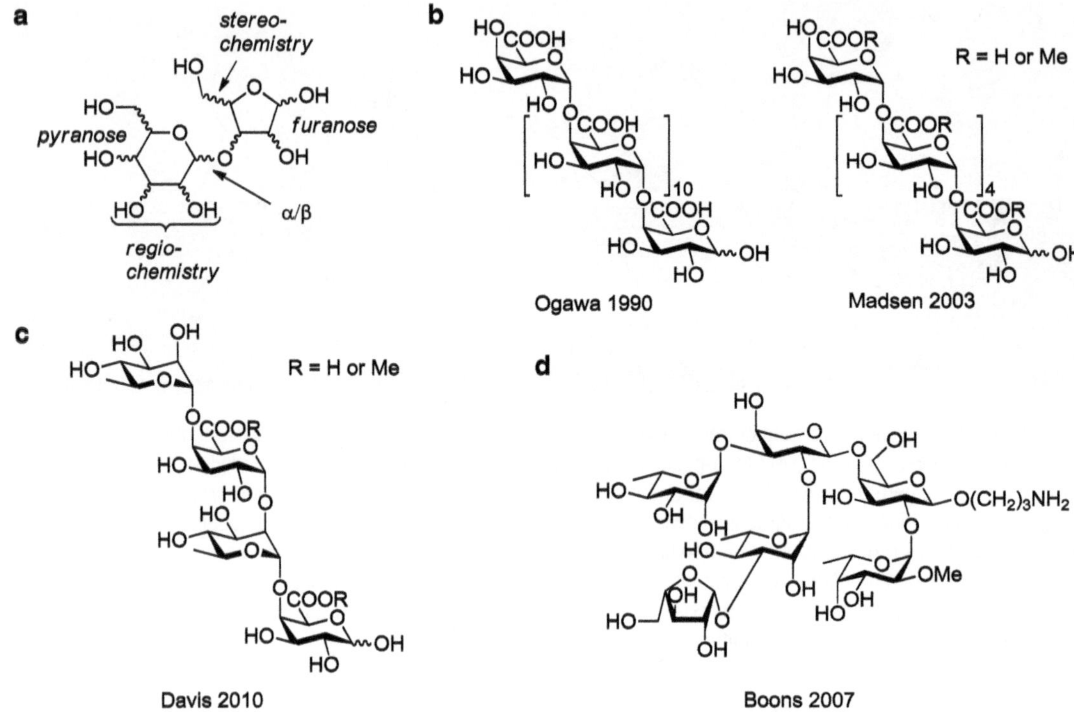

Fig. 3. Synthesis of plant cell wall oligosaccharides. (**a**) The challenges faced by synthetic chemist when assembling glycans. (**b**) Homogalacturonans. (**c**) Rhamnogalacturonan-I backbone tetrasaccharide. (**d**) Rhamnogalacturonan-II hexasaccharide. See text for details.

and stereochemistry (glucose/galactose, etc. and α/β-configuration, Fig. 3a, (36)). For chemists, the execution of a successful synthesis of an oligosaccharide hinges on controlling regiochemistry, i.e. linkages, which despite recent advances (37, 38) is still done most effectively by applying protective groups for functionalities that are not involved in the assembly of the oligosaccharide. However, protective groups are not inconsequential in a synthesis—rather, they have a profound influence on both the reactivity and stereoselectivity of a given glycosylation reaction (39, 40). Therefore matching the protective groups with the synthetic route is a substantial challenge that can often only be solved through trial-and-error (41). A further important parameter in oligosaccharide synthesis is the stereochemical outcome of a given glycosylation reaction. This is under stereoelectronic control, but can also be influenced by factors such as solvent, temperature, protective groups, the nature of the glycosyl acceptor and donor, and the method used to activate the latter (42). Although these factors render chemical synthesis of oligosaccharides a significant scientific challenge, it also provides unique opportunities, compared to isolating relevant oligosaccharides from plant material after chemical and/or enzymatic degradation of the polymers and a subsequent tedious isolation step. Synthetic oligosaccharides can typically be made in significant amounts (50–100 mg is not unusual); they have a high degree of purity and are not contaminated with similar structures, such as can be seen after chromatographic purification of plant extract mixtures; and the synthetic chemist has the option of functionalizing the final products, making it possible to couple them to proteins, surfaces or fluorophores and thus enabling biological investigations.

In spite of the challenges in chemical synthesis mentioned above, impressive examples of oligosaccharide synthesis with relevance to plant biologists have been disclosed. Ogawa and coworkers prepared a dodecasaccharide of α-1,4-GalAp as early as 1990 (43) and more recently, homogalacturonan oligosaccharides with defined patterns of methylation have been published (44–46) (Fig. 3b). Rhamnogalacturonan-I has been the subject of papers from several groups (47–49), so far culminating in the synthesis of a backbone tetrasaccharide by the group of Davis (50) (Fig. 3c). Arabinans (51), galactans (52) and arabinogalactans (53) are other RG-I structures that have been prepared chemically, although more detailed studies of these will benefit the plant biology field. The highly complex RG-II structure, consisting of a number of unusual carbohydrates, has also attracted significant attention (54–56), albeit without succumbing to a total synthetic effort yet (Fig. 3d). In addition to the pectic oligosaccharides mentioned, efforts have also been devoted to hemicelluloses, for example xylans (57).

Recent advances in solid-phase oligosaccharide synthesis have raised the hope that 1 day these important structures will be as easily assessable as, e.g. peptides are today (58). Nonetheless, despite the achievements of, for example, the groups of Seeberger (59) and Boons (60), the routine preparation of oligosaccharides is still a distant goal—and one that will require the chemical community to further address the problems of regio- and stereoselectivity, which makes carbohydrate chemistry so challenging. In the meantime, we will continue to rely on classical, solution-phase synthesis, typically of one oligosaccharide at a time—a proven method that hopefully makes the population of microarrays with well-defined oligosaccharides less of a bottleneck and in turn increases our knowledge of plant biology.

1.4. Future Directions

So far the CoMPP technique has been applied exclusively to plant cell wall research, but in principle this approach can be applied to plant material from which polysaccharides, or indeed proteins can be extracted and for which there is a suitable set of probes (mAbs, CBMs, lectins, or other binding markers). The extracted molecules must be large enough to adhere by passive adsorption to appropriate slides or membranes, but this requirement includes virtually all proteins and most polysaccharides. With this in mind there is no reason why CoMPP, or a version of it, could not be more widely applied in plant research where it may be of considerable value.

The plant kingdom is a hugely diverse and abundant source of complex polysaccharides, many of which may have hitherto unrecognized industrial or medical applications. At present it is not possible to produce synthetic analogs of these polysaccharides, so exploiting this bio-resource depends upon discovery in nature and this in turn requires massively HTP screening platforms. As has been shown to a limited extent already, microarrays are ideals tools for this technology, together with the analysis of CBPs and carbohydrate active enzymes are likely to be important areas of growth for plant carbohydrate microarrays.

References

1. Ekins R, Chu FW (1999) Microarrays: their origins and applications. Trends Biotechnol 17:217–218
2. Schena M, Shalon D, Davis RW, Brown PO (1995) Quantitative monitoring of gene expression patterns with a complementary DNA microarray. Science 270:467–470
3. McWilliam I, Chong Kwan M, Hall D (2011) Inkjet printing for the production of protein microarrays. Methods Mol Biol 785:345–361
4. Park S, Lee M-R, Shin I (2008) Carbohydrate microarrays as powerful tools in studies of carbohydrate-mediated biological processes. Chem Commun:4389–4399
5. Willats WG, Rasmussen SE, Kristensen T, Mikkelsen JD, Knox JP (2002) Sugar-coated microarrays: a novel slide surface for the high-throughput analysis of glycans. Proteomics 2:1666–1671
6. Fukui S, Feizi T, Galustian C, Lawson AM, Chai W (2002) Oligosaccharide microarrays for high-throughput detection and specificity assignments of carbohydrate-protein interactions. Nat Biotechnol 20:1011–1017
7. Feizi T (2000) Progress in deciphering the information content of the 'glycome'—a crescendo in the closing years of the millennium. Glycoconj J 17:553–565
8. Carpita NC, Gibeaut DM (1993) Structural models of primary cell walls in flowering plants: consistency of molecular structure with the physical properties of the walls during growth. Plant J 3:1–30
9. Fry SC (2004) Primary cell wall metabolism: tracking the careers of wall polymers in living plant cells. New Phytol 161:641–675
10. Willats WGT, McCartney L, Mackie W, Knox JP (2001) Pectin: cell biology and prospects for functional analysis. Plant Mol Biol 47:9–27
11. Bacic A, Harris AJ, Stone BA (1988) In Preiss J (ed) The biochemistry of plants. Academic, New York
12. Lee KJ, Marcus SE, Knox JP (2011) Cell wall biology: perspectives from cell wall imaging. Mol Plant 4:212–219
13. Willats WGT, Knox JP, Mikkelsen JD (2006) Pectin: new insights into an old polymer are starting to gel. Trends Food Sci Technol 17:97–104
14. Albersheim P, Darvill A, Roberts K, Sederoff R, Staehelin A (2011) In Masson S (ed) Plant cell walls. Garland Science, Taylor and Francis Publishing Group, New York
15. Gilbert HJ (2010) The biochemistry and structural biology of plant cell wall deconstruction. Plant Physiol 153:444–455
16. Smith DF, Song X, Cummings RD (2008) Use of glycan microarrays to explore specificity of glycan-binding proteins. Methods Enzymol 480:417–444 (Chapter 19)
17. Larsen K, Thygesen MB, Guillaumie F, Willats WG, Jensen KJ (2006) Solid-phase chemical tools for glycobiology. Carbohydr Res 341:1209–1234
18. Blixt O, Head S, Mondala T, Scanlan C, Huflejt ME, Alvarez R, Bryan MC, Fazio F, Calarese D, Stevens J, Razi N, Stevens DJ, Skehel JJ, van Die I, Burton DR, Wilson IA, Cummings R, Bovin N, Wong CH, Paulson JC (2004) Printed covalent glycan array for ligand profiling of diverse glycan binding proteins. Proc Natl Acad Sci USA 101:17033–17038
19. Moller I, Marcus SE, Haeger A, Verhertbruggen Y, Verhoef R, Schols H, Ulvskov P, Mikkelsen JD, Knox JP, Willats WGT (2008) High-throughput screening of monoclonal antibodies against plant cell wall glycans by hierarchical clustering of their carbohydrate microarray binding profiles. Glycoconj J 25:37–48
20. Feizi T, Chai W (2004) Oligosaccharide microarrays to decipher the glyco code. Nat Rev Mol Cell Biol 5:582–588
21. Feizi T, Fazio F, Chai W, Wong CH (2003) Carbohydrate microarrays—a new set of

technologies at the frontiers of glycomics. Curr Opin Struct Biol 13:637–645

22. Shipp M, Nadella R, Gao H, Farkas V, Sigrist H, Faik A (2008) Glyco-array technology for efficient monitoring of plant cell wall glycosyltransferase activities. Glycoconj J 25:49–58
23. Kosík O, Auburn RP, Russell S, Stratilová E, Garajová S, Hrmova M, Farkaš V (2010) Polysaccharide microarrays for high-throughput screening of transglycosylase activities in plant extracts. Glycoconj J 27:79–87
24. Moller I, Sørensen I, Bernal AJ, Blaukopf C, Lee K, Øbro J, Pettolino F, Roberts A, Mikkelsen JD, Knox JP, Bacic A, Willats WG (2007) High-throughput mapping of cell-wall polymers within and between plants using novel microarrays. Plant J 50:1118–1128
25. Øbro J, Sørensen I, Derkx P, Madsen CT, Drews M, Willer M, Mikkelsen JD, Willats WG (2009) High-throughput screening of Erwinia chrysanthemi pectin methylesterase variants using carbohydrate microarrays. Proteomics 9:1861–1868
26. Øbro J, Sørensen I, Moller I, Skjøt M, Mikkelsen J, Willats WGT (2007) High-throughput microarray analysis of pectic polymers by enzymatic epitope deletion. Carbohydr Polym 70:77–81
27. Singh B, Avci U, Eichler Inwood SE, Grimson MJ, Landgraf J, Mohnen D, Sørensen I, Wilkerson CG, Willats WG, Haigler CH (2009) A specialized outer layer of the primary cell wall joins elongating cotton fibers into tissue-like bundles. Plant Physiol 150:684–699
28. Moller IE, De Fine Licht HH, Harholt J, Willats WG, Boomsma JJ (2011) The dynamics of plant cell-wall polysaccharide decomposition in leaf-cutting ant fungus gardens. PLoS One 6:e17506
29. Sørensen I, Pettolino FA, Wilson SM, Doblin MS, Johansen B, Bacic A, Willats WGT (2008) Mixed linkage (1 3), (1 4)-β-D-glucan is not unique to the Poales and is an abundant component of *Equisetum arvense* cell walls. Plant J 54:510–521
30. Chae K, Lord EM (2010) Pollen tube growth and guidance: roles of small, secreted proteins. Ann Bot 108:627–636
31. Mollet J-C, Park S-Y, Nothnagel EA, Lord EM (2000) A lily stylar pectin is necessary for pollen tube adhesion to an in vitro stylar matrix. Plant Cell 12:1737–1750
32. Cid M, Pedersen HL, Kaneko S, Coutinho PM, Henrissat B, Willats WG, Boraston AB (2010) Recognition of the helical structure of beta-1,4-galactan by a new family of carbohydrate-binding modules. J Biol Chem 285: 35999–36009
33. Mahendran T, Williams PA, Phillips GO, Al-Assaf S, Baldwin TC (2008) New insights into the structural characteristics of the arabinogalactan-protein (AGP) fraction of gum arabic. J Agric Food Chem 56:9269–9276
34. Guillaumie F, Sterling JD, Jensen KJ, Thomas OR, Mohnen D (2003) Solid-supported enzymatic synthesis of pectic oligogalacturonides and their analysis by MALDI-TOF mass spectrometry. Carbohydr Res 338:1951–1960
35. Spadiut O, Ibatullin FM, Peart J, Gullfot F, Martinez-Fleites C, Ruda M, Xu C, Sundqvist G, Davies GJ, Brumer H (2011) Building custom polysaccharides in vitro with an efficient, broad-specificity xyloglucan glycosynthase and a fucosyltransferase. J Am Chem Soc 133:10892–10900
36. Davis BG (2000) Recent developments in oligosaccharide synthesis. J Chem Soc Perkin Trans 1:2137–2160
37. Kaji E, Nishino T, Ishige K, Ohya Y, Shirai Y (2010) Regioselective glycosylation of fully unprotected methyl hexopyranosides by means of transient masking of hydroxy groups with arylboronic acids. Tetrahedron Lett 51:1570–1573
38. Lawandi J, Rocheleau S, Moitessier N (2011) Directing/protecting groups mediate highly regioselective glycosylation of monoprotected acceptors. Tetrahedron 67:8411–8420
39. Carmona AT, Moreno-Vargas AJ, Robina I (2008) Glycosylation methods in oligosaccharide synthesis. Part 1. Curr Org Synth 5:33–60
40. Carmona AT, Moreno-Vargas AJ, Robina I (2008) Glycosylation methods in oligosaccharide synthesis. Part 2. Curr Org Synth 5:81–116
41. Demchenko AV (ed) (2008) Handbook of chemical glycosylation. Wiley-VCH, Weinheim
42. Fraser-Reid B, López JC (eds) (2011) Reactivity tuning in oligosaccharide assembly. Topics in current chemistry, vol 301. Springer, Berlin
43. Nakahara Y, Ogawa T (1990) Stereoselective total synthesis of dodecagalacturonic acid, a phytoalexin elicitor of soybean. Carbohydr Res 205:147–159
44. Clausen MH, Madsen R (2003) Synthesis of hexasaccharide fragments of pectin. Chem Eur J 9:3821–3832
45. Clausen MH, Madsen R (2004) Synthesis of oligogalacturonates conjugated to BSA. Carbohydr Res 339:2159–2169
46. Petersen BO, Meier S, Duus JØ, Clausen MH (2008) Structural characterization of homogalacturonan by NMR spectroscopy—assignment of reference compounds. Carbohydr Res 343:2830–2833
47. Reiffarth D, Reimer KB (2008) Synthesis of two repeat units corresponding to the backbone

of the pectic polysaccharide rhamnogalacturonan I. Carbohydr Res 343:179–188

48. Maruyama M, Takeda T, Shimizu N, Hada N, Yamada H (2000) Synthesis of a model compound related to an anti-ulcer pectic polysaccharide. Carbohydr Res 325:83–92
49. Nemati N, Karapetyan G, Nolting B, Endress HU, Vogel C (2008) Synthesis of rhamnogalacturonan I fragments by a modular design principle. Carbohydr Res 343:1730–1742
50. Scanlan EM, Mackeen MM, Wormald MR, Davis BG (2010) Synthesis and solution-phase conformation of the RG-I fragment of the plant polysaccharide pectin reveals a modification-modulated assembly mechanism. J Am Chem Soc 132:7238–7239
51. Du Y, Pan Q, Kong F (2000) An efficient and concise regioselective synthesis of α-(1→5)-linked L-arabinofuranosyl oligosaccharides. Carbohydr Res 329:17–24
52. El-Shenawy H, Schuerch C (1984) Synthesis and characterization of propyl *O*-β-D-galactopyranosyl-(1→4)-*O*-β-D-galactopyranosyl-(1→4)-α-D-galactopyranoside. Carbohydr Res 131:239–246
53. Fekete A, Borbás A, Antus S, Lipták A (2009) Synthesis of 3,6-branched arabinogalactan-type tetra and hexasaccharides for characterization of monoclonal antibodies. Carbohydr Res 344:1434–1441
54. Chauvin AL, Nepogodiev SA, Field RA (2005) Synthesis of a 2,3,4-triglycosylated rhamnoside fragment of rhamnogalacturonan-II side chain A using a late stage oxidation approach. J Org Chem 70:960–966
55. Rao Y, Boons GJ (2007) A highly convergent chemical synthesis of conformational epitopes of rhamnogalacturonan II. Angew Chem Int Ed 46:6148–6151
56. Rao Y, Buskas T, Albert A, O'Neill MA, Hahn MG, Boons GJ (2008) Synthesis and immunological properties of a tetrasaccharide portion of the B side chain of rhamnogalacturonan II (RG-II). Chembiochem 9:381–388
57. Takeo K, Ohguchi Y, Hasegawa R, Kitamura S (1995) Synthesis of (1→4)-β-D-xylo-oligosaccharides of dp 4–10 by a blockwise approach. Carbohydr Res 278:301–313
58. Seeberger PH, Haase WC (2000) Solid-phase oligosaccharide synthesis and combinatorial carbohydrate libraries. Chem Rev 100:4349–4394
59. Weishaupt M, Eller S, Seeberger PH (2010) In Fukuda M (ed) Solid phase synthesis of oligosaccharides. Methods Enzymol 478: 463–484
60. Boltje TJ, Kim JH, Park J, Boons GJ (2010) Chiral-auxiliary-mediated 1,2-cis-glycosylations for the solid-supported synthesis of a biologically important branched alpha-glucan. Nat Chem 2:552–557

INDEX

Jennifer Normanly (ed.), *High-Throughput Phenotyping in Plants: Methods and Protocols*, Methods in Molecular Biology, vol. 918, DOI 10.1007/978-1-61779-995-2,

Q

R

S

T

V

W

X

MIX
Papier aus verantwortungsvollen Quellen
Paper from responsible sources
FSC® C105338

If you have any concerns about our products,
you can contact us on
ProductSafety@springernature.com

In case Publisher is established outside the EU,
the EU authorized representative is:
Springer Nature Customer Service Center GmbH
Europaplatz 3, 69115 Heidelberg, Germany

Printed by Libri Plureos GmbH
in Hamburg, Germany